Y0-AAY-449

RECENT ADVANCES IN CELLULAR AND MOLECULAR ASPECTS OF ANGIOTENSIN RECEPTORS

ADVANCES IN EXPERIMENTAL MEDICINE AND BIOLOGY

Editorial Board:

NATHAN BACK, *State University of New York at Buffalo*

IRUN R. COHEN, *The Weizmann Institute of Science*

DAVID KRITCHEVSKY, *Wistar Institute*

ABEL LAJTHA, *N. S. Kline Institute for Psychiatric Research*

RODOLFO PAOLETTI, *University of Milan*

Recent Volumes in this Series

Volume 389
INTRACELLULAR PROTEIN CATABOLISM
Edited by Koichi Suzuki and Judith S. Bond

Volume 390
ANTIMICROBIAL RESISTANCE: A Crisis in Health Care
Edited by Donald L. Jungkind, Joel E. Mortensen, Henry S. Fraimow, and Gary B. Calandra

Volume 391
NATURAL TOXINS 2: Structure, Mechanism of Action, and Detection
Edited by Bal Ram Singh and Anthony T. Tu

Volume 392
FUMONISINS IN FOOD
Edited by Lauren S. Jackson, Jonathan W. DeVries, and Lloyd B. Bullerman

Volume 393
MODELING AND CONTROL OF VENTILATION
Edited by Stephen J. G. Semple, Lewis Adams, and Brian J. Whipp

Volume 394
ANTIVIRAL CHEMOTHERAPY 4: New Directions for Clinical Application and Research
Edited by John Mills, Paul A. Volberding, and Lawrence Corey

Volume 395
OXYTOCIN: Cellular and Molecular Approaches in Medicine and Research
Edited by Richard Ivell and John A. Russell

Volume 396
RECENT ADVANCES IN CELLULAR AND MOLECULAR ASPECTS OF ANGIOTENSIN RECEPTORS
Edited by Mohan K. Raizada, M. Ian Phillips, and Colin Sumners

Volume 397
NOVEL STRATEGIES IN THE DESIGN AND PRODUCTION OF VACCINES
Edited by Sara Cohen and Avigdor Shafferman

A Continuation Order Plan is available for this series. A continuation order will bring delivery of each new volume immediately upon publication. Volumes are billed only upon actual shipment. For further information please contact the publisher.

RECENT ADVANCES IN CELLULAR AND MOLECULAR ASPECTS OF ANGIOTENSIN RECEPTORS

Edited by

Mohan K. Raizada, M. Ian Phillips, and
Colin Sumners

University of Florida
Gainesville, Florida

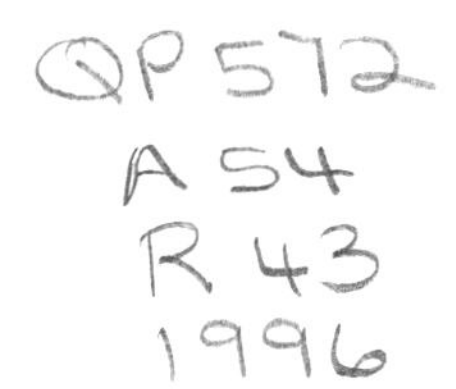

PLENUM PRESS • NEW YORK AND LONDON

Library of Congress Cataloging-in-Publication Data

Recent advances in cellular and molecular aspects of angiotensin receptors / edited by Mohan K. Raizada, M. Ian Phillips, and Colin Sumners.
p. cm. -- (Advances in experimental medicine and biology ; 396)
Includes bibliographical references and index.
ISBN 0-306-45209-X
1. Angiotensin--Receptors. I. Raizada, Mohan K. II. Phillips, M. Ian. III. Sumners, Colin. IV. Series.
QP572.A54R43 1996
612'.015--dc20
96-3812
CIP

Proceedings of the First International Symposium on the Cellular and Molecular Aspects of Angiotensin Receptors, held December 9 – 11, 1994, in Gainesville, Florida

ISBN 0-306-45209-X

© 1996 Plenum Press, New York
A Division of Plenum Publishing Corporation
233 Spring Street, New York, N. Y. 10013

10 9 8 7 6 5 4 3 2 1

All rights reserved

No part of this book may be reproduced, stored in a retrieval system, or transmitted in any form or by any means, electronic, mechanical, photocopying, microfilming, recording, or otherwise, without written permission from the Publisher

Printed in the United States of America

FOREWORD

Scientific advances over the past two decades have afforded unprecedented opportunities to understand the structure and function of receptors, receptor–ligand interactions, and receptor signaling. The extent of progress in this area is underscored by the recent Nobel Prize for Medicine and Physiology to Alfred Gilman and Martin Rodbell, both of whose work in understanding receptor/G-protein interactions has redefined the way in which we think of how hormones and neurochemicals exert their activity on cellular function.

This book is replete with examples of current research approaches to help us better understand the cellular roles in which the renin–angiotensin system and the angiotensin receptors participate. Clearly, defining the structure of angiotensin receptor subtypes is an important first step in clarifying the mechanisms by which these receptors take part in cellular function. However, the chapters within this book range far beyond structural studies and encompass research on tissue specific expression of the angiotensin receptor subtypes, the genetic regulation of these receptors, and the unique function of various angiotensin subtypes in different organ systems, such as the brain, the reproductive system, adipose tissue, the heart, and the kidneys.

Continued and increased multidisciplinary efforts that combine molecular, genetic, cellular, and physiological approaches will be crucial for our quest to understand fully how the renin–angiotensin system functions in both health and disease. The use of genetically altered animals to correlate molecular and physiological events is a key development in this regard. The knowledge provided by such endeavors will lead to innovative and novel diagnostic, preventive, and therapeutic strategies, and will also yield insight into key fundamental mechanisms that are important to all biology.

Claude Lenfant, M.D.
Director, Heart, Lung and
Blood Institute

PREFACE

There have been so many recent advances in cellular and molecular aspects of angiotensin receptors that it is timely to put them together in a single volume. Since the discovery of the role of the renin–angiotensin system in hypertension by several independent pioneers from Argentina, Germany, and the United States in the 1960s, there has been continuous interest in developing knowledge of the system and how to inhibit its overactivity. For many years, angiotensin was considered to be a blood-borne peptide, produced by the combined actions of the kidney, the lung, and the liver. The discoveries of angiotensin in the brain by Ganten[1] and Phillips[2] and angiotensin receptors in the brain by Snyder,[3] and their distribution by Mendelsohn[4] altered how we view the system. Since angiotensin II (Ang II) receptors in the brain are isolated from blood-borne angiotensin, there must be more than a single renin–angiotensin system. There had to be a brain renin–angiotensin system.

What has developed in the past decade is the realization that in addition to the blood-borne renin–angiotensin system, there are multiple renin–angiotensin systems localized in specific tissues, probably independent of the blood-borne angiotensin. Evidence for components of the renin–angiotensin system and receptors for Ang II and for fragments of Ang II have been found in many tissues, including not only brain, but also heart, blood vessels, kidney, pancreas, adrenal, gut, and gonads.[5,6,7,8] Effective inhibitors of the renin–angiotensin system have been designed by attacking the angiotensin-converting enzyme, and somewhat less successfully antagonizing the renin enzyme. The effector mechanism of all the components of the renin–angiotensin system, however, lies in the receptor. Early in the 1970s a peptide antagonist for the angiotensin receptor known as saralasin (Sar^1Ile^8 Ang II) was developed. It did not turn out to be a clinically useful inhibitor because of agonistic effects. Nevertheless, research on receptors proceeded with modified peptides and the first angiotensin II type 2 (AT_2) receptor ligand, developed by deGasparo and colleagues, was a peptide.[9] The big breakthroughs in angiotensin receptor advances came from two different sources. The first was the development of novel nonpeptide antagonists in 1989 (Chiu *et al.*[10]), which revealed that there are two receptor subtypes of angiotensin receptors, a type-1 subtype (AT_1-R) and a type-2 subtype (AT_2-R). The second was the use of molecular biology to resolve the structure of the subtypes. For many years, experimenters tried to isolate and purify angiotensin receptors, but these attempts were dogged by the paradox that the mechanism of binding and transduction was lost in the process of purification. This difficulty was finally overcome in 1991 by expression cloning the angiotensin II type 1 (Ang II AT_1) receptor from vascular smooth muscle cells (Murphy *et al.*[11] and by Inagami et al.[12]) from adrenal cells. The expression cloning method did not require purification of a receptor protein, and this has proven to be the most productive way to isolate and identify the molecular structure of angiotensin receptors. Since the AT_2 receptor was first cloned (Kambayashi *et al.*[13] and by Mukoyama *et al.*[14]), at least 20 different angiotensin receptor subtypes have been identified in different species and in different tissues. In the rat there appear to be AT_{1A} and AT_{1B} receptors, in addition to AT_2 receptors, and a new subtype, the AT_4 receptor.[15] In human tissue there appears to be only the AT_{1A} receptor and the AT_2 receptor. In many other species there are receptors that differ from both the AT_1 and the AT_2 receptors of the rat. Eventually, through the accumulation of this new data, we will see the evolution of the angiotensin receptor and discover its functional origins.

At the more practical level, the novel antagonists developed for inhibition of AT_1 receptors have been developed so far that they are now available for the treatment of hypertension. The Merck Pharmaceutical Company introduced Cozaar© in June of 1995, and several other companies are completing trials with their own AT_1 antagonists. The AT_1 receptor appears to be, by far, the most ubiquitous functional receptor for cardiovascular disease and growth in all tissues. The AT_2 receptor by contrast remains something of a mystery. There are active debates on the transductional changes that occur when receptors are stimulated and on their physiological function. Recent evidence indicates that AT_2 receptors inhibit the actions of AT_1 receptors in certain tissues. This suggests that AT_2 receptors put the brakes on AT_1 receptor activation in a number of physiological functions.

In this book you will find 26 chapters by leading investigators in angiotensin receptor research from around the world. The book is derived from an international symposium held December 9–11, 1994, at the University of Florida in Gainesville. Chapters were solicited after the symposium to bring readers up-to-date and complete coverage of current work on angiotensin receptors. The work extends from the regulation of the genes for the receptors to the interaction between the receptors with neurotransmitters and peptides. The distribution of the receptors in various tissues is covered and the variety of receptors, including atypical receptors, is described. Features of the book are regulation of expression, cloning, intracellular signaling and functions of AT_1 and AT_2 receptors. The book should have lasting value because it includes both knowledge about the angiotensin receptors, as well as hypotheses for future knowledge. Therefore the reader will find a source for reference and a source for stimulation.

M. Ian Phillips
Mohan K. Raizada
Colin Sumners
Gainesville, Florida

REFERENCES

1. Ganten, D., Fuxe, J., Phillips, M.I., and Mann, J.F.E. The brain isorenin angiotensin system: Biochemistry, localization and possible role in drinking and blood pressure regulation, *In.*: Frontiers in Endocrinology (Ganong, W.F. and Martini, L., eds.) Raven Press, New York, pg. 61, 1978.
2. Phillips, M.I., Mann, J.F.E., Haebara, H., Hoffman, W.E., Dietz, R., Schelling, P. and Ganten, D. Lowering of hypertension by central saralasin in the absence of plasma renin, *Nature* 270:445, 1977.
3. Bennett, J.P., Jr. and Snyder, S.H. Angiotensin II binding to mammalian brain membranes, *J. Biol. Chem.* 251:7423, 1976.
4. Mendelsohn, F.A.O., Quirion, R., Saavedra, J.M., Aguilera, G. and Catt, K.J. Autoradiographic localization of angiotensin II receptors in rat brain, *Proc. Natl. Acad. Sci. U.S.A.* 81:1575, 1984.
5. Raizada, M.K., Phillips, M.I., Crews, F.T. and Sumners, C. Distinct angiotensin II receptor in primary cultures of glial cells from rat brain, *Proc. Natl. Acad. Sci. U.S.A.* 84:4655, 1987.
6. Phillips, M.I., Speakman, E. and Kimura, B. Levels of angiotensin and molecular biology of the tissue renin–angiotensin system, *Reg. Pep.* 43:1–20, 1992.
7. Saavedra, J.M. and Timmermans, P.B.M.W.M. Angiotensin Receptors, Plenum Press, New York, 1994.

8. Campbell, D.J. and Habener, J.G. Cellular localization of angiotensinogen gene expression in brown adipose tissue and mesentery: Quantification of messenger ribonucleic acid abundance using hybridization *in situ, Endocrinology* 12:1616, 1987.
9. Whitebread, S., Mele, M., Kamber, B. and deGasparo, M. Preliminary biochemical characterization of two angiotensin II receptor subtypes, *Biochem. Biophys. Res. Comm.* 163:284, 1989.
10. Chiu, A.T., Herblin, W.F., McCall, D.E., Ardecky, R.J., Carini, D.J., Duncia, J.V., Pease L.J., Wong, P.C., Wexler, R.R., Johnson, A.L and Timmermans, P.B.M.W.M. Identification of angiotensin II receptor subtypes, *Biochem. Biophys. Res. Commun.* 165:196, 1989.
11. Murphy, T.J., Alexander, R.W., Griendling, K.K., Runge, M.S. and Bernstein, K.E. Isolation of a cDNA encoding a vascular type-1 angiotensin II receptor, *Nature* 351:233, 1991.
12. Sasaki, K., Yamano, Y., Bardham, S., Iwai, N., Murray, J.J., Hasegawa, M., Matsuda, Y., and Inagami, T. Cloning and expression of a complementary DNA encoding a bovine adrenal angiotensin II type-1 receptor, *Nature* 351:230, 1991.
13. Kambayashi, Y., Bardhan, S., Takahashi K., Tsuzuki, S., Inhui H., Hamakubo, T. and Inagami, T. Molecular cloning of a novel angiotensin II receptor isoform involved in phosphotyrosine phosphatase inhibition, *J. Biol. Chem.* 168(33):24543–6, 1993.
14. Mukoyama, M., Nakajima, M., Horiuchi, M., Sasamura, H., Pratt, R.E., Dzau, V.J. Expression of cloning type-2 angiotensin II receptor reveals a unique class of seven-transmembrane receptors, *J. Biol. Chem.* 268(33):24539–42, 1993.
15. Swanson, G.N., Hanesworth, J.M., Sardinia, M.F., Coleman, J.K.M., Wright, J.W., Hall, K.L., Miller-Wing, A.V., Stobb, J.W., Cook, V.I., Harding, E.C., and Harding, J.W. Discovery of a distinct binding site for angiotensin II (3–8), a putative angiotensin IV receptor, *Reg. Pep.* 40(3)409–419, 1992.

ACKNOWLEDGMENT

The editors wish to acknowledge the expert secretarial assistance of Ms. Jennifer Brock in the preparation of manuscripts and all aspects of putting together this proceedings. We would also like to acknowledge generous financial support from the Division of Sponsored Research; College of Medicine; the ICBR; the Hypertension Center; Departments of Pharmacology and Physiology at the University of Florida; Bristol-Myers Squibb Company; Ciba-Geigy Ltd.; Basel; Dupont Merck Pharmaceutical Company; Cadus Pharmaceutical Corporation; The Upjohn Company; and Glaxo Research. Without their support, the symposium and this volume would not have been possible.

CONTENTS

1

CHARACTERIZATION OF A *CIS*-REGULATORY ELEMENT AND *TRANS*-ACTING PROTEIN THAT REGULATES TRANSCRIPTION OF THE ANGIOTENSIN II TYPE 1A RECEPTOR GENE

Satoshi Murasawa, Hiroaki Matsubara, Yasukiyo Mori, Kazuhisa Kijima, Katsuya Maruyama, and Mitsuo Inada

Second Department of Internal Medicine
Kansai Medical University
Fumizonocho 1, Moriguchi
Osaka 570, Japan

INTRODUCTION

Angiotensin II (Ang II) has multiple physiological effects in the cardiovascular, endocrine, and nervous systems that are initiated by binding to specific receptors located on the plasma membrane (1). Two major subtypes (type 1 and type 2) of Ang II receptors have been revealed by their differential affinity for nonpeptide drugs (2). Ang II type 1a receptor (AT1a-R) cDNAs have been cloned from rat vascular smooth muscle cells (3), bovine adrenal zona glomerular cells (4) and rat kidney (5). Ang II mRNA is expressed in a variety of cells and tissues including vascular smooth muscle cells, liver, kidney and spleen, while the mRNA abundance is low in other tissues such as heart, brain, thymus and testis. AT1a-R gene expression is regulated in an ontogenic manner (6). Thus, the rat AT1a-R gene is cell-specifically and developmentally regulated.

We characterized one negative and three positive *cis*-regulatory elements in the 5′-flanking region of the rat AT1a-R gene (7). The negative *cis*-regulatory element (NRE) was located between -489 and -331 and inhibited the promoter activity of the 590 bp 5′-flanking region by a factor of ten. The *trans*-acting factor that binds to the element was present in PC12, not in vascular smooth muscle and glial cells. This suggested that the *trans*-acting factor is a major determinant which regulates the expression of the rat AT1a-R gene in PC12 cells. However, the NRE was located within 159 nucleotides (nt) from -489 to -331 and the core sequence has not been mapped in detail.

Here, we identified the core sequence to clarify the negative *cis*-regulation, using the gel retardation assay and the DNase I foot print analysis. We showed that the core sequence

Recent Advances in Cellular and Molecular Aspects of Angiotensin Receptors
Edited by Mohan K. Raizada et al., Plenum Press, New York, 1996

is A+T-rich (AATCTTTTATTTTA, nt -455 to -442). Southwestern blotting revealed that a nuclear protein of about 53-kDa bound to the NRE in PC12 cells and the rat brain, but not in vascular smooth muscle cells, glial cells, kidney, spleen, adrenal gland and liver.

METHODS

The gel retardation, DNase footprint and Southwestern analyses were performed as described in "Current Protocol in Molecular Biology" (8). The 159 bp NRE fragment (nt -489 to -331) was used as a probe in these experiment. A mutation in the NRE was created by PCR overlap extension mutagenesis. All plasmids were transfected by calcium-phosphate method into PC12, A10, glial and primary culture of rat vascular smooth muscle cells. CAT activity was determined by a dual phase diffusion assay, and normalized for transfection efficiency by means of β-galactosidase activity and for cell density by the protein concentration (7).

RESULTS

Gel Retardation Analysis of Negative cis-Regulatory Element. Our previous study demonstrated that transient expression analysis using the CAT reporter gene system, and competitive transfection using the *cis*-regulatory elements and the 5′-deletion mutants fused to CAT constructs, located three positive *cis*-regulatory elements (P1, P2 and P3) and one strong NRE in the 980 bp 5′-flanking region of the rat AT1a-R gene (7). The latter studies suggested that the *trans*-acting factor bound to the NRE is expressed in PC12 cells, but not

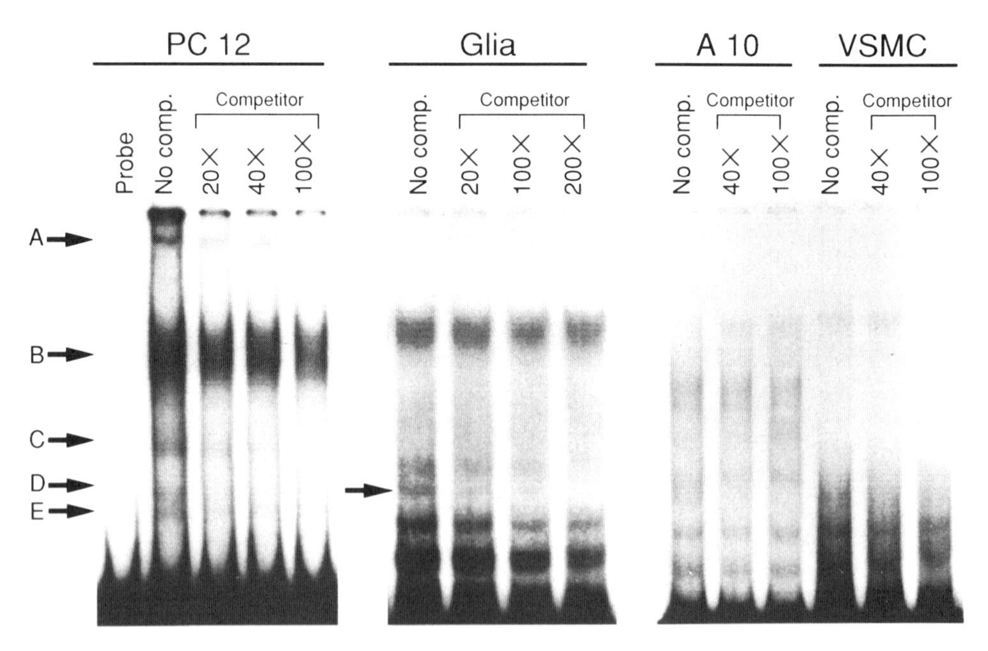

Figure 1. Gel retardation analysis of the NRE fragment. The NRE fragment (nt -489 to -331) was labeled and used in gel retardation analyses with nuclear extract from PC12 (2 μg), glial cells (10 μg) and A10 cells (2 μg) and primary cultured vascular smooth muscle cells of rat aorta (VSMC) (2 μg). The unlabeled NRE fragment at a 20× to 200× molar excess was the competitor.

in vascular smooth muscle (A10) or glial cells. However, the NRE core sequence has not been mapped and the *trans*-acting factor has not been identified.

Figure 1 shows the result of gel retardation analyses using 159 bp of the NRE fragment (nt -489 to -331) as a probe. The NRE fragment formed five retarded bands (arrows A, B, C, D and E in Fig. 1) upon incubation with cellular nuclear extract from PC12 cells. Although these bands differed in mobility, the addition of a 100-fold molar excess of the same unlabeled NRE fragment completely competed with the slowest band (arrow A). Other retarded bands (arrows C and D) also competed with the unlabeled NRE fragment, whereas the inhibition was not complete even with a 100-fold molar excess of the competitor. The

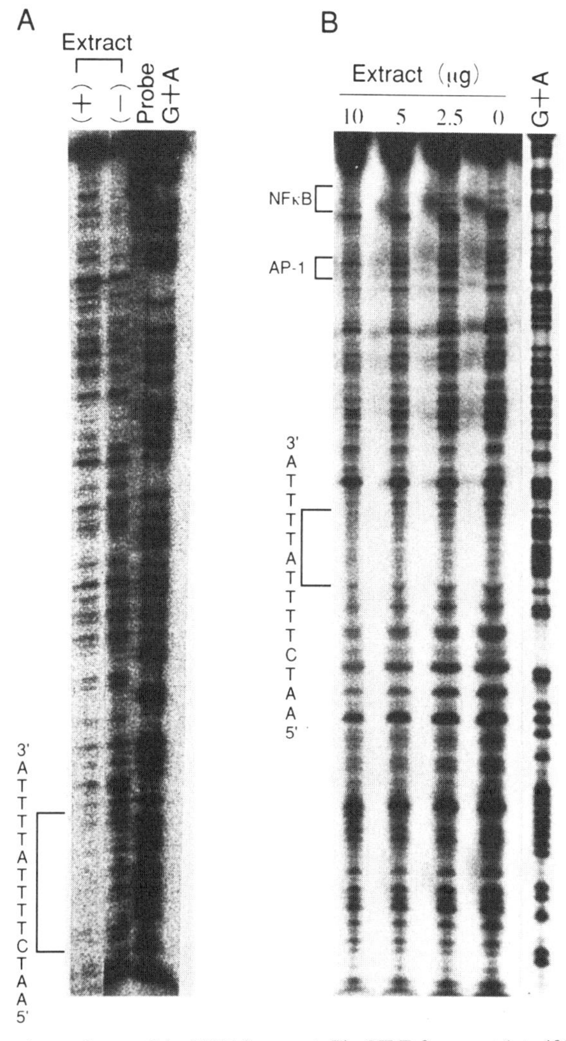

Figure 2. DNase I footprint analyses of the NRE fragment. The NRE fragment (nt -489 to -331) was labeled at the XhoI site (A) or the BssHII site (B) with 32P, incubated in the presence or absence of nuclear extract from PC12 cells, and partially digested with DNase I as described in "Methods". Twenty μg of nuclear extract was used in panel A and increasing amounts of nuclear extract were used in panel B. G+A lane indicates the sequence ladder by the Maxam-Gilbert reactions. The protected portion is shown in the left side of panel A and B. The NF B and AP-1 indicate the proposed regulatory elements by these transcriptional factors.

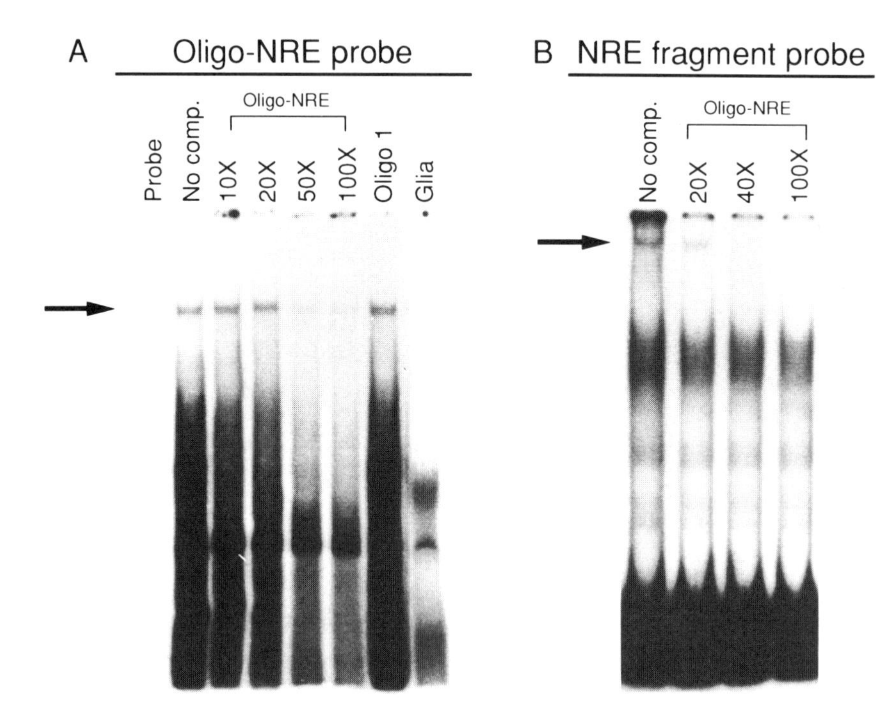

Figure 3. Gel retardation analyses of the oligo-NRE. A, the oligo-NRE (nt -456 to -442) was labeled and used in gel retardation analyses with nuclear extract (10 μg) from PC12 cells. Glia refers to gel retardation using nuclear extracts (10 μg) from glial cells. Oligo I shows gel retardation when oligo I encompassing the more upstream region (nt -489 to -457) was the competitor (molar ratio, 100×). B, the NRE fragment (nt -489 to -331) and the oligo-NRE (nt -456 to -442) were used as the probe and the competitor, respectively, and incubated with nuclear extract (2 μg) from PC12 cells. Arrows indicate the retarded products specifically bound to the probe.

arrow B and E bands were not inhibited with an excess of the competitor. The retarded band corresponding to the arrow A was not detected in glial cells, A10 cells and VSMC. Several retarded bands were detected in glial cells and one band indicated by the arrow was effectively inhibited by the competitor. No specific band complex was observed when the nuclear extract from A10 cells or VSMC was incubated with the NRE fragment. Since there are some transcriptional regulatory elements in the NRE fragment, such as an AP-1 site (TGAGTCA, located between -387 to -381), and an NF B site (GGGAGTTCC, located between -364 to -356), these transcriptional factors may interact with the NRE fragment and formed specific retarded bands in these cells.

Mapping the Core Sequence of the NRE. To better define the protein-binding site in the region responsible for the transcriptional inhibition, the NRE fragment (nt -489 to -331) was analysed by DNase I footprinting using a nuclear extract from PC12 cells. A protected region between nt -456 and -442 (5′-TAATCTTTTATTTTA-3′) was detected when a sense strand of the NRE fragment was labelled. When an antisense strand of the NRE fragment was used as a probe and partially digested with DNase I in the absence of nuclear extract, no or few sequence ladders were detected in the region between nt -455 and -442. Thus, since an antisense probe was not suitable for the foot print analysis, we extended the 5′ end of the NRE fragment by 52 bp using the vector nucleotide sequences and move the protected region to the middle of the gel. As shown in Fig. 2B, a protected region was identical to that observed in Fig. 2A, and other protected regions were not clearly characterized in this study. The regions corresponding to AP-1 and NF B sequences were not protected (Fig. 2B) (Fig. 2A).

An oligonucleotide (oligo-NRE) was designed to encompass the protected 15 bp sequences. With oligo- NRE as the labeled probe, the nuclear extract from PC12 cells produced a single strong band that was eliminated with an excess of the same oligo, but not by an unrelated oligo (oligo I) encompassing from nt -489 to -457 of the more uprestream NRE region (Fig. 3A). The nuclear extract from glial cells did not form a retarded band (Fig. 3A). Thus, oligo-NRE contains a binding site for a sequence-specific DNA binding protein.

Gel retardation analyses were performed using the promoter NRE fragment as a probe, to determine whether or not the retarded band could be competed by the oligo-NRE. Excess oligo-NRE interfered with the protein-NRE binding and reduced the amount of a slowest retarded product (Fig. 3B). These findings suggest that a slowest retarded band (arrow A in Fig. 1) is due to the interaction between nuclear protein and oligo-NRE sequence.

Site Directed Mutagenesis of the NRE. Two sets of mutations were designed within the NRE core sequence; one (oligo-NRE mutation 1) was mutated at -452 and -447 and the other (oligo-NRE mutation 2) at -450 and -449 (Fig. 5A). Gel retardation analyses using oligo-NRE mutation 1 as a probe, showed that the mutation had no effect on the protein/oligo binding, whereas an oligo-NRE mutation 2 probe efficiently reduced it (Fig. 4B). Next, we

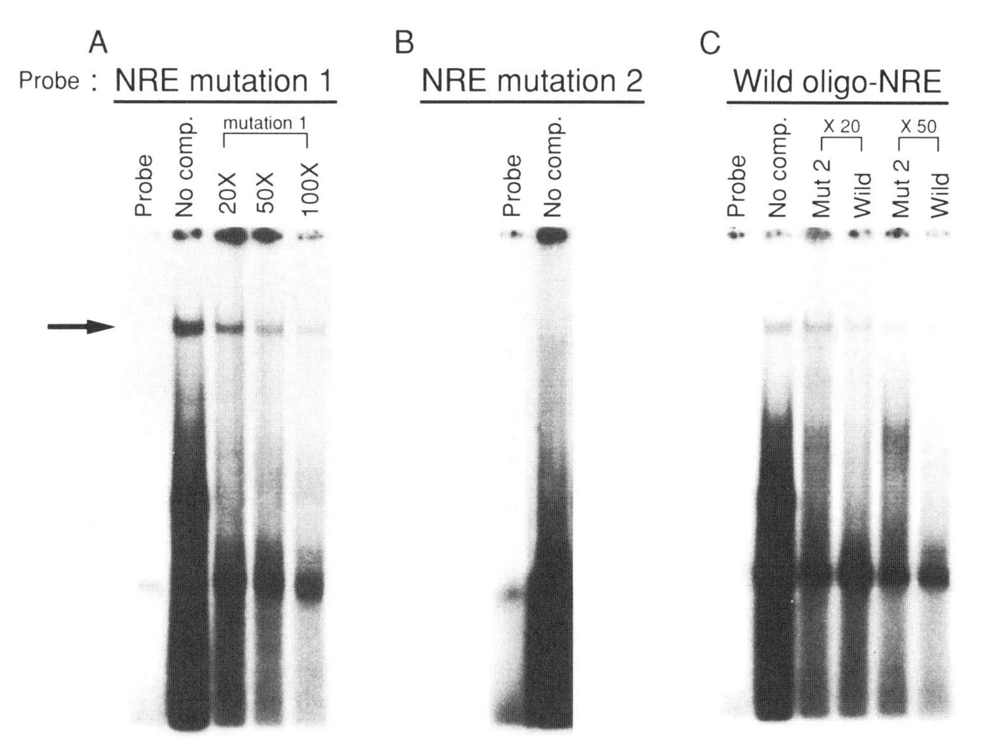

Figure 4. Gel retardation analyses of the oligo-NRE mutations. A, the NRE mutation 1 (-452 and -447) was labeled and competed with the unlabeled NRE mutation 1. B, the NRE mutation 2 (-450 and -449) was labeled and reacted with nuclear extract. C, the wild oligo-NRE was the probe and competed with wild oligo-NRE or unlabeled NRE mutation 2. Nuclear extract (10 μg) was prepared from PC12 cells. The arrow indicates the retarded band. Sequences for the wild oligo-NRE and the mutations are shown in Figure 5.

examined whether the oligo-NRE mutation 2 interferes with the wild-type oligo-NRE binding to the nuclear protein. As shown in Fig. 4C, mutation 2 slightly inhibited the protein/wild-oligo interaction, whereas the competition was weaker than that by the wild-oligo NRE. These results confirmed the assignment of this band to the oligo-NRE, and demonstrated that the T^{450} T^{449} sequences in the middle of the NRE core sequence are important for the protein/DNA binding, but this mutation does not completely lose an ability to bind to the trans-acting factor.

Functional Significance of the NRE in Transcriptional Regulation. We found that the NRE fragment (nt -489 to -331) inhibits the transcriptional activity of the 489 bp upstream region containing the proximal two positive cis-regulatory elements about 10-fold (P2: nt -331 to -201, P3: nt -201 to -61) (7). Since the relatively strong positive cis-regulatory element was located between nt -560 to -489 (7), we examined the inhibitory effect of the NRE fragment on the 980 bp of promoter region using the NRE fragment-deleted CAT construct. The CAT activity was normalized for transfection efficiency by means of the co-transfected β-Gal gene and for cell density by the protein concentration, and expressed as a relative value to that of the promoterless CAT construct transfected in PC12 or glial cells (7). The experiment was repeated six times and the analysis of variance and the Dunnet's test were used for statistical comparisons. The obtained relative CAT value of the wild AT1a980 CAT construct in PC12 cells was arbitrarily assigned a value of 1.0 for quantitative comparison. The results showed that deleting the NRE fragment from the promoter region yielded about a 2.7-fold increase ($p<0.01$, n=6) in the relative CAT activity and reached a level similar to the promoter activity of glial cells (Fig. 5B). In addition, we constructed an oligo-NRE mutation 2 in the 980 bp of promoter region, and fused it to the CAT gene. As shown in Fig. 5B, the relative CAT activity significantly ($p<0.01$, n=6) increased about

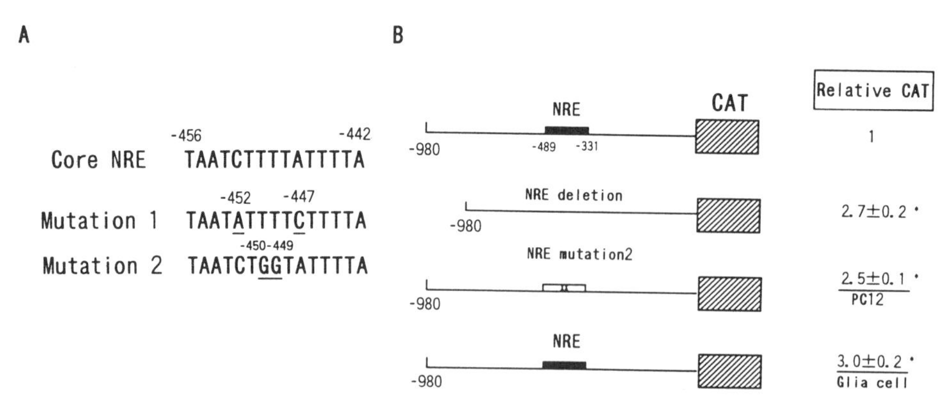

Figure 5. Nucleotide sequences of NRE mutations and functional analyses of NRE in the promoter activity. A, nucleotide sequences of NRE mutations 1 (-452 and -447) and 2 (-450 and -449) are shown in comparison with wild NRE sequences. B, the NRE fragment (nt -489 to -331) was deleted from 980 bp of 5′-flanking region and fused to the CAT reporter gene as described in "Methods". A mutation corresponding to NRE mutation 2 was designed by means of PCR overlap extension mutagenesis. These CAT fusion genes (15 μg) were co-transfected with 5 μg of β-Gal genes into PC12 or glial cells. The CAT activity levels were normalized with β-Gal activities and protein contents, and expressed as relative values to those of the promoterless CAT construct. For quantitative comparison, the relative value of the wild AT1a 980-CAT construct in PC12 cells is assigned a value of 1.0. The mean activities and the standard error of the mean from six separate assays are presented. Analysis of variance and the Dunnet's test were used for multigroup comparisons. *$p<0.01$ vs value of wild CAT construct in PC12 cells.

2.5-fold compared with the CAT construct containing the wild NRE, indicating the functional significance of the NRE core sequence in the transcriptional inhibition of PC12 cells.

Southwestern Blots of the trans-Acting Factor in PC12 Cells. To characterize the binding factor, the nuclear extract from PC12 cells was resolved by SDS-PAGE and Southwestern blotted with the 32P-labeled NRE 159 bp fragment or oligo-NRE probes. These probes bound to three proteins with the same intensity. The molecular mass of these bands was almost same between the NRE fragment and oligo-NRE probes, which disappeared in the presence of excess cold probe (Fig. 6) (data using oligo-NRE probe not shown). When the oligo-NRE was used as a competitor against the nuclear proteins bound to the NRE fragment probe, the highest molecular weight (~53-kDa) protein signal disappeared and other two signals were not affected (Fig. 6). On the other hand, the nuclear extract from glial, A10 cells or VSMC did not contain the protein bound to the NRE fragment and oligo-NRE. We also examined the presence of this nuclear protein in the rat brain, liver, spleen, adrenal gland and kidney. As shown in Fig. 7, a single band corresponding to the 53-kDa protein was found in the nuclear extract from the rat brain, but not in the liver, spleen, adrenal gland and kidney (data for PC12 cells and rat brain are shown). Thus, we concluded that the 53-kDa protein specifically binds to the NRE core sequence, whereas the other two bands non-specifically bind to the NRE core sequence in PC12 cells. Although the condition of protein-DNA binding quite differs between the gel shift assay and the Southwestern blot,

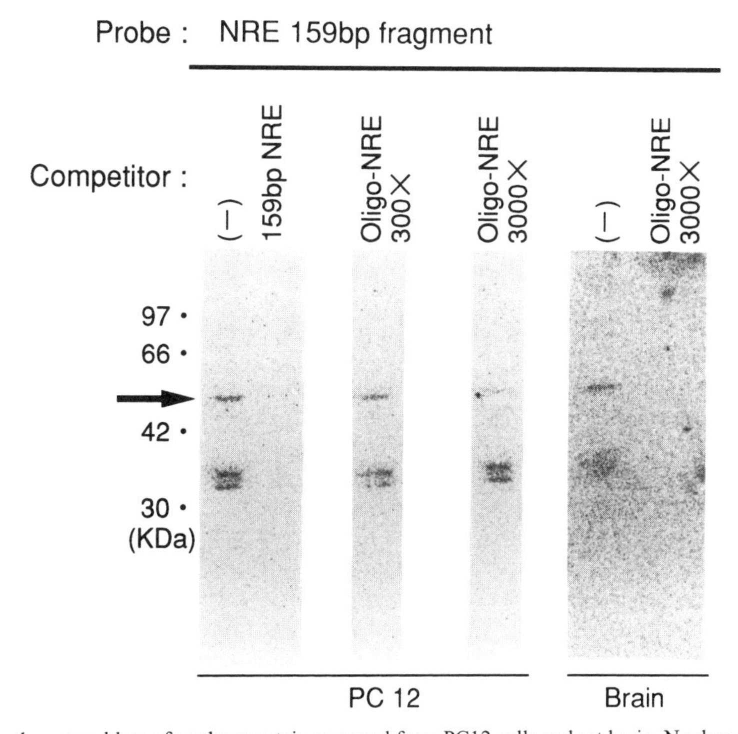

Figure 6. Southwestern blots of nuclear protein prepared from PC12 cells and rat brain. Nuclear extracts (20 μg of protein per lane) were separated by SDS-PAGE, blotted onto a nitrocellulose membrane, then detected with the 32P-labeled NRE fragment (159 bp, nt -489 to -331). The unlabeled NRE and wild oligo-NRE were used as competitors. The arrow indicates the 53-kDa protein specifically bound to the probe. Films were exposed for 1 (PC12 cells) or 14 days (rat brain) at -80°C with intensifying screens. Size markers are indicated in kDa on the left side of gels.

the finding that the oligo-NRE inhibits a single retarded band alone as shown in Fig. 3B, may also support this contention.

Action of the NRE Fragment upon a Heterologous Promoter. To test whether the NRE fragment could regulate a heterologous promoter, the fragment was subcloned upstream of the thymidine kinase (TK) promoter fused to the CAT gene. In transient assays using PC12 cells, the expression of the fusion gene containing the TK promoter alone appreciably increased compared with the TK-less CAT gene. When the 159 bp NRE fragment was added to the TK-CAT construct and transfected into PC12 cells, the CAT activity was unaffected, although the NRE fragment was placed immediately upstream of the TK gene (data not shown). These results suggested that the NRE sequence acts as a negative regulatory element of the rat AT1a gene, rather than as a silencer element, and that it works in junction with the native, rather than the heterologous promoter.

DISCUSSION

We demonstrated that three positive cis-regulatory elements and one strong NRE are present in the 5′-flanking region of rat AT1a-R gene, and suggested that this NRE is one of major determinants that regulate the level of gene expression in PC12 cells (7). Here, we discovered and mapped the core sequence of the NRE (5′-TAATCTTTTATTTTA-3′) between -456 and -442. This element reduced AT1a-R gene expression by a factor of 2.7 in transient transfection assays using PC12 cells. Site-directed mutagenesis of two base pairs at -450 and -449 affected the specific DNA-protein interaction and eliminated the suppression of AT1a-R gene expression in the transient transfection assays. These data demonstrated that specific protein-DNA interaction at this sequence down-regulates the gene expression in PC12 cells.

Although NREs have been detected in a variety of genes (9), the molecular mechanisms by which they exert their effects remain obscure. While the NREs can exhibit enhancer-like qualities and function on heterologous promoters in a distance and orientation-independent fashion (10), they often reduce rather than abolish the activity of heterologous promoters and demonstrate a preference for a specific promoter (11,12). The NRE in the rat AT1a-R gene, when transferred to the TK promoter, had no significant effect on the transcriptional activity, suggesting that the NRE in the AT1a-R gene works more effectively in conjunction with the native AT1a-R gene promoter than with the heterologous promoter.

Recently, many cis-acting, negative transcriptional elements in mammals have been identified (9), some of which include A+T rich sequences. These include human Ig heavy chain (AATATTTT) (13), human MHC class I (AAAATTATCTGAAAAAGGTTATTAAAA) (14), rat prolactin (AAATAAA, TATAATTTTATA) (15), and mouse Ig (ATTAATTTAT) (16). However, the A+T rich NRE sequence identified in the rat AT1a-R gene does not match these A+T rich sequences, indicating that this NRE is a novel negative element.

The rat AT1a-R gene has a subtype gene, AT1b-R, which has very high homology (96 %) in the coding sequence (17-20). Although the receptor-mediated second signal observed in the expression study was similar to that in the AT1a-R, the profile for the tissue distribution between the AT1a-R and AT1b-R mRNAs was quite distinct. The AT1a-R is the dominant form expressed in the liver, kidney, vasculature, lung, ovary, testis and heart, whereas the AT1b-R is expressed in greater quantities in the adrenal gland, anterior pituitary, and uterus (17-19). Given the high similarity of nucleotides in the coding region between the AT1a-R and AT1b-R, the 5é-flanking region may regulate the tissue-specific gene expression. Very recently, Guo and Inagami have sequenced the rat AT1b-R promoter region (21), in which the homology between both subtypes was low and the A+T rich NRE sequence observed in the AT1a-R gene was not detected. Since the abundant expression of AT1b-R gene is restricted in a few organs compared with that of AT1a-R gene, a much stronger NRE may be present and regulate a tissue-specific expression. In the 5é-flanking region of the human AT1a-R gene, Guo et al. (22) and Takayanagi et al. (23) found using vascular smooth muscle cells, that the *cis*-regulatory region inhibits the gene transcription between -881 to -642, and -962 to -114, respectively. However, the A+T rich NRE sequence observed in the rat AT1a-R gene was not located in these regions, and a similar element (AAATTTATTTTA) was present more upstream. This study demonstrated that the *trans*-acting protein bound to the NRE of rat AT1a-R gene is present in PC12, but not in primary cultured vascular smooth muscle cells of rat aorta, A10 cells and glial cells. Thus, whether the A+T rich sequence in the human AT1a-R gene can function as a NRE depends on a suitable cell model containing an abundant amount of the specific *trans*-acting protein.

Southwestern blots suggested the involvement of a 53-kDa protein factor in the AT1a-R promoter function in PC12 cells. PC12 cells are derived from the clonal isolation of a rat adrenal chromaffin cell tumor, and in the presence of nerve growth factor, the cells cease to multiply and differentiate into neurons (24). We showed in the previous study (7) that the expression of the AT1-R is very low in PC12 cells: the mRNA level was detectable only by the reverse transcriptase-PCR method, not by the Northern blot, and the AT1-R protein was not quantified by the binding assay. Although the binding of Ang II to the receptor in the brain alters the activity of the central neuronal pathway and secretion of vasopressin and adrenocorticotropic hormone (25), the AT1-R is expressed in highly restricted regions in the rat brain (26), in contrast with the finding that the AT1-R is expressed in a variety of peripheral tissues (27). Sumners et al. have reported using the neonatal rat brain, that glial cells predominantly express AT1-R, whereas neurons contain a small amount of AT1-R (28). We also found that the promoter region of rat AT1a-R gene contains a positive *cis*-regulatory element active only in glial and PC12 cells (7), suggesting that the regulatory mechanism of rat AT1a-R gene expression may differ between the central and peripheral tissues. The finding that a nuclear protein bound to the NRE in the brain, but not in vascular smooth muscle cells, glial cells, the spleen, kidney, adrenal gland and liver, also supports this contention. The expression of AT1a-R gene is also inhibited in the adrenal gland and the pituitary gland, in which the AT1b-R subtype is predominant. The 53-KDa protein was not detected in the adrenal gland, suggesting that other regulatory mechanisms to suppress the AT1a-R gene transcription may function in these tissues. Although the transcriptional regulation of the NRE region is complicated by the presence of several binding protein, the 53-kDa nuclear protein may be one of major *trans*-acting factors that determine the neuron-specific regulation of AT1a-R gene. The cloning of a gene encoding the 53-kDa protein bound to the NRE may help to elucidate the exact profile of tissue distribution of this protein and the cell-type specific expression mechanisms of the AT1a-R gene, as well as the isolation of a novel nuclear protein that inhibits the expression of the Ang II receptor.

ACKNOWLEDGMENTS

This study was supported in part by research grants from the Ministry of Education, Science and Culture, Japan, the Study Group of Molecular Cardiology, the Naito Foundation, the Clinical Pharmacology Foundation in Japan.

REFERENCES

1. Peach, M.J. (1981). Biochemical Pharmacol. *30*: 2745-2751
2. Timmermans, P.B.M.W.M., P.C. Wong, A.T. Chiu, and W.F. Herblin. (1991). Trend. Pharmacol. Sci. *12*: 55-61
3. Murphy, T.J., R.W. Alexander, K.K. Griendling, M.S. Runge, and K.E. Bernstein. (1991). Nature. *351*: 233-236
4. Sasaki, K., Y. Yamano, S. Bardham, N. Iwai, J.J. Murray, M. Hasegawa, Y. Matsuda, and T. Inagami. (1991). Nature. *351*: 230-233
5. Iwai, N., Y. Yamano, S. Chaki, F. Konishi, S. Bardhan, C. Tibbetts, K. Sasaki, M. Hasegawa, Y. Matsuda, and T. Inagami. (1991). Biochem. Biophys. Res. Commun. *177*: 299-304
6. Suzuki, J., H. Matsubara, M. Urakami, and M. Inada. (1993). Circ. Res. *73*: 439-447
7. Murasawa, S., H. Matsubara, M. Urakami and M. Inada. (1993). J Biol Chem. *268*: 26996-27003
8. Abmayr, S.M., L.A. Chodosh, A.S. Baldwin, M. Brenowitz, D.F. Senear, and R.E. Kingston, DNA-protein interactions, In Current Protocol in Molecular Biology. F.M. Ausubel, R. Brent, R.E. Kingston, D.D. Moore, J.G. Seidman, J.A. Smith, and K. Struhl, editors. Greene Publishing Associates, Brooklyn, NY. 12.0.1-12.9.4
9. Levine, M. and J. Manley. (1989). Cell. *59*: 405-408
10. Muglia, L., and L.B. Rothman-Denes. (1986). Proc. Natl. Acad. Sci. USA. *83*: 7653-7657
11. Nakamura, N., D.W. Burt, M. Paul, and V.J. Dzau. (1989). Proc. Natl. Acad. Sci. USA. *86*: 56-59
12. Maue, R.A., S.D. Kraner, R.H. Goodman, and G. Mandel. (1990). Neuron. *4*: 223-231
13. Imler, J.L., C. Lemaire, C. Wasylyk, and B. Wasylyk. (1987). Mol. Cell. Biol. *7*: 2558-2567
14. Weissman, J.D., and D.S. Singer. (1991). Mol. Cell. Biol *11*: 4217-4227.
15. Zhang, Z.H., V. Kumar, R.T. Rivera, J. Chisholm, and D.K. Biswas. (1990). J. Biol. Chem. *265*: 4785-4788
16. Pierce, J.W., A.M. Gifford and D. Baltimore. (1991). Mol. Cell. Biol. *11*: 1431-1437
17. Iwai, N., and T. Inagami. (1992). FEBS Lett. *298*: 257-260
18. Sandberg, K., H. Ji, A.J.I. Clark, H. Shapira and K.J. Catt. (1992). J. Biol Chem. *267*: 9455-9458
19. Kakar, S.S., J.C. Sellers, D.C. Devor, L.C. Musgrove, and J.D. Neil. (1992). Biochem. Biophys Res. Commun. *183*: 1090-1096
20. Elton, T.S., C.C. Stephan, G.R. Taylor, M.G. Kimball, M.M. Martin, J.N. Durand and S. Oparil. (1992). Biochem. Biophys. Res. Commun. *184*: 1067-1073
21. Guo, D-F and T. Inagami. (1994). Biochem. Biophys. Acta. *1218*: 91-94
22. Guo, D-F., H. Furuta, M. Mizukoshi and T. Inagami. (1994). Biochem. Biophys. Res. Commun. *200*: 313-319
23. Takayanagi, R., K Ohnaka, Y. Sakai, S. Ikuyama, and H. Nawata. (1994). Biochem. Biophys. Res. Commun. *200*: 1264-1270
24. Greene, L.A., and A.S. Tischler. (1976). Proc. Natl. Acad. Sci. USA *73*: 2424-2428
25. Phillips, M.I. (1987). Annu. Rev. Physiol. *49*: 413-435
26. Millan, M.A., D.M. Jacobowitz, G. Aguilera and K.J. Catt. (1991). Proc. Natl. Acad. Sci. USA. *88*: 11440-11444
27. Inagami, T., and Y. Kitami. (1994). Hypertens. Res. *17*: 87-97
28. Summers, C., W. Tang. B. Zelezna, and M.K. Raizada. (1991). Proc. Natl. Acad. Sci. USA. *88: 7567-7571*

2

HUMAN AT_1 RECEPTOR GENE REGULATION

Baogen Su, Mickey M. Martin, and Terry S. Elton

Brigham Young University
Department of Chemistry and Biochemistry
Benson Science Building
Provo, Utah 84602

INTRODUCTION

The peptide hormone angiotensin II (AII), the biologically active component of the renin-angiotensin system, regulates a variety of physiological responses including fluid homeostasis, aldosterone production, renal function and contraction of vascular smooth muscle (1). In addition, AII has actions in the central nervous system (2) and may play a role in development (3). AII binds to specific receptors that mediate intracellular Ca^{2+} mobilization through stimulation of phospholipase C and production of inositol trisphosphate (4), activation of Ca^{2+} channels (5), and inhibition of adenylate cyclase through a pertussis toxin-sensitive G-protein (6). Given the diversity of AII-mediated events and the differential signal transduction mechanisms, it has been proposed that multiple AII receptor subtypes exist (1,7).

Recently, two major pharmacologically distinct AII receptor subtypes have been described by using specific antagonists (8-13). The AII type 1 receptor (AT_1R) has a high affinity for the nonpeptide antagonist DuP 753 (losartan), whereas the AII type 2 receptor (AT_2R) has a high affinity for the nonpeptide antagonist PD123319 and the peptide agonist CGP42112A and a low affinity for losartan (8-13). To date, all of the effects currently associated with AII seem to be mediated by the AT_1R. In contrast, the physiological function of the AT_2R, and its mechanism(s) of signal transduction, are poorly understood. Since all of the known biochemical functions elicited by AII are blocked by AT_1R specific antagonists, this suggests the existence of subtypes of the AT_1R.

Several laboratories have succeeded in cloning the cDNA for the AT_1R from bovine and human tissues (14-17). The cloned AT_1R has the typical features of G-protein coupled receptors (i.e., seven transmembrane spanning region topology) and is linked to the phospholipase C/IP_3 signal transduction pathway. Interestingly, Southern blot analysis utilizing rat genomic DNA suggested the existence of another closely related AT_1R (18). We and others have isolated and characterized distinct AT_1R subtypes, now designated as $AT_{1A}R$ and $AT_{1B}R$ (18-21). The rat $AT_{1A}R$ has been localized to chromosome 17 and the $AT_{1B}R$ to chromosome 2 (22). The open reading frames of both the $AT_{1A}R$ and $AT_{1B}R$ encode proteins of 359 amino acids with 95% homology (18-21). Although it has been suggested that both

Recent Advances in Cellular and Molecular Aspects of Angiotensin Receptors
Edited by Mohan K. Raizada et al., Plenum Press, New York, 1996

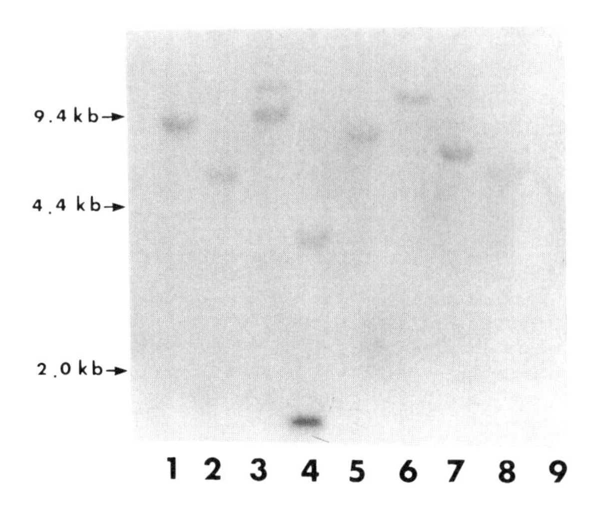

Figure 1. Southern blot containing 8 μg of genomic DNA per lane from nine eukaryotic species. DNA is restricted with Eco RI. Lanes 1-9 contain, in order, genomic DNA from human, monkey, rat, mouse, dog, cow, rabbit, chicken and yeast. The ZOO blot was probed with a random-prime labeled hAT_1R fragment.

receptor subtypes are pharmacologically and functionally identical (19,23,24) they are expressed in a tissue-specific manner (19-21) and seem to be regulated differentially (20,25-27).

Currently, it is thought that only rodents have multiple AT_1R subtypes. However, recently it was suggested by Konishi et al. (28) that multiple human AT_1R (hAT_1R) subtypes also exist. In this paper, we will discuss the existence of multiple AT_1R subtypes, alternatively spliced hAT_1R mRNA transcripts, the organization of the hAT_1R gene and some of the mchanisms by which this gene can be regulated.

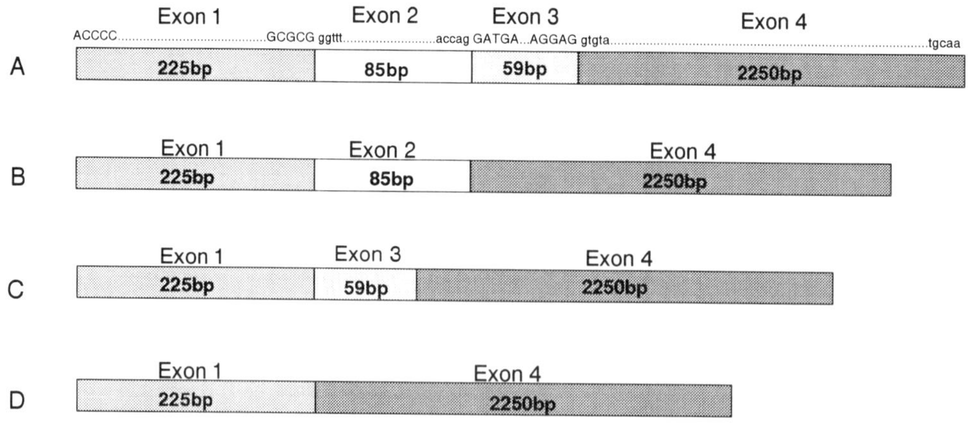

A: Exon 1 (225bp), Exon 2 (85bp), Exon 3 (59bp), Exon 4 (2250bp)
ACCCC.......GCGCG ggttt.......accag GATGA...AGGAG gtgta.......tgcaa

B: Exon 1 (225bp), Exon 2 (85bp), Exon 4 (2250bp)

C: Exon 1 (225bp), Exon 3 (59bp), Exon 4 (2250bp)

D: Exon 1 (225bp), Exon 4 (2250bp)

Figure 2. A schematic representation of the 5′-RACE experiments. Human lung 5′-RACE-Ready cDNA was PCR amplified using an anchor and hAT_1R specific primers. PCR products were subcloned and sequenced. Four distinct hAT_1R cDNAs were characterized (A-D). (From Su et al., Biochem. Biophys. Res. Commun., 204, 1041, 1994. With permission.)

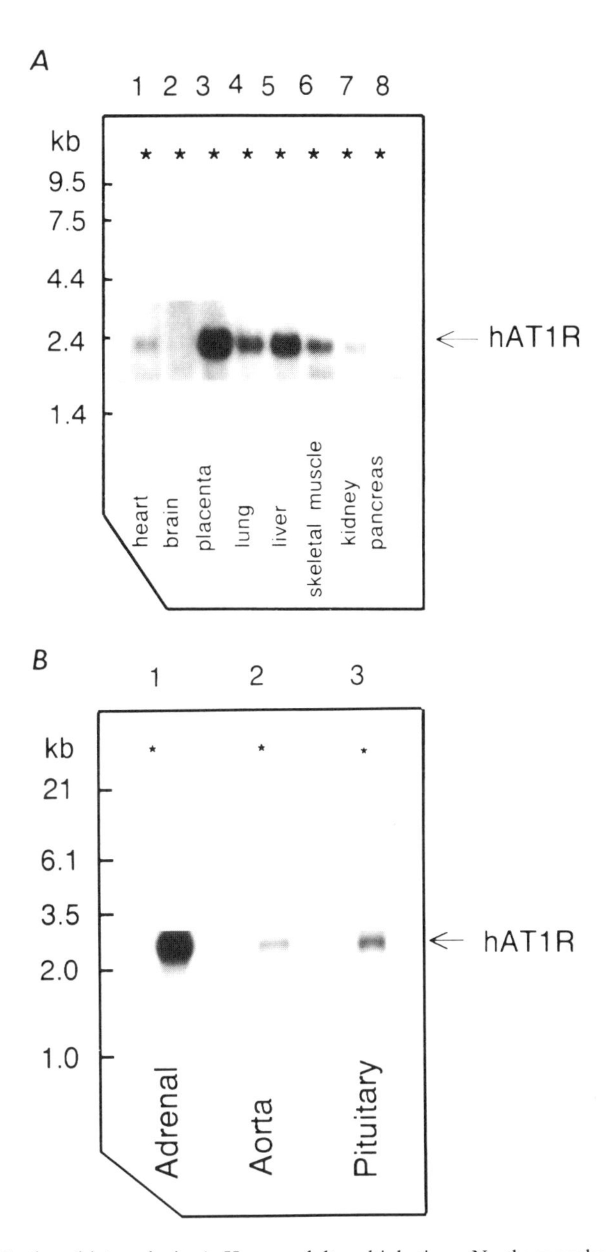

Figure 3. Human Northern blot analysis. A. Human adult multiple tissue Northern probed with a radiolabeled hAT_1R cDNA probe. Lanes 1-8 contained, in order, two μg of poly A^+ mRNA isolated from the tissues as listed. B. Human adult multiple tissue Northern probed as above. Lanes 1-3 contained, in order, five μg of poly A^+ mRNA isolated from the tissues as listed.

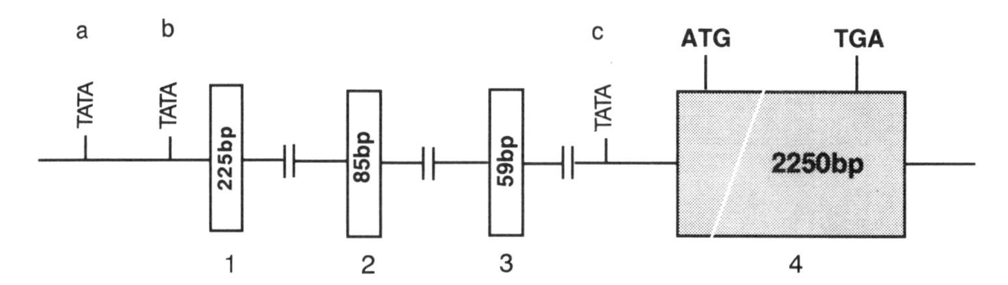

Figure 4. Schematic representation of the hAT_1R gene. Exons 1-3 code for 5′ untranslated sequences, while exon 4 harbors the entire open reading frame. Sequence analysis suggests that there are three putative TATA boxes (A-C) in this gene. (From Su et al., Biochem. Biophys. Res. Commun., 204, 1042, 1994. With permission.)

HUMAN AT_1R SUBTYPES AND ALTERNATIVE SPLICING

Although in rodents it has clearly been demonstrated that AT_1R subtypes exist (18-21), in humans this is a very controversial topic. Until recently, only a single AT_1R subtype had been reported. This conclusion was based on two independent reports. First, Southern blot analyses suggest that the human genome contains a single AT_1R gene since only one restriction fragment hybridized to the probe (29,30). Second, Curnow et al. (31) have mapped the hAT_1R gene to chromosome 3q and found no evidence of hAT_1R subtype genes on other chromosomes. In contrast to these results, Konishi et al. (28) reported on the isolation and characterization of a novel hAT_1R which they designated as $hAT_{1B}R$. This cDNA clone is greater than 97% identical at the nucleotide level when compared with other published hAT_1R cDNA clones (i.e., when comparing the open reading frame and the 3′ untranslated region (UTR) sequence). Interestingly, the 5′ UTR sequences of these clones are 80% homologous. Konishi et al. (28) speculate that the $hAT_{1B}R$ is encoded by a novel AT_1R gene. Unfortunately, the human genomic Southern data does not support this conclusion. Figure 1 shows a "ZOO" blot which is a Southern blot utilizing genomic DNA from various animal sources. This data indicates that of the animal species investigated, only rat and mouse have multiple AT_1R genes (lanes 3 and 4). Our laboratory has performed a number of human genomic Southern blots utilizing a wide variety of restriction enzymes and have never seen any evidence of multiple hAT_1R genes.

If there is only one hAT_1R gene, how can the apparent discrepancy with Konishi's results be explained? One possible explanation may be alternative splicing. The characterization of several distinct hAT_1R cDNA clones suggests that these receptors can be encoded by multiple mRNA transcripts sharing an identical open reading frame, but differing in the 5′ UTRs (29-31). Additionally, it has been demonstrated that the entire coding region of the hAT_1R is harbored on a single uninterrupted exon (32,33). Taken together, these data suggest that alternative splicing events combine various 5′ UTR exons with the exon that contains the hAT_1R coding region.

To decipher how these multiple transcripts arise and to begin to determine whether this gene has multiple promoter regions (i.e., determine whether there are hAT_1R mRNA transcripts that are comprised of distinct first exons), 5′-rapid amplification of cDNA ends (RACE) experiments were performed utilizing human lung 5′ RACE-ready cDNA (Clontech, Palo Alto, CA). Multiple PCR products were amplified, subcloned and characterized. Figure 2 summarizes our 5′-RACE results. These results suggest that the 5′ UTR of hAT_1R

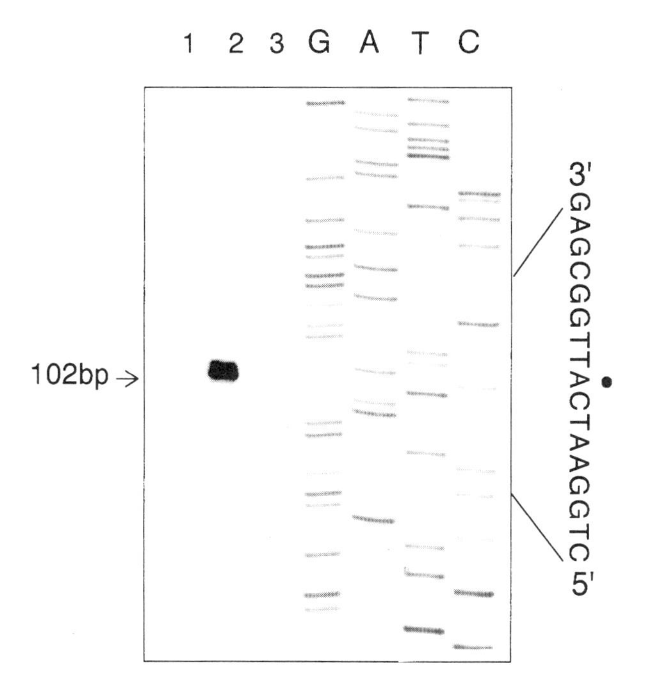

Figure 5. Primer extension analysis of the transcription start site of the hAT$_1$R gene. Lane 1: 40 μg of E. coli tRNA, lane 2: 40 μg of total RNA from human adrenal. Marker lanes G, A, T and C indicate sequencing ladder of the human gene using the same primer. Arrow indicates the extended product and the corresponding "A" nucleotide. (From Su et al., Biochem. Biophys. Res. Commun., 204, 1043, 1994. With permission.)

mRNA is comprised of at least three exons (i.e., 201 bp, 85 bp and 59 bp; now designated as exons 1, 2 and 3 with the entire coding region of the hAT$_1$R mRNA as exon 4). Our results also indicate that at least four distinct alternatively spliced hAT$_1$R mRNA transcripts are transcribed from the hAT$_1$R gene (e.g., hAT$_1$R mRNA transcripts are comprised of exons 1, 2, 3, 4 or exons 1, 2, 4 or exons 1, 3, 4 or exons 1 and 4). Importantly, all of the alternatively spliced hAT$_1$R mRNA transcripts encode for the same receptor since all of these transcripts contain exon 4 (i.e., the entire coding region). These results also suggest that, at least in

Table 1. A summary of the tissue distribution of each hAT$_1$R exon. Northern blots were sequentially probed with exon-specific probes. Data was normalized for loading and/or transferring efficiency using a human β-actin probe. (From Su et al., Biochem. Biophys. Res. Commun., 204, 1044, 1994. With permission.)

	Adrenal	Aorta	Brain	Heart	Kidney	Liver	Lung	Pancreas	Placenta	Pituitary	Skeletal Muscle
Exon 1	+++++	+		+	+	+++	+		+++++	+	+
Exon 2	+++++	+		+	+	++++	+++		+++++	+	+
Exon 3	+++++	+		+	+	+++++	+		++++	+	+
Exon 4	+++++	+		+	+	++++	+		+++++	+	+

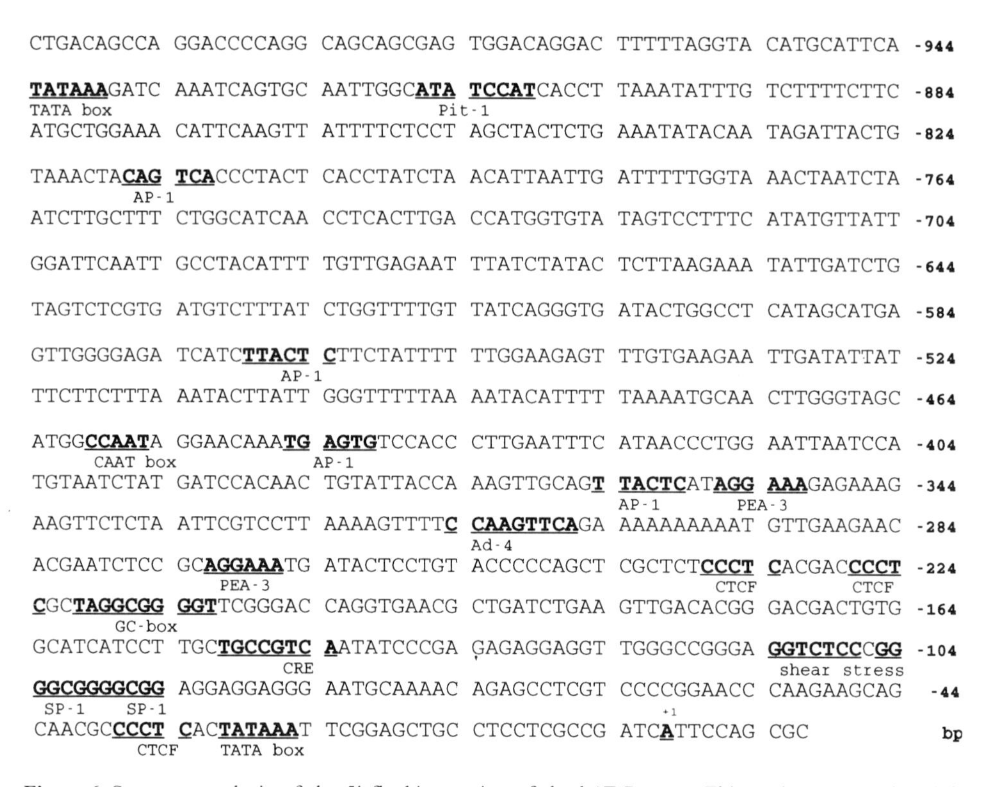

Figure 6. Sequence analysis of the 5′ flanking region of the hAT_1R gene. This region was analyzed for consensus regulatory sequences (35). Putative transcription factor recognition sites are underlined and labeled. The transcription initiation site is labeled +1 and an arrow marks the start and direction of transcription. (From Su et al., Biochem. Biophys. Res. Commun., 204, 1044, 1994. With permission).

human lung, there is only one functional promoter in the hAT_1Rgene since all hAT_1R mRNA transcripts characterized contain the same first exon.

To investigate whether the various exons were expressed in a tissue-specific manner, probes specific for each exon were generated by PCR and used as probes for Northern analysis. Figure 3 demonstrates that the hAT_1R mRNA is abundantly expressed in the adrenal, liver, lung, pituitary and placenta, when exon 4 was utilized as a probe. Identical Northern results were obtained when exon 1, 2 or 3 specific probes were used (summarized in Table 1). These data show that the individual hAT_1R 5′ UTR exons are not expressed differentially when compared to each other. However, from this data we cannot determine whether these human tissues are expressing all of the possible hAT_1R mRNA transcripts or the relative quantities of each transcript.

CLONING OF THE hAT_1R GENE

To determine the genomic organization and to continue to investigate putative promoter regions of the hAT_1R gene, a human genomic library was screened with probes specific for each exon. Four distinct clones which harbored specific exons were characterized

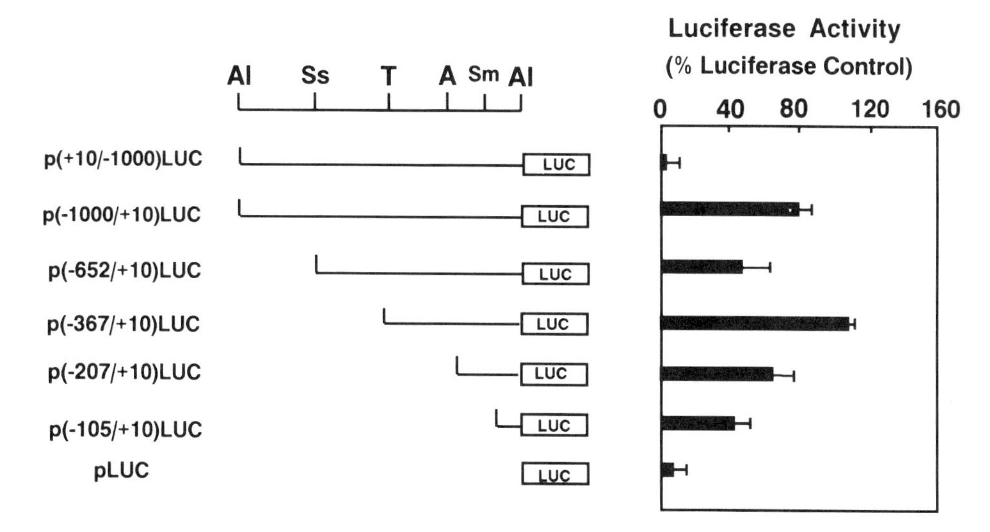

Figure 7. Expression of hAT_1R promoter/luciferase constructs in human adrenal cells. The luciferase activity levels, normalized with β-gal activities, are expressed as a percent of the positive control (SV40 promoter/enhancer construct), which is assigned a value of 100%. The mean activities and standard error of the mean from three separate transfections are shown. Al indicates Alw NI; SS, Ssp I; T, Taq I; A, Ava II; Sm, Sma I. (From Su et al., Biochem. Biophys. Res. Commun., 204, 1045, 1994. With permission.)

by Southern analysis, restriction mapping and sequencing. The genomic organization of the hAT_1R gene is shown schematically in Figure 4. The gene is comprised of at least four exons and spans at least 60 kb. Each intron encompasses at least 10 kb. Sequence analysis suggests that there are at least three putative TATA boxes (A-C). Primer extension analysis was performed to determine if TATA boxes B and/or C were functional. We utilized a 5′ end-labeled antisense primer (primer C) which anneals approximately 160 bp downstream of the third putative TATA (i.e., exon 4) and one (primer B) which anneals approximately 120 bp downstream of the second putative TATA (exon 1). No evidence of primer-mediated extension was observed with primer C when human adrenal RNA was utilized as template (data not shown). When a PCR generated DNA fragment, specific to the downstream region of the putative TATA box C region, was utilized to probe the multiple tissue Northern no hybridization signal was detected (data not shown). Furthermore, when a probe specific for the region downstream from putative TATA box A was utilized in Northern analysis, again no hybridization signal was detected (data not shown). Taken together, these data suggest that, at least in the tissues we investigated, TATA boxes A and C are not functional. In contrast, however, when primer B was hybridized to human adrenal RNA, a 102 bp band was detected (Figure 5); this major extended DNA product corresponds to the "A" residue (designated +1) located 25 bp downstream of the second TATA. Since eukaryotic polymerase II transcription generally initiates at ~25-35 bp downstream from the TATA box at an "A" nucleotide (34), our primer extension data correlates with this observation perfectly. Additionally, positive results were obtained when Northern analysis was performed utilizing a probe specific to the region downstream from TATA box B (see Table 1, exon 1). Taken

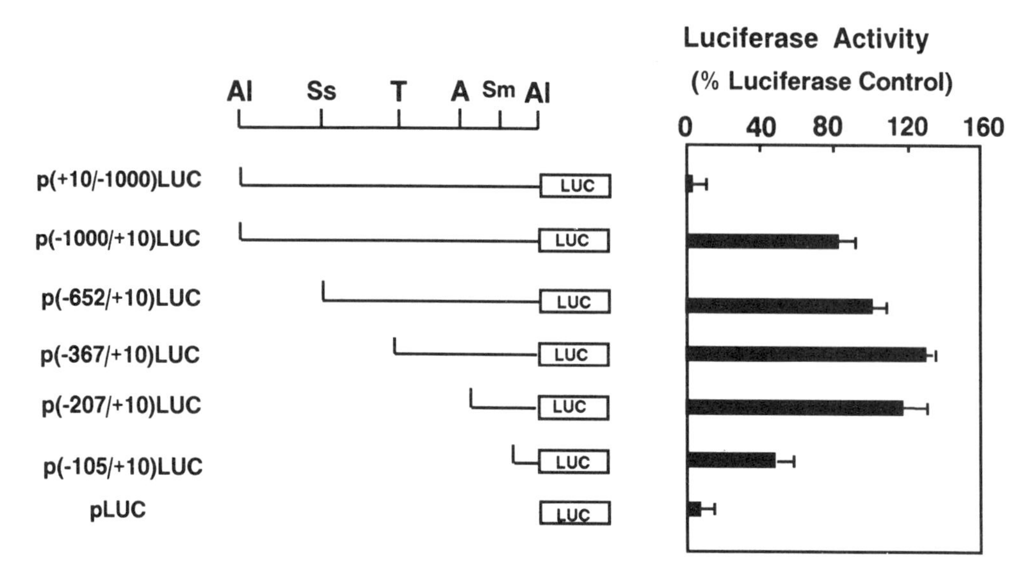

Figure 8. Expression of hAT_1R promoter/luciferase constructs in rat aortic smooth muscle cells. The luciferase activity levels, normalized with β-gal activities, are expressed as a percent of the positive control (SV40 promoter/enhancer construct), which is assigned a value of 100%. The mean activities and standard error of the mean from three separate transfections are shown. Al indicates Alw NI; SS, Ssp I; T, Taq I; A, Ava II; Sm, Sma I.

together, these results clearly demonstrate that TATA box B is functional in all the tissues investigated, suggesting againn that the hAT_1R gene has at least one functional promoter.

REGULATION OF THE hAT_1R GENE

To further characterize the promoter region upstream from exon 1, a 1000 bp Alw NI 5′ flanking region fragment was sequenced and analyzed for consensus regulatory sequences (35) (Figure 6). In addition to the TATA box (at -25 bp), several potential binding sites for transcription factors were identified. There is one binding site for the CTF (5′ CCAAT 3′ at -459 bp) transcription factor at a CCAAT-box. The CCAAT-box element seems to be related to the level of basal expression and functions in either orientation (36). There are two binding sites for the PEA-3 (5′ AGGAAA 3′ at -271 bp and -356 bp). The PEA-3 motif has been shown to mediate transcriptional activation by the c-ets-1 and -2 proto-oncogenes, v-src, phorbol ester and serum components (37,38). There is also one binding site for Pit-1 (5′ (A/T) TAT (T/C) CAT 3′ at -916 bp) transcription factor (39). The pituitary-specific transcription factor Pit-1 is a member of the POU family of transcription factors which play a critical role in both the proliferation of specific cells and their expression of specific genes that define their phenotype within a tissue (40). The hAT_1R promoter also has one cyclic-AMP inducible response element (CRE) (5′ TGCCGTCA 3′ at -150 bp) (41) and an adrenal

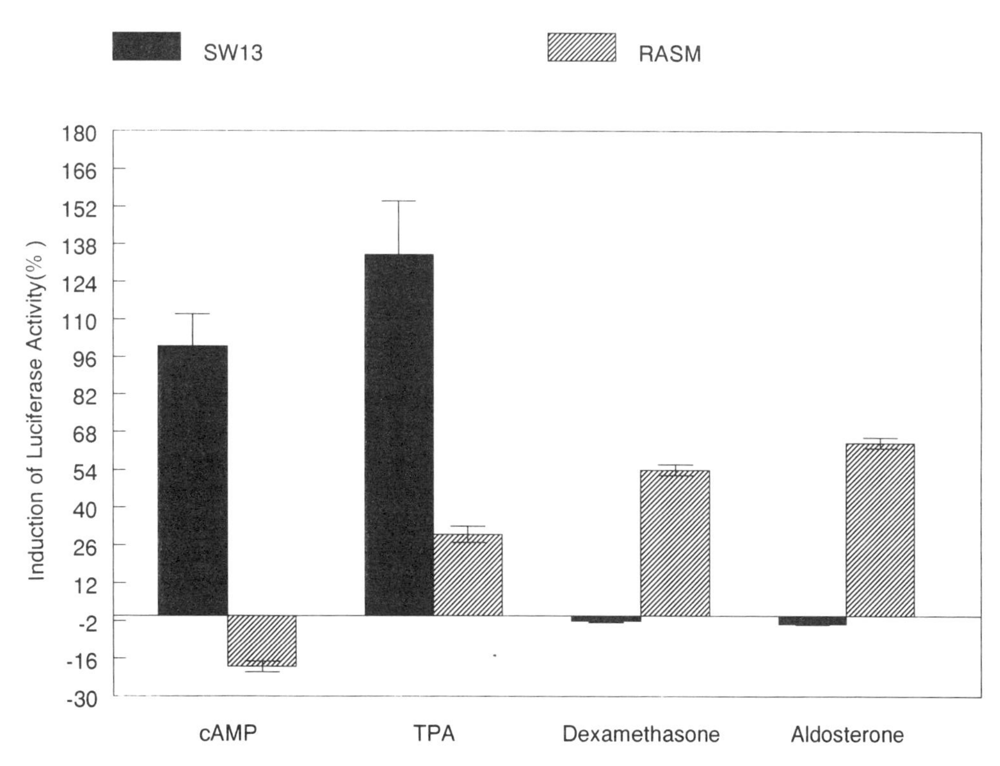

Figure 9. Effects of various regulators on luciferase activities of phAT$_1$R(-1000/+10)Luc transfected SW-13 or RASM cells. Transfected cells were treated with the regulators for 24 h. Relative luciferase activity indicates the ratio of the regulator-stimulated activity to the basal activity (N=3).

specific response element (Ad-4) (5′ CCAATTCA at -314 bp) (42). Both of these response elements are activated cAMP (41,42). Finally, there is one binding site for a shear-stress inducible transcription factor (5′ GGTCTCC 3′ at -113 bp) (43). The presence of this sequence suggests that the hAT$_1$R gene may respond to shear stress *in vivo.*.

To test whether this hAT$_1$R 5′ flanking region fragment had promoter activity, it was subcloned immediately upstream to a promoterless, enhancerless luciferase gene. This construct and several 5′ deletion promoter/luciferase constructs were transfected into SW-13 cells (i.e., a human adrenal cell line) and rat aortic smooth muscle (RASM) cells. The luciferase results (Figure 7 and 8) clearly demonstrate that this 5′ flanking region fragment has promoter activity in both cell lines. Deletion constructs phAT$_1$R(-652/+10)Luc and phAT$_1$R(-207/+10)Luc showed a pronounced decrease in luciferase activity in adrenal cells but not RASMs, suggesting that the deleted promoter sequence plays an important role in the regulation of the hAT$_1$R gene in adrenal cells only. Deletion construct phAT$_1$R(-105/+10)Luc showed a large decrease in luciferase activity in RASMs, indicating that the promoter sequence located between -207 and -105 is critical for the hAT$_1$R to be expressed in vascular smooth muscle.

To investigate whether the hAT$_1$R promoter region could differentially respond to various physiological stimuli, the largest construct, pAT$_1$R(-1000/+10)Luc was transfected into adrenal and RASM cells. Twenty-four hours after transfection, the cells were incubated with dibutyryl cAMP ([1 mM]), Phorbol 12-myristate 13-acetate (TPA) ([30 ng/ml]), dexamethasone ([1 μM]) or aldosterone ([10 μM]). Interestingly, the hAT_1R promoter region responded in a unique fashion that was dependent upon the cellular environment that the promoter was placed (Figure 9). For example, dibutyryl cAMP and TPA significantly increased luciferase activity only in adrenal cells. In contrast, only dexamethasone and aldosterone stimulated hAT_1R promoter activity in RASM cells. These results suggest that multiple transcription factors can govern the regulation of the hAT_1R gene.

CONCLUSION

Currently, it is thought that only rodents have multiple AT_1R genes. Recently, Konishi et al. (28) suggested that multiple hAT_1R genes also exist. Unfortunately, the appearance of only one hybridizable restriction fragment on genomic Southern blots suggests the presence of a single hAT_1R gene. This apparent discrepancy may result from the extensive alternative splicing that the hAT_1R heterogenous nuclear RNA undergoes (i.e., unprocessed hAT_1R mRNA). We have demonstrated that the hAT_1R can be encoded by at least four distinct mRNA transcripts. These transcripts share the same exon that encodes for the open reading frame, but differ in the organization of their 5′ UTR sequeqnces. Therefore, even though multiple alternatively spliced forms of hAT_1R are transcribed, they are all translated into identical AT_1 receptors.

By comparing the hAT_1R cDNA with the hAT_1R genomic sequence, the organization of the hAT_1R gene was determined. The hAT_1R gene is comprised of at least four exons and spans more than 60 kb. The promoter region was sequenced and analyzed for putative transcription factor binding sites. This analysis demonstrated that the hAT_1R has the capability of responding to a number of physiological stimuli. More importantly, we have demonstrated that this promoter region can respond in a distinctive manner depending upon the cellular environment that the promoter finds itself. Now that the promoter region of the hAT_1R gene has been cloned, we can continue to elucidate the mechanisms that govern the transcriptional regulation of this important gene.

ACKNOWLEDGMENTS

This work was supported by research grants NIH/HL 48848 and HL 48848-52.

REFERENCES

1. Peach, M.J. and Dostal, D.E. (1990). J. Cardiovasc. Pharmacol. *16*:525-530.
2. Phillips, M.J.A. (1987). Annu. Rev. Physiol. *49*:413-435.
3. Millan, M.A., Carvallo, P., Izumi, S.I., Zemel, S., Catt, K.J. and Aguilera, G. (1989). Science *244*:1340-1342.
4. Catt, K.J., Carson, M.C., Hausdorff, W.P., Leach-Harper, C.M., Baukal, A.J., Guillemette, G., Balla, T. and Aguilera, G. (1987). J. Steroid Biochem. *27*:915-927.
5. Hausdorff, W.P. and Catt, K.J. (1988). Endocrinology *123*:2818-2826.
6. Pobiner, B.F., Hewlett, E.L. and Garrison, J.C. (1985). J. Biol. Chem. *260*:16200-16209.
7. Douglas, J. (1987). Am. J. Physiol. *22*:F1-F7.
8. Chiu, A.T., Herblin, W.F., McCall, D.E., Ardecky, R.J., Carini, D.J., Dunica, J.V., Pease, L.J., Wong, P.C., Wexler, R.C., Johnson A.L. and Timmermans, P.B.M.W.M. (1989). *165*:196-203.
9. Whitebread, S., Mele, M., Kamber, B. and de Gasparo, M. (1989). Biochem. Biophys. Res. Commun. *163*:284-291.
10. Dudley, D.T., Panek, R.L., Major, T.C., Lu, G.H., Bruns, R.F., Klinkefus, B.A., Hodges, J.C. and Weishaar, R.E. (1990). Mol. Pharmacol. *38*:370-377.

11. Chang, R.S.L., Lotti, V.J., Chen, T.B. and Faust, K.A. (1990). Biochem. Biophys. Res. Commun. *171*:813-817.
12. Zemel, S., Millan, M.A. and Aguilera, G. (1989). Endocrinology *124*:1774-1780.
13. Grady, E.F., Sechi, L.A., Griffin, C.A., Schambelan, M. and Kalinyak, J.E. (1991). J. Clin. Invest. *88*:921-933.
14. Sasaki, K., Yamano, Y., Bardhan, S., Iwai, N., Murray, J.J., Hasegawa, M., Matsuda, Y. and Inagami, T. (1991). Nature *351*:230-233.
15. Murphy, T.J., Alexander, R.W., Griendling, K.K., Runge, M.S. and Bernstein, K.E. (1991). Nature *351*:233-236.
16. Iwai, N., Yamano, S., Chaki, S., Konishi, F., Bardhan, S., Tibbets, C., Sasaki, K., Hasegawa, M., Matsuda, Y. and Inagami, T. (1991). Biochem. Biophys. Res. Commun. *177*:299-304.
17. Takayanagi, R., Ohnaka, K., Sakai, Y., Nakao, R., Yanase, T., Haji, M., Inagami, T., Furata, H., Guo, D.F. (1992). *183*:910-916.
18. Elton, T.S., Stephan, C.C., Taylor, G.R., Kimball, M.G., Martin, M.M., Durand, J.N. and Oparil, S. (1992). *184*:1067-1073.
19. Iwai, N. and Inagami, T. (1992). FEBS Letters *298*:257-260.
20. Kakar, S., Sellers, J., Devor, D., Musgrove, L. and Neill, J. (1992). Biochem. Biophys. Res. Commun. *183*:1090-1096.
21. Sandberg, K., Ji, H., Clark, A.J., Shapira, H. and Catt, K. (1992). J. Biol. Chem. *267*:9455-9458.
22. Lewis, J.L., Serikawa, T. and Warnock, D.G. (1993). Biochem. Biophys. Res. Commun. *194*:677-682.
23. Chiu, A.T., Dunscomb, J., Kosierowski, J., Burton, C.R., Santomenna, L.D., Corjay, M.H. and Benfield, P. (1993). Biochem. Biophys. Res. Commun. *197*:440-449.
24. Conchon, S., Monnot, C., Teutsch, B., Corvol, P. and Clausser, E. (1994). FEBS Letters *349*:365-370.
25. Iwai, N., Inagami, T., Ohmichi, N., Nakamura, Y., Saeki, Y. and Kinoshita, M. (1992). Biochem. Biophys. Res. Commun. *188*:298-303.
26. Kitami, Y., Okura, T., Marumoto, K., Wakamiya, R. and Hiwada, K. (1992). Biochem. Biophys. Res. Commun. *188*:4469-452.
27. Llorens-Cortes, C., Greenberg, B., Huang, H. and Corvol, P. (1994). Hypertension *24*:538-548.
28. Konishi, H., Kuroda, S., Inada, Y. and Fujisawa, Y. (1994). Biochem. Biophys. Res. Commun. *199*:467-474.
29. Takayanagi, R., Ohnaka, K., Sakai, Y., Nakao, R., Yanase, T., Haji, M., Inagami, T., Furuta, H., Gou, D., Nakamuta, M. and Nowata, H. (1992). Biochem. Biophys. Res. Commun. *183*:910-916.
30. Bergsma, D., Ellis, C., Kumar, C., Griffin, E., Stadel, J. and Aiyar, N. (1992). Biochem. Biophys. Res. Commun. *183*:989-995.
31. Curnow, K.M., Pascoe, L. and White, P.C. (1992). Mol. Endocrin. *6*:1113-1118.
32. Furuta, H., Guo, D. and Inagami, T. (1992). Biochem. Biophys. Res. Commun. *183*:8-13.
33. Mauzy, C.A., Hwang, O., Egloff, A.M., Wu, L.H. and Chung, F.Z. (1992). Biochem. Biophys. Res. Commun. *186*:277-284.
34. Xu, L., Thali, M. and Schaffner, W. (1991). Nucleic Acids Res. *19*:6699-6704.
35. Ghosh, D. (1990). Nucleic Acids Res. *18*:1749-1756.
36. Morgan, W.D., Williams, G.T., Morimoto, R.I., Greene, J., Kingston, R.E. and Tjian, R. (1987). Mol. Cell. Biol. *7*:1129-1138.
37. Wasylyk, C., Flores, P., Gutman, A. and Wasylyk, B. (1989). EMBO J. *8*:3371-3378.
38. Wasylyk, B., Wasylyk, S., Flores, P., Beque, A.I., Leprinc, D. and Stehelin, D. (1990). Nature *346*:191-193.
39. Elsholtz, H.P., Albert, V.R., Treacy, M.N. and Rosenfeld, M.G. (1990). Genes & Dev. *4*:43-51.
40. Rosenfeld, M.G. (1991). Genes & Dev. *5*:897-907.
41. Fink, J.S., Verhave, M., Kasper, S., Tsukuda, T., Mandel, G. and Goodman, R.H. (1988). Proc. Natl. Acad. Sci. USA *85*:6662-6666.
42. Honda, S., Morobashi, K., Nomura, M., Takeya, H., Kitajima, M. and Omura, T. (1993). J. Biol. Chem. *268*:7494-7502.
43. Resnick, N., Collins, T., Atkinson, W., Bonthron, D.T., Dewey, C.F. and Gimbrone, M.A. (1993). Proc. Natl. Acad. Sci. USA *90*:4591-4595.

3

REGULATION OF GENE TRANSCRIPTION OF ANGIOTENSIN II RECEPTOR SUBTYPES IN THE HEART

Hiroaki Matsubara, Yutaka Nio, Satoshi Murasawa, Kazuhisa Kijima, Katsuya Maruyama, Yasukiyo Mori, and Mitsuo Inada

Second Department of Internal Medicine
Kansai Medical University, Fumizonocho 1
Moriguchi, Osaka 570, Japan

INTRODUCTION

The process of left ventricular remodeling after acute myocardial infarction involves alterations in the topography of both infarcted and noninfarcted ventricular regions (1). In the infarcted area, infarct expansion with regional dilation and thinning of the infarct zone occur within 1 day after myocardial infarction (2). The myocardium remote from the area of infarction is subjected to increased diastolic wall stress (2,3), resulting in myocyte slippage (3) as well as myocyte hypertrophy (4). The myocardial hypertrophy exhibits characteristics of combined pressure and volume overload (4,5).

The renin-angiotensin system is activated after acute myocardial infarction and angiotensin I-converting enzyme (ACE) inhibitors improve the clinical condition of patients after myocardial infarction (6,7). In rat models of heart failure, ACE inhibitors cause hemodynamic improvement or increase survival after long-term therapy (8,9). The mechanism(s) of action of the benefical effect of ACE inhibition has generally been attributed to afterload reduction (1). However, recent studies have suggested an important role for the renin-angiotensin system in the development of cardiac hypertrophy (10,11), and for the renin-angiotensin system in the production of myocardial fibrosis (12). Of interest, there is now evidence that cardiac hypertrophy is associated with the induction of gene expression for the ACE (13), angiotensinogen (14) and angiotensin II (AngII) receptors (10), and increased local synthesis of AngII within the ventricular myocardium (13). In addition, it has been demonstrated that mRNA or protein levels for angiotensinogen (15), cardiac ACE (16) and AngII receptor (17,18) are increased in rat hearts that have undergone remodeling after experimental myocardial infarction.

At least two main AngII receptor subtypes, AT1-R and AT2-R, have been identified by the use of receptor subtype-specific nonpeptide antagonists (19,20). Two subvariants of the AT1-R have been identified in the rat through the cDNA sequencing and are identified as AT1a-R and AT1b-R (21-24). The cDNA clone encoding AT2-R has been recently isolated

Recent Advances in Cellular and Molecular Aspects of Angiotensin Receptors
Edited by Mohan K. Raizada et al., Plenum Press, New York, 1996

and it has 32% homology with AT1a-R protein (25,26). We found that AT1-R and AT2-R expression is increased in response to the development of cardiac hypertrophy (10), and that in the rat heart, AngII receptors are predominantly expressed in cardiac fibroblasts rather than in myocytes (27). In addition, AT1a-R is dominant in cardiac fibroblasts, whereas AT1b-R gene expression is upregulated in myocytes (27). Anversa et al. found using isolated ventricular myocytes after myocardial infarction, that the expression of AT1-R, not AT2-R, is exclusively upregulated in myocytes (17,18). Since myocytes as well as cardiac fibroblasts undergo constitutive and functional remodeling after myocardial infarction, it is important to determine the changes in gene expressions of AngII-R subtypes and the effects of AngII-R antagonists on the infarcted heart. Here, we demonstrated that (1) AT1a-R and AT2-R mRNA levels are increased in rat ventricles with myocardial infarction, whereas the AT1b-R mRNA level is unaffected, (2) the gene transcriptional mechanism induces an increase in AT1a-R and AT2-R mRNAs and (3) therapy with AT1-R antagonist effectively reduces the increased expression of AT1-R and AT2-R in infarcted myocardium, while AT2-R antagonist does not affect it.

METHODS

Myocardial infarction was surgically induced in rats by ligation of the left anterior coronary artery. Quantification of AT1a-R, AT1b-R and AT2-R mRNAs was was performed by means of the competitive reverse-transcription (RT)-polymerase chain reaction (PCR). Gene transcription was evaluated by nuclear run off assay. The protein levels of AT1-R and AT2-R were determined by the antagonist-specific inhibition in the ^{125}I-AngII saturation bindings. These were described in detail in a recent publication (28).

RESULTS

Changes of Weight, Blood Pressure and Infarct Size. Body and heart weights in the untreated, TCV116-treated, and PD123319-treated rats are shown in Table 1. Although the body weight of the untreated myocardial infarction group was decreased as compared with that of the sham-operated control group, the heart weight and heart/body weight ratios were similar in both. Neither TCV116 nor PD123319 changed any of these parameters. There

Table 1. Changes of hemodynamic parameters, heart weight and infarct size in myocardial infarction

	Control	MI	TCV116	PD123319
Systolic blood pressure (mmHg)	123 ± 8	119 ± 5	95 ± 4**	108 ± 2
Heart rate (bpm)	426 ± 13	441 ± 29	470 ± 20	458 ± 15
Body weight (g)	312 ± 6	288 ± 5*	279 ± 4**	284 ± 5**
Heart weight (g)	0.75 ± 0.05	0.77 ± 0.03	0.71 ± 0.02	0.78 ± 0.02
Heart/Body weight	0.24 ± 0.01	0.26 ± 0.01	0.25 ± 0.01	0.26 ± 0.01
Infarct size (%)		38 ± 4.5	38 ± 4.2	40 ± 4.1
Numbers	5	6	6	6

All data were obtained from rats 7 days after onset of myocardial infarction. Administration of angiotensin II receptor antagonists was maintained for 7 days. MI: myocardial infarction. All data are means SE. *p, **p vs sham-operated controls

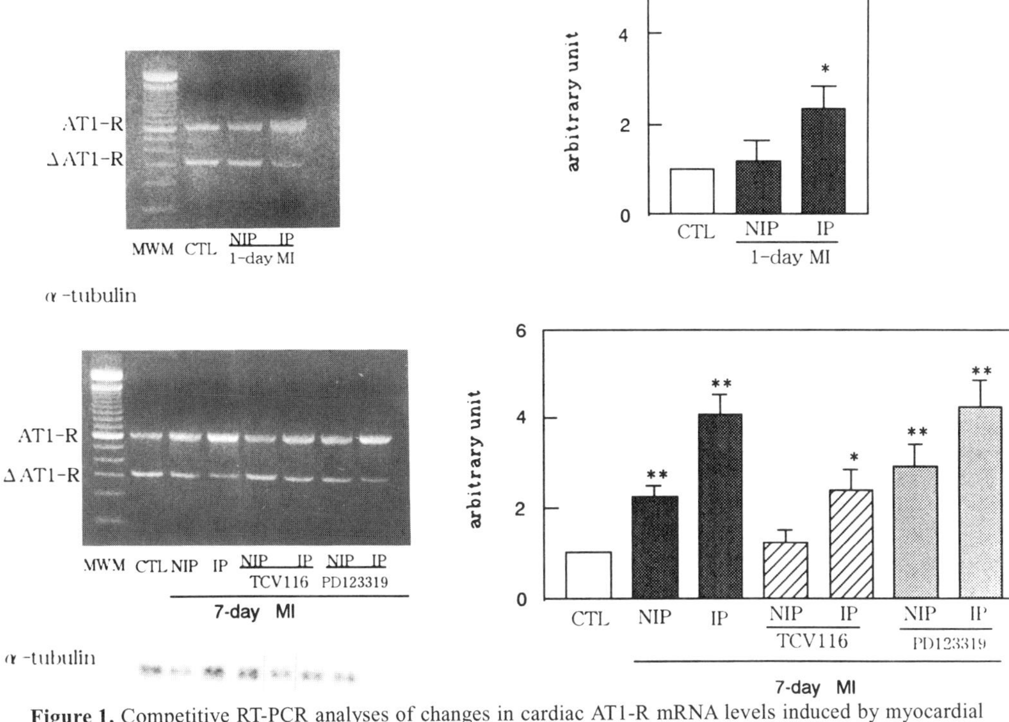

Figure 1. Competitive RT-PCR analyses of changes in cardiac AT1-R mRNA levels induced by myocardial infarction. Total RNA (1μg) was reverse-transcribed with deletion mutated cRNA (ΔAT1-R, 1pg) and the resultant cDNA mixtures were amplified by PCR in the presence of [32P]dCTP. The PCR products were loaded onto a 2% agarose gel and autoradiographic signals (7μg of total RNA) from Northern blots using α-tubulin are shown. The incorporated 32P counts in AT1-R signals were normalized to those in ΔAT1-R and α-tubulin autoradiographic counts measured using a densitometry. The normalized value in sham-operated controls is arbitrarily expressed in 1 unit. Panels A and B show data from rat hearts 1 and 7 days after myocardial infarction, respectively. TCV116 (n=6) and PD123319 (n=6) were given for 7 days after myocardial infarction. The values given are the means ± SE of separate rat ventricular samples. CTL: sham-operated controls (n=5). MI: myocardial infarction (n=6). NIP: non-infarcted portion. IP: infarcted portion. MWM: molecular weight marker. *p<0.05, **p<0.01 vs controls.

were no significant differences in infarct size among the three groups of rats with myocardial infarction (Table 1).

The heart rates in rats treated with TCV116 or PD123319 did not differ from those in the control rats. There was a significant ($p<0.01$) decrease in systolic blood pressure in TCV116-treated group in comparison with that in the control group, whereas a significant change was not evident in the PD123319-treated group.

Myocardial Infarction Causes an Increase in Cardiac AT1-R mRNA Accumulation. We examined cardiac AT1-R mRNA levels 1 and 7 days after the onset of myocardial infarction. The infarcted myocardium was dissected into the non-infarcted and the infarcted portions based upon planimetry as described in "Methods". The borderline area between non-infarct and infarct was included in the infarct area. As shown in Fig. 1A, 1-day infarction caused a 2.4-fold increase ($p<0.05$) in AT1-R mRNA levels in the infarcted portion as compared with those in sham-operated rat hearts, whereas the AT1-R mRNA levels in the non-infarcted portion did not significantly differ from those in sham-operated rats. In the 7-day infarction (Fig. 1B), the AT1-R mRNA levels in the non-infarcted portion were

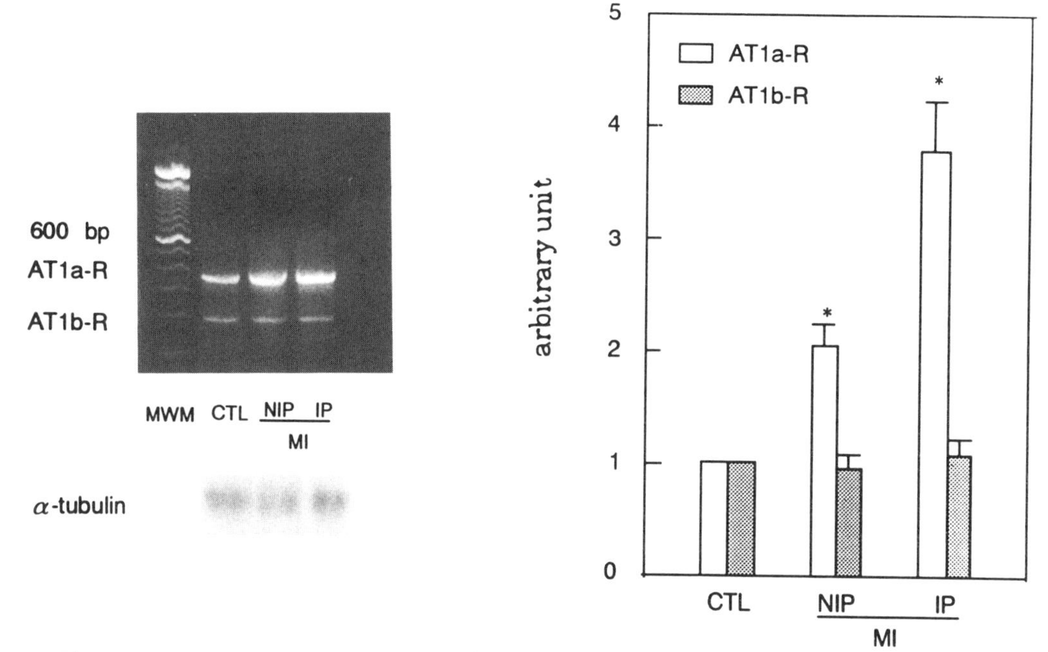

Figure 2. RT-PCR analyses of changes in cardiac AT1a-R and AT1b-R mRNA levels induced by myocardial infarction. Total RNA (1μg) was reverse-transcribed and amplified with AT1a-R and AT1b-R specific primers in the presence of [32P]dCTP. The resultant PCR products were loaded onto a 2% agarose gel. The normalized value in sham-operated controls is arbitrarily expressed as 1 unit. The values given are the means ± SE of separate rat ventricular samples. All samples were obtained from the rat heart 7 days after myocardial infarction. CTL: sham-operated controls (n=5). MI: myocardial infarction (n=6). NIP: non-infarcted portion. IP: infarcted portion. MWM: molecular weight marker. *p<0.01 vs controls

increased up to 2.2-fold in comparison with those in sham-operated controls, and the increase in the infarcted portion reached 4.2-fold.

Myocardial Infarction Induces the Accumulation of AT1a-R mRNA, But Not That of AT1b-mRNA. AT1-R mRNA levels quantified by competitive RT-PCR (Fig. 1) are composed of mRNAs from both AT1a-R and AT1b-R genes, because the PCR primers used in the assay were designed from sequences common to AT1a-R and AT1b-R cDNAs. To examine the specific regulation of these genes in response to myocardial infarction, we used PCR primers specific for AT1a-R or AT1b-R cDNA (27). As shown in Fig. 2, we found that only AT1a-R mRNA levels are increased in the infarcted heart, whereas AT1b-R mRNA levels are not significantly different from those in the sham-operated rat heart. The increased ratio of AT1a-R mRNA levels in the 7-day infarcted portion was 3.8-fold relative to those in sham-operated controls, in agreement with the increase observed in AT1-R mRNA levels (Fig. 1B). The ratio of AT1a-R mRNA to AT1a-R+AT1b-R mRNAs in the sham-operated rat heart was approximately 71±2%. In the 7-day infarcted portion, the ratio was increased up to about 93±3%.

Myocardial Infarction Stimulates AT2-R mRNA Accumulation. The cDNA sequence coding for AT2-R has been described by two groups (25,26), in which no AT2-R mRNA signal was detected in heart in Northern blot using a cDNA probe. Here, we designed the

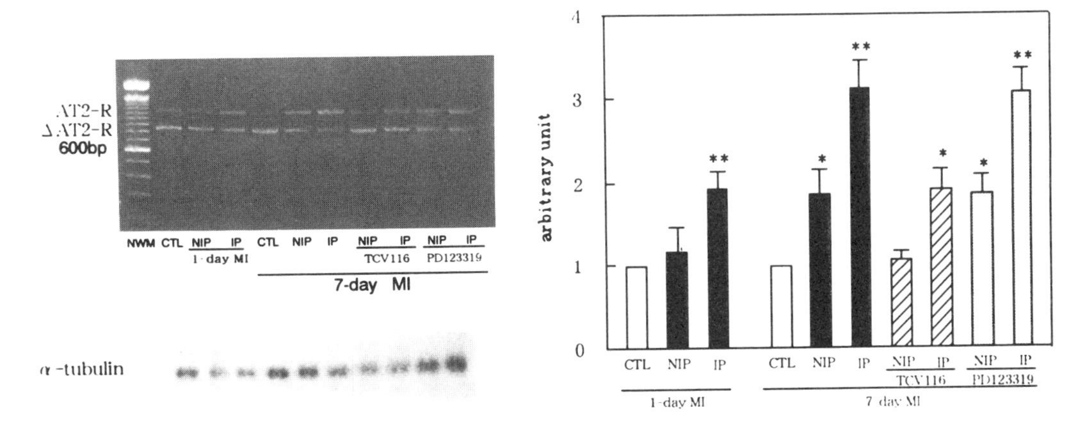

Figure 3. Competitive RT-PCR analyses for changes in cardiac AT2-R mRNA levels induced by myocardial infarction. Total RNA (3μg) was amplified with PCR primers specific for AT2-R, loaded onto a 2% agarose gel, and normalized as described in the Legend to Fig. 1. The normalized value in sham-operated controls is arbitrarily expressed as 1 unit. TCV116 (n=6) and PD123319 (n=6) were given for 7 days after myocardial infarction. The values are the means ± SE of separate rat ventricular samples. CTL: sham-operated controls (n=5). MI: myocardial infarction (n=6). NIP: non-infarcted portion. IP: infarcted portion. MWM: molecular weight marker. *p<0.05, **p<0.01 vs controls.

the PCR primers from the non-coding regions of AT2-R cDNA sequences and attempted to measure the AT2-R mRNA level in the rat heart. As shown in Fig. 3, a signal for AT2-R mRNA was detectable in the sham-operated rat heart. One-day post infarction induced a 1.9-fold increase in AT2-R mRNA levels in the infarcted portion compared with those in the controls. However, there was no significant changes in the non-infarcted portion. In the 7-day infarction, AT2-R mRNA levels in both non-infarcted and infarcted portions were significantly increased. The increased ratios relative to the AT2-R mRNA level of the sham-operated control were 1.8 and 3.1-fold in the non-infarcted and infarcted portions, respectively (Fig. 3). The total RNA was extracted by means of CsCl centrifugation followed by RNase-free DNase digestion, and no message was obtained in the RT-PCR assay in the absence of reverse-transcriptase, suggesting that contamination by genomic DNA was negligible. In addition, partial sequencing of the PCR product confirmed that the amplified product was the AT2-R transcript (data not shown).

An AT1-R, But Not the AT2-R Antagonist Reduces the Increased Level of AT1-R and AT2-R mRNAs. Treating infarcted rats with the AT1-R antagonist TCV-116, resulted in a reduction in systolic arterial blood pressure, whereas the AT2-R antagonist, PD123319, did not affect the blood pressure (Table 1). We reported that antihypertensive therapy with TCV-116 normalizes the increased level of AT1-R mRNA in hypertrophic rat ventricles together with the complete regression of hypertrophy (10). Here, we treated the animals with TCV-116 or PD123319 for 7 days after myocardial infarction and examined the effects on the AT1-R and AT2-R mRNA levels. As shown in Fig.1B, TCV-116 reverted the increase in AT1-R mRNA levels in the non-infarcted portion to the control level, while in the infarcted portion they were significantly (p<0.05) decreased in comparison with those in the untreated infarcted heart. However, they did not revert to the levels in sham-operated controls. On the other hand, PD123319 did not significantly change the AT1-R mRNA levels in either the infarcted or non-infarcted portions as compared with those in the controls. Similar findings were also evident in the AT2-R mRNA levels; exposure to TCV-116 for 7 days normalized the increased levels of AT2-R mRNA in the non-infarcted portion alone, but PD123319 did not (Fig. 3).

Myocardial Infarction Stimulates Gene Transcription of AT1a-R and AT2-R Genes. The effect of myocardial infarction on AT1a-R and AT2-R gene transcription was assessed by the nuclear run-off assay. As shown in Fig. 4, the transcriptional rate of AT1a-R gene relative to that of ∝-tubulin gene was increased 5.8-fold in the infarcted portion of the 7-day infarcted heart as compared with that in sham-operated control, whereas that of AT1b-R gene was unchanged, confirming the data shown in Fig. 2. The rate of transcription of ∝-tubulin gene was not altered in the infarcted heart. The AT1a-R probe used in the run-off assay was designed from the 3'non-coding region of the AT1a-R gene and was less likely to hybridize with the AT1b-R gene transcripts. In fact, when the AT1b-R probe was used in the run-off assay, no significant signals were detected in either the infarcted or control myocardium (Fig. 4). No significantly hybridized signal for AT2-R transcript was detected in sham-operated heart (background level) possibly because of the low transcription level of the AT2-R gene in the rat heart, whereas a detectable band was obviously increased (4.6-fold) in the infarcted heart (Fig. 4). The increases in the gene transcriptional levels (5.8-fold in AT1a-R and 4.6-fold in AT2-R) were slightly higher than those of the mRNA determined by RT-PCR (4.2-fold in AT1a-R and 3.2-fold in AT2-R). Although changes in the half-lives of AT1a-R and AT2-R mRNAs were not measured in this study, these results suggest that the gene transcriptional mechanism plays an important role in the increased mRNA levels of AT1a-R and AT2-R genes in myocardial infarction, and furthermore, the half-lives of these mRNAs may be altered under this pathological state.

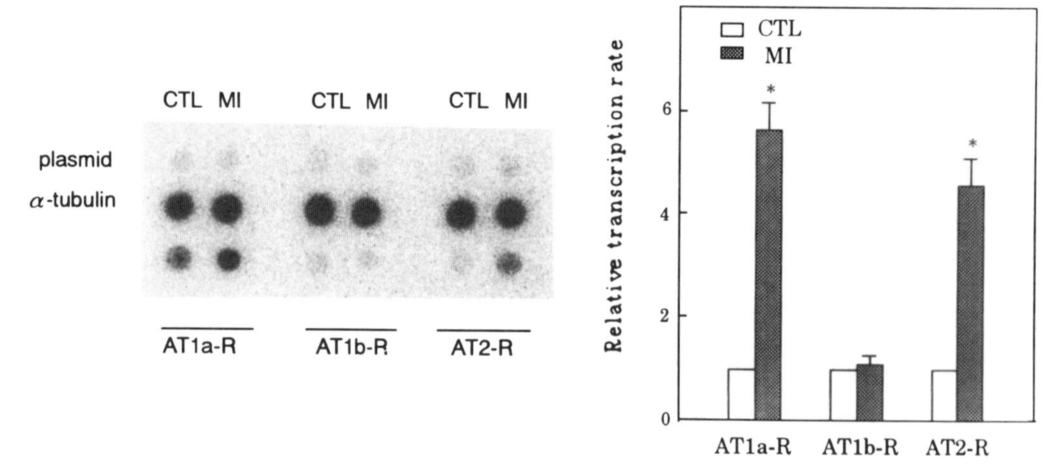

Figure 4. Nuclear run-off analyses of changes in cardiac AT1a-R, AT1b-R and AT2-R transcriptional levels induced by myocardial infarction. The nuclei were isolated and incubated in the presence of [32P]UTP as described in Methods. The [32]RNA was isolated and hybridized to pBluescript II KS(-) DNAs (5μg/dot) containing AT1a-R, AT1b-R, AT2-R, ∝-tubulin or pBluescript II KS(-) plasmid alone. The transcription rates were expressed relative to the ∝-tubulin transcription rate and the values in sham-operated controls (CTL, n=5) were normalized to 1 arbitrary unit. The values are the means ± SE of separate rat ventricular samples. Myocardial infarction (MI, n=6) samples were obtained from infarcted portion 7 days after onset of myocardial infarction. *p<0.01 vs CTL.

Myocardial Infarction Induces an Increase in AT1-R and AT2-R Protein Expression. As shown in Fig. 5, the AngII receptor density estimated by the Bmax (35±4 fmol/mg protein) in the infarcted myocardium increased 2.9-fold as compared with those in the sham-operated controls (12±2 fmol/mg protein). To characterize the AT1-R and AT2-R subtypes, competition binding was performed using their respective antagonists (10,27). The proportion of receptor subtypes were 62±3% for AT1-R and 39±2% for AT2-R in the ventricles of sham-operated rats. In the infarcted myocardium, the expression of both AT1-R (3.3-fold) and AT2-R (2.3-fold) was significantly increased as compared with those of the controls; the subtype proportions were changed to 70±5% for AT1-R and 32±3% for AT2-R.

Exposure to TCV116 for 7 days effectively reduced the increased density of the AngII receptor and subtypes ($p<0.05$ compared with those in the untreated infarcted heart). However, statistical analyses showed that the reduced levels were significantly higher than those in the sham-operated controls. On the other hand, the treatment with PD123319 did not affect the increased density of AT1-R and AT2-R. These results indicated the effectiveness of the AT1-R antagonist in inhibiting myocardial infarction-induced increases in cardiac AngII receptor expression.

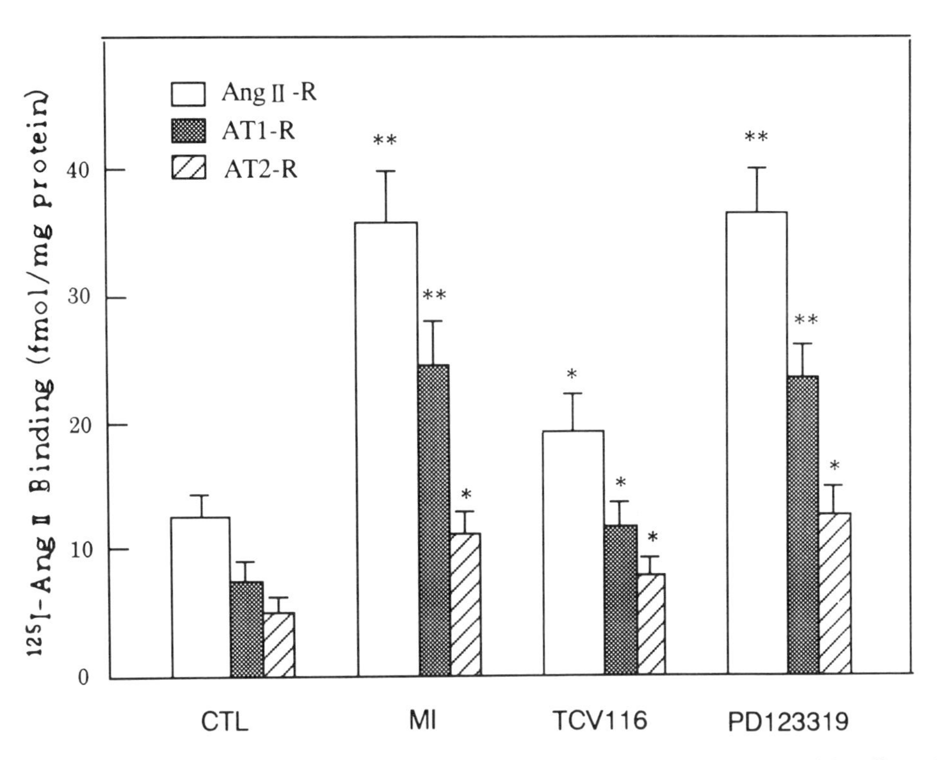

Figure 5. Changes in cardiac AngII receptor densities induced by myocardial infarction and the effects of TCV116 or PD123319. AngII receptor densities were calculated from Scatchard analyses of the competitive binding curves. Specific 12@I-AngII binding sensitive to TCV116 or PD123319 was estimated as AT1-R or AT2-R, respectively. The membrane fraction was prepared from tissue samples including both non-infarcted and infarcted left ventricles 7 days after myocardial infarction. The values are the means ± SE of separate rat ventricular samples. CTL: sham-operated controls (n=5), MI: myocardial infarction (n=6). *$p<0.05$, **$p<0.01$ vs CTL.

DISCUSSION

We examined the gene regulation of AngII receptor subtypes (AT1a-R, AT1b-R and AT2-R) in rat hearts with myocardial infarction and the effects of AngII receptor antagonists on their gene expression. The major findings of the present study were as follows: 1) myocardial infarction causes increases in mRNAs and proteins encoding cardiac AT1a-R and AT2-R, whereas AT1b-R expression is not affected, 2) induction of gene transcriptional activity plays an important role in the increased accumulation of AT1a-R and AT2-R mRNAs, 3) therapy with AT1-R antagonist effectively reduced the increased expression of AT1a-R and AT2-R, whereas therapy with AT2-R antagonist had no effect on these levels. Anversa et al. have reported that acute myocardial infarction enhances the message of the AngII receptor (AT1a-R) in the surviving myocytes, in which the gene regulation of AT1b-R and AT2-R was not analyzed (4,5). Our present data with respect to AT1a-R mRNA levels are in agreement with their results. Previous evidence indicates that the increased DNA synthesis levels in interstitial cells after myocardial infarction was inhibited by an ACE inhibitor independent of its effect on afterload changes (29), and that AngII causes fibrosis and increased collagen deposition in the cardiac interstitium, the effect of which is prevented by an ACE inhibitor (30). Sadoshima and Izumo have reported that AngII has a mitogenic effect on cardiac fibroblasts and induces a hypertrophic change on cardiomyocytes (31). Baker et al. also found that AngII is mitogenic in cardiac fibroblasts (32). These findings suggest that cardiac fibroblasts are a target of AngII, and that AngII plays a role in the structural remodeling of the cardiac interstitium after infarction by exerting a proliferative effect on fibroblasts.

Recent evidence (40) including ours (27) has shown that the expression of AngII receptor is more predominant in cardiac fibroblasts of rats than in myocytes. In addition, we found that AT1a-R is exclusively present in cardiac fibroblasts of neonatal rats, whereas in myocytes the ratio of AT1b-R expression is higher than that of AT1a-R (27). This study demonstrate that the gene expression of AT1a-R subtype is increased in the myocardium with infarction, whereas the AT1b-R mRNA level is unchanged. Anversa et al. reported that acute myocardial infarction causes a reactive hypertrophy of viable myocytes in which AT1-R expression is significantly increased, while in viable fibroblasts the amount of AT1-R gene transcripts does not change (17,18). We also showed that cardiac hypertrophy induces an increase in mRNA and protein levels of cardiac AT1a-R in the experimental rat model (10). These findings suggest that the overexpression of AT1-R in myocardial infarction can be ascribed to the enhancement of AT1a-R gene expression, whereas the AT1b-R gene that is expressed mainly in myocytes is in the static state even under such conditions and it does not contribute to the changes of AngII receptors observed in the infarcted heart. Thus, although AT1b-R exhibits high similarity to AT1a-R in amino acid sequence (95% identity), the binding of AngII analogues and utilization of Ca^{++} as an intracellular second message (23,24), the regulatory mechanism between AT1a-R and AT1b-R genes appears to be quite distinct. Indeed, we found that the expression of AT1a-R gene in cardiac fibroblasts is stimulated by glucocorticoid, whereas that of AT1b-R gene is unresponsive (27).

In contrast to the AT1-R genes, much less was known about the structure and function of the AT2-R. Mukoyama et al. (25) and Kambayashi et al. (26) have described the cDNA sequences for AT2-R, in which Northern blots of poly(A)$^+$ RNA from rat hearts did not generate detectable AT2-R transcripts. Here, RT-PCR using total RNA prepared from rat hearts, revealed a detectable transcript for the AT2-R gene. Because no message was detectable in the RT-PCR assay in the absence of reverse-transcriptase, and since RNA was extracted with CsCl centrifugation followed by DNase digestion, contamination of the genomic DNA was considered negligible. Partial sequencing of the PCR product confirmed that the amplified product is the AT2-R transcript. Our ligand binding studies, using membrane fraction (10) or viable myocytes prepared from rat hearts (27), proved the presence of AT2-R protein, which is consistent with the observation in the rabbit heart (33). Since previous studies (25,26) have used Sprague-Dawley rats and we used Wistar rats, the mRNA levels for cardiac AT2-R in the static state may be distinct in rat species or other animals species.

As shown in Figs. 3 and 5, the process of cardiac remodeling after myocardial infarction induces not only AT1a-R, but also AT2-R expression at the mRNA and protein levels. The AT1-R mediates many of the biological responses hitherto attributed to AngII (34). Kambayashi et al. have identified the ability of AT2-R to inhibit protein phosphotyrosine phosphatase activity through a pertussis toxin-sensitive G-protein (26). The abundant expression of AT2-R has been found in the mesenchymal tissues of a developing rat fetus (35) and in rat aortic smooth muscle cells during embryonic and immediate postnatal development (36), indicating an important role of AT2-R in growth and development. A sudden occlusion of a major coronary artery leads to an acute loss of contractile function in the supplied myocardium and to a redistribution of cardiac loading on the remaining viable tissue (1). This mechanical stimulus has been coupled with reactive hypertrophic process in surviving myocytes (17,18). Since most of the AT2-R is present in cardiac myocytes, rather than in cardiac fibroblasts (27), it is suggested that the induction of cardiac AT2-R expression observed in myocardial infarction is accompanied with the reactive hypertrophic process of surviving cardiac myocytes. The increase of AT2-R expression in the hypertrophic myocardium of experimental hypertensive rats is also in good agreement with this contention (10). Although the role of AT2-R in the physiological and pathological states of the heart remains to be determined, these findings indicate that the signal of AngII mediated through the increased AT2-R contributes to the remodeling process of the heart after myocardial infarction.

The results of the nuclear run-off transcription assay indicated that the increase in AT1a-R and AT2-R mRNA levels after myocardial infarction is involved in an increase in the reactive transcription rate of the genes. However, the increased rates in AT1a-R and AT2-R gene transcriptions appear to be higher than those of the mRNA levels determined by RT-PCR analysis. Although changes in the half-lives of AT1a-R and AT2-R transcripts could not be measured in the present study, the results suggested that gene transcription as well as mRNA turnover are enhanced in the infarcted myocardium.

In conclusion, this study demonstrated myocardial infarction induced increases in AT1a-R and AT2-R gene transcriptions and the inhibitory effect of an AT1-R antagonist on the increased receptor expression, suggesting that a direct action of AngII, mediated through its increased receptor expression, is involved in the remodeling process after myocardial infarction via a hypertrophic action upon myocytes and a mitogenic action upon cardiac fibroblasts. Further studies are needed to elucidate the exact molecular mechanism(s) responsible for the increased expression of AngII receptor subtype genes in the infarcted heart.

REFERENCES

1. Pfeffer, M.A. and E. Braunwald. (1990). Circulation. *81*: 1161-1172
2. Weisman, H.F., D.E. Bush, J.A. Mannisi, and G.H. Bulkley. (1985). J. Am. Coll. Cardiol. *5*: 1355-1362
3. Olivetti, G., J.M.Capasso, and E.H. Sonnenblick. (1990). Circ. Res. *67*: 23-34
4. Anversa, P., A.V. Loud, V. Levicky, and G. Guideri. (1985). Am. J. Physiol. *248*: H876-H882
5. Grossman, W.,D. Jones, and L.P. McLaurin. (1975). J. Clin. Invest. *56*: 56-64
6. Pfeffer, M.A., G.A. Lamas, D.E. Vaughan, A.F. Parisi, and E. Braunwald. (1988). N. Engl. J. Med. *319*: 80-86

7. Pfeffer, M, E. Braunwald, L. Moye, et al. (1992). N. Engl. J. Med. *327*: 669-677
8. Eaton, L.W., J.L. Weiss, B.H. Bulkley, J.B. Garrison, and M.L. Weisfeldt. (1979). N. Engl. J. Med. *300*: 57-62
9. Erlebacher, J.A., J.L. Weiss, M.L. Weisfeldt, and B.H. Bulkley. (1984). J. Am. Coll. Cardiol. *4*: 201-208
10. Suzuki, J., H. Matsubara, M. Urakami, and M. Inada. (1993). Circ. Res. *73*: 439-447
11. Matsubara, H., J. Yamamoto, Y. Hirata, Y. Mori, S. Oikawa, and M. Inada. (1990). Circ. Res. *66*: 176-184
12. Brilla, C.G., R. Pick, L.B. Tan, J.S. Janicki, and K.T. Weber. (1990). Circ. Res. *67*: 1355-1364
13. Schunkert, H., V.J. Dzau, S.S. Tahn, A.T. Hirsch, C.S. Apstein, and B.H. Lorell. (1990). J. Clin. Invest. *86*: 1913-1920
14. Baker, K.M., M.I. Chernin, S.K. Wixson and J.F. Aceto. (1990). Am. J. Physiol. *259*: H324-H332
15. Lindpainter, K., L. Wenyan, N. Niedermajer, B. Schieffer, H. Just, D. Ganten and H. Drexler. (1993). J. Mol. Cell. Cardiol. *25*: 133-143
16. Hirsch, A.T., C.E Talsness, H. Schunkert, M. Paul and V.J. Dzau. (1991). Circ. Res. *69*: 475-482
17. Meggs, L.G., J. Coupet, H. Huang, W. Cheng, P. Li, J.M. Capasso, C.J. Homcy and P. Anversa. (1993). Circ. Res. *72*: 1149-1162
18. Reiss, K., J.M. Capasso, H. Huang, L.G. Meggs, P. Li and P. Anversa. (1993). Am. J. Physiol. *264*: H760-H769.
19. Chiu, A.T., W.F. Herblin, D.E. McCall, R.J. Ardecky, D.J. Carini, J.V. Duncia, L.J. Pease, P.C. Wong, R.R. Wexler, A.L. Johnson, and P.B.M.W.M. Timmermans. (1989). Biochem. Biophys. Res. Commun. *165*: 196-203
20. Wong, P.C., S.D. Hart, A.M. Zaspel, A.T. Chiu, R.J. Ardecky, R.D. Smith, and P.B.M.W.M. Timmermans. (1990). J. Pharmacol. Exp. Ther. *255*: 584-592
21. Murphy, T.J., R.W. Alexander, K.K. Griendling, M.S. Runge, and K.E. Bernstein. (1991). Nature (Lond). *351*: 233-236
22. Sasaki, K., T. Yamano, S. Bardham, N. Iwai, J. Murray, M. Hasegawa, Y. Matsuda and T. Ingami. (1991). Nature (Lond). *351*: 230-233
23. Iwai, N. and T. Inagami. (1991). FEBS Lett. *177*: 299-304
24. Kakar, S.S., J.C. Sellers, D.C. Devor, L.C. Musgrove and J.D. Neill. (1992). Biochem. Biophys. Res. Commun. *183*: 1090-1096
25. Mukoyama, M., M. Nakajima, M. Horiuchi, H. Sasamura, R.E. Pratt, and V.J. Dzau. (1993). J. Biol. Chem. *268*: 24539-24542
26. Kambayashi, Y., S. Bardhan, K. Takahashi, S. Tsuzuki, H. Inui, T. Hamakubo, and T. Inagami. (1993). J. Biol. Chem. *268*: 24543-24546
27. Matsubara, H., M. Kanasaki, S. Murasawa, Y. Tsukaguchi, Y. Nio and M. Inada. (1994). J. Clin. Invest. *93*: 1592-1601
28. Nio, Y., H. Matsubara, S. Murasawa, M. Kanasaki and M. Inada. (1995). J. Clin. Invest. *95*: 46-54
29. van Krimpen, C., J.F.M. Smits, J.P.M. Cleutjens, J.J.M. Debets, R.G. Schoemaker, H.A.J. Struyker, F.T. Bosman and M.J.A.P. Saemen. (1991). J. Mol. Cell. Cardiol. *23*: 1245-1253
30. Weber, K.T. and C.G. Brilla. (1991). Circulation. *83*: 1849-1865
31. Sadoshima, J., and S. Izumo. (1993). Circ. Res, *73*: 413-423
32. Schorb, W., G.W. Booz, D.E. Dostal, K.M. Conrad, K.C. Chang, and K.M. Baker. (1993). Circ. Res. *72*: 1245-1254
33. Rogg, H., A. Schmid, and M. de Gasparo. (1990). Biochem. Biophys. Res. Commun. *173*: 416-422
34. Timmermans, P.B.M.W.M., P.C. Wong, A.T. Chiu, W.F. Herblin, P. Benfield, D.J. Carini, R.J. Lee, R.R. Wexler, J.A.M. Saye, and R.D. Smith. (1993). Pharmacol. Rev. *45*: 205-251
35. Grady, E.F., L.A. Sechi, C.A. Griffen, M. Schambelan, and J.E. Kalinyak. (1991). J. Clin. Invest. *88*: 921-933
36. Vinswanathan, M.K., K. Tsutsumi, F.M.A. Correa, and J.M. Saavedra. (1991). Biochem. Biophys. Res Commun. *179*: 1361-1367

4

SODIUM INDUCED REGULATION OF ANGIOTENSIN RECEPTOR 1A AND 1B IN RAT KIDNEY

Donna H. Wang, Yong Du, and Aqing Yao

Department of Internal Medicine
Hypertension and Vascular Research
University of Texas Medical Branch
Galveston, Texas 77555-1065

INTRODUCTION

Two AT1 receptor subtypes, AT1A and AT1B, have been cloned and sequenced in the rat.[1-5] Although AT1A and AT1B share a high degree (96%) of amino acid homology,[3,5] they may be functionally different. In studies of AT_{1A} and AT_{1B} receptors transfected into COS-7 cells,[4] Sanderberg et al have shown that there are dose-related differences during angiotensin II (Ang II)-induced Ca^{2+} signalling and differences in binding affinity for Ang II agonists and antagonists between AT1A and AT1B. Although it has been noted that AT1 receptor subtypes in the rat brain can be differentially regulated by hormones or changes in dietary sodium,[5,6] gene regulation of the AT1 receptor subtypes in the kidney has not been fully investigated. Considering the key role of the kidney in the control of fluid homeostasis and blood pressure, it is important to understand the regulation of these two distinct receptor genes in the kidney. Such studies could lead to further insights into the relevance of these receptor genes in hypertensive disease states and possibly could lead to the development of new and more specific therapies. In this study, we used Northern blot and *in situ* hybridization analysis to test the hypothesis that dietary sodium intake differentially regulates gene expression of the AT_{1A} and AT_{1B} receptor subtypes in the kidney.

METHODS

Treatment Groups and Tissue Preparation. Seven-week-old male Wistar rats weighing between 150-200g were randomly divided into two groups and pair-fed a low (0.07%, n=9) or a normal (0.5%, n=9) sodium diet for 14 days. The rat was then anesthetized and the left kidney was removed, frozen in liquid nitrogen and processed for RNA extraction. The right kidney was perfusion-fixed and harvested for the *in situ* hybridization studies.

Recent Advances in Cellular and Molecular Aspects of Angiotensin Receptors
Edited by Mohan K. Raizada et al., Plenum Press, New York, 1996

cDNA Probes for Northern Blot Analysis. Because AT_{1A} and AT_{1B} cDNAs exhibit a minimun sequence homology in their 5′ and 3′ untranslated regions, a 235-bp (+1142-+1377) fragment derived from 3′ non-coding region of AT_{1A} cDNA (a generous gift from Dr. Tadashi Inagami, Vanderbilt University, Nashville)[1] was used as a template for making AT1A cDNA probes. A 3′ non-coding region fragment (395 bp, +1246-+1641) of the rat AT1B cDNA (a generous gift from Dr. Tadashi Inagami)[3] was used as a template for making AT_{1B} cDNA probes. AT1A and AT1B cDNA probes were labeled with ^{32}P-dCTP and used in Northern blot analysis.

cRNA Probes for in Situ Hybridization. A 395 bp AT1B cDNA template described above was subcloned into the Hind III and EcoR I site of pGEM-3Z. Transcription from the T7 promoter (digested with Hind III) yielded the antisense probe. Transcription from the SP6 promoter (digested with EcoR1) yielded the sense probe. Digoxigenin-labeled cRNA probes were synthesized using *in vitro* transcription.

RNA Extraction and Northern Blots. Total kidney RNA was extracted using the guanidine thiocyanate-phenol-chloroform extraction protocol.[7] Thirty μg of denatured RNA from each preparation was used for electrophoresis. RNA was transferred to a positively charged nylon membrane and hybridized with the ^{32}P-labeled probes. The membrane was washed successively in 2x, 1x and 0.5x SSC containing 0.1% SDS at 65°C. Blots were exposed to X-ray film and autoradiographic signals were scanned with a laser densitometer.

In Situ Hybridization. Five micron tissue sections from the paraffin embedded kidney were cut, mounted on superfrost-plus slides. After deparaffinization, rehydration, and prehybridization, tissue sections were hybridized by addition of 300 ng/ml of digoxigenin-labeled AT1B cRNA probes at 55°C overnight. After post-hybridization washes, immunological detection was accomplished with an anti-digoxigenin antibody conjugated to alkaline phosphatase, followed by calorimetric reaction.

Quantitative Image Analysis. Using a image analysis system, the positive staining was automatically marked and converted to optical area units by the computer, and the percent positive staining was calculated as positive stained area divided by circumscribed area multiplied by 100%.

Statistical Analysis. Results were expressed as mean ± SEM. The data from the normal and low sodium treated groups was analyzed by Students unpaired *t* test. Differences were considered statistically significant at the $p<0.05$ level.

RESULTS

The body weight (BW), mean arterial pressure (MAP), plasma renin activity (PRA) and urinary sodium of the rats are listed below:

	BW(g)	MAP(mmHg)	PRA(ng AI/h/ml)	Urinary Na^+(mmol)
Control	253±5	116±6	28.8±3.7	218.5±11.6
Low Na^+	250±4	125±4	27.0±4.3	60.8±10.3*

* $p<0.05$ vs control.

Northern blot analysis of renal AT_{1A} mRNA content of the low sodium treated and control rats is shown in Figure 1. Densitometric analysis indicated that the ratio of AT1A mRNA-to-18s rRNA was significantly elevated by sodium restriction (1.28±0.11) compared with control (0.69±0.11).

Since no AT1B mRNA was detected by Northern blot analysis, we performed *in situ* hybridization analysis of AT1B mRNA in the kidney of the rats fed a low sodium and a normal sodium diet (Figure 2). AT_{1B} mRNA was expressed in the proximal and collecting tubules and glomeruli of the renal cortex, and in the collecting tubules of the renal medulla in the rats fed a normal sodium diet (Figure 2). Although AT_{1B} mRNA was still detectable after the treatment (Figure 2), the low sodium diet markedly attenuated AT_{1B} mRNA expression in the both cortex and medulla of the kidney. Quantitative image analysis indicated that low sodium treatment significantly decreased the percent positive staining area of AT1B mRNA in the renal cortex (2.73±0.35% vs the control 5.51±0.77%, $p<0.05$) and medulla (2.01±0.43% vs the control 4.76±0.70%, $p<0.05$).

DISCUSSION

The major new findings of these studies are that the message for the AT_{1A} subtype is far more abundant than for AT_{1B} subtype, suggesting that the AT_{1A} is the predominant renal subtype of the AT_1 receptor, and that sodium depletion increased renal AT_{1A} mRNA levels but decreased AT_{1B} mRNA content, indicating that there is differential regulation of two receptor subtypes.

A strength of the present study is the fact that we used probes from the 3′-untranslated regions of the cDNAs encoding the AT_{1A} and AT_{1B} receptor subtypes which allowed us to distinguish between mRNA expression for the two receptor subtypes. Using these probes we found that the AT_{1A} is the predominant subtype of the AT_1 receptor in the kidney. This conclusion is based on the fact that AT_{1A} mRNA was readily detected with Northern blot analysis whereas AT_{1B} could only be detected with the more sensitive *in situ* hybridization. The results from *in situ* hybridization indicate that AT_{1B} mRNA is present in the tubules and glomeruli of the kidney in rats fed with a normal diet. The localization of AT_{1B} mRNA appears to be different from that of the AT_{1A}. It has been reported that AT_{1A} mRNA localized in mesangial areas, predominantly at the vascular pole[8]. The differential distribution of AT_1 receptor subtypes may suggest that the AT_{1A} and AT_{1B} play different roles in the kidney under physiological conditions.

Our finding that there is an inverse relationship between the expression of the AT_{1A} and AT_{1B} in the kidney in response to a low sodium diet deserves comment. These data

LS **NS**

AT_{1A}

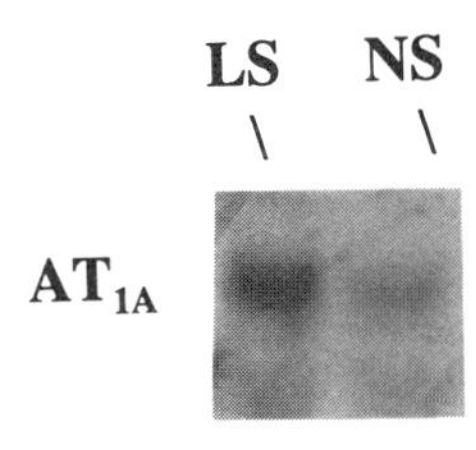

18S

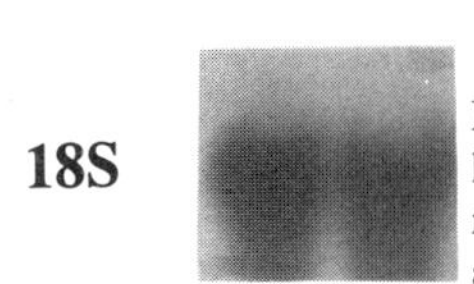

Figure 1. Representative Northern blot using ^{32}P-dCTP labeled AT_{1A} cDNA to hybridize RNA isolated from kidney of normal sodium and low sodium treated rats. Thirty μg of total renal RNA were used in each lane. The blots were stripped and rehybridized with ^{32}P-dCTP labeled 18S ribosome RNA cDNA.

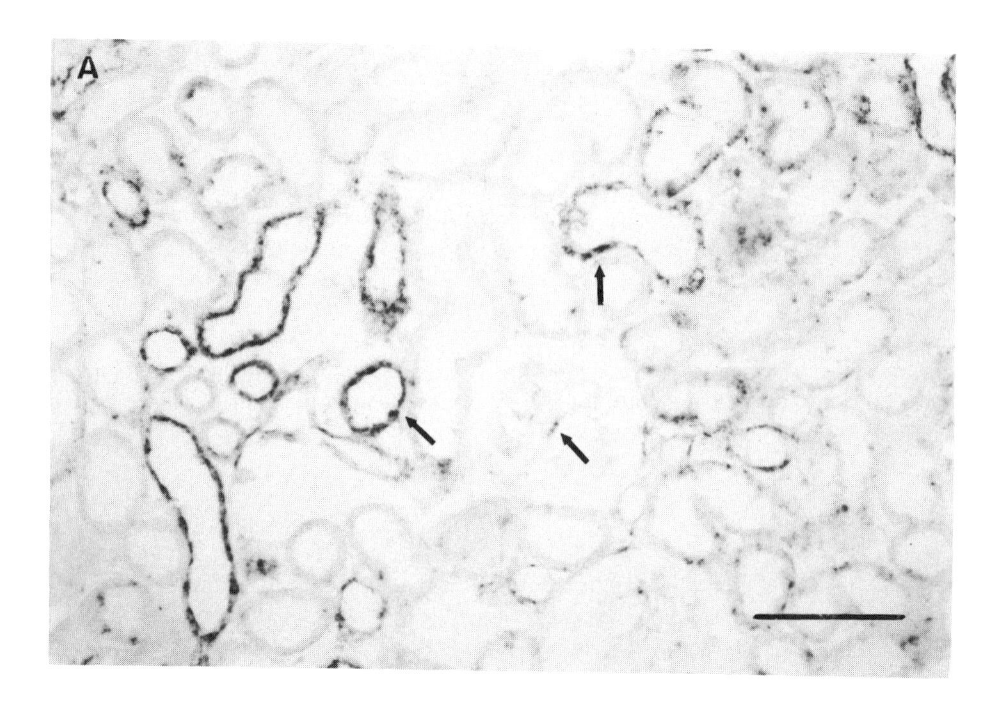

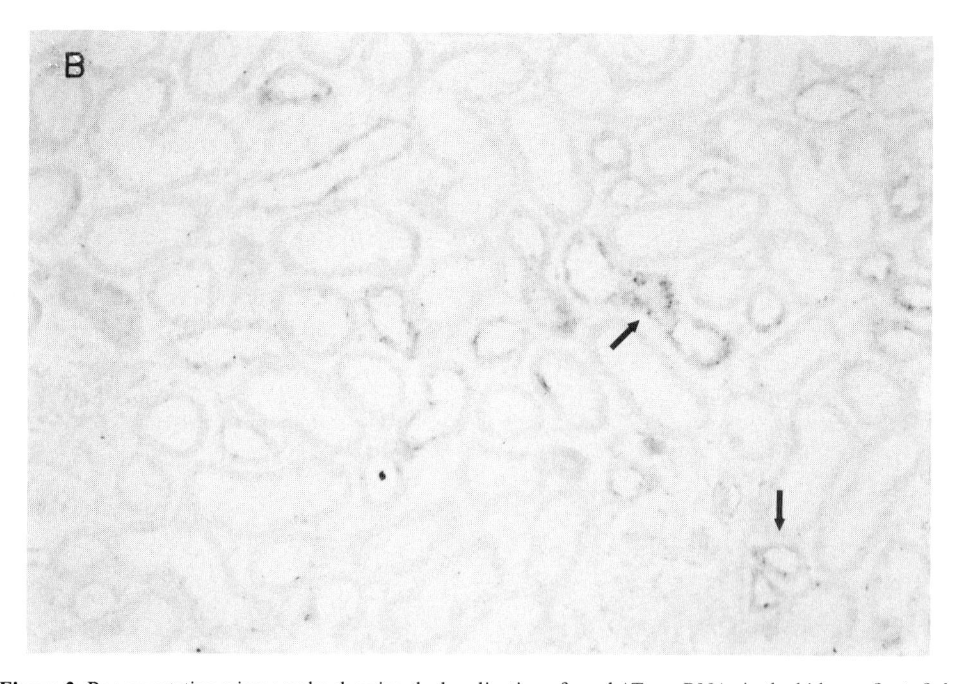

Figure 2. Representative micrographs showing the localization of renal AT_{1B} mRNAs in the kidney of rats fed a normal sodium diet (A) and a low sodium diet (B) by in situ hybridization. Micrographs showed that renal tubular and glomerular cells of the control rats hybridized to digoxigenin-labeled antisense AT_{1B} cRNA probe (arrows in A). Renal AT_{1B} mRNA expression was markedly decreased by low sodium treatment (arrows in B). The black bars represent 100 μm.

suggest that dietary sodium differentially regulates the genes encoding AT_1 receptor subtypes. Interestingly, an opposite effect of sodium diet on gene expression of the AT_1 receptor subtype in the brain has been reported[6] where it was shown that sodium deprivation enhances the expression of AT_{1B} mRNA and high sodium intake increases expression of AT_{1A} mRNA[6]. Considering that differences in the 5′ regulatory regions of the AT_{1A} and AT_{1B} genes exist[9,10], differential regulation of AT_1 receptor subtypes in different tissues in response to sodium may indicate distinct functional specificities.

A question that remains to be answered is how sodium restriction induces changes in mRNA expression of AT_1 receptor subtypes. In the present study, sodium restriction decreased urinary sodium content, but it had no effect on plasma renin activity. Thus it is likely that the circulating renin angiotensin system was not activated. One possible explanation is that another compensatory mechanism, such as the release of aldosterone or local renal production of Ang II, was activated in response to the low sodium diet and contributed to the differential regulation of AT_1 receptor subtypes. This deserves additional investigation.

In conclusion, the present studies demonstrate that sodium depletion significantly increases the gene expression of the AT_{1A} receptor subtype, and decreases AT_{1B} mRNA levels in the kidney. Such differential regulation of kidney angiotensin receptor subtypes by sodium load suggests that they play unique roles in the maintenance of fluid homeostasis. An understanding of the individual roles of these Ang II receptor subtypes in the physiology and pathophysiology of the renin angiotensin system may provide new information regarding the regulation of blood pressure and the pathogenesis of arterial hypertension.

ACKNOWLEDGMENT

This study was supported in part by National Institutes of Health grant HL-52279 (D.H. Wang).

REFERENCES

1. Iwai N, Yamano S, Chaki S, Konishi F, Bardhan S, Tibbetts C, Sasaki K, Hasegawa M, Matsuda Y, Inagami T: Rat angiotensin II receptor: cDNA sequence and regulation of the gene regulation. *Biochem Biophys Res Commun.* 1991;177:299-304.
2. Murphy TJ, Alexander RW, Griendling KK, Runge MS, Bernstein KE: Isolation of cDNA encoding the vascular type 1 angiotensin II receptor. *Nature.* 1991;351:233-236.
3. Iwai N and Inagami T: Identification of two subtypes in the rat type I angiotensin II receptor. *FEBS Letters.* 1992;298;257-260.
4. Sanderberg K, Ji H, Clark AJL, Shapira H, Catt KJ: Cloning and expression of novel angiotensin II receptor subtype. *J Biol Chem.* 1992;267:9455-9458.
5. Kakar SS, Sellers JC, Devor DC, Musgrove LC, Neill JD: Angiotensin II type-1 receptor subtype cDNAs: Differential tissue expression and hormonal regulation. *Biochem Biophys Res Commun.* 1992;185:688-692.
6. Sandberg K, Ji H, Catt KJ: Regulation of Angiotensin II receptors in rat brain during dietary sodium changes. *Hypertension.* 1994;23 [suppl I]:I-137-I-141.
7. Chomczynski P, Sacchi N: Single-step method of RNA isolation by acid guanidinium thiocyanate-phenol-chloroform extraction. *Anal Biochem.* 1987;162:156-159.
8. Kakinuma Y, Fogo A, Inagami T, Ichikawa I. Intrarenal localization of angiotensin II type 1 receptor mRNA in the rat. *Kidney International.* 1993;43:1229-1235.
9. Takeuchi K, Alexander W, Nakamura Y, Tsujino T, Murphy TJ. Molecular structure and transcriptional function of the rat vascular AT1A angiotensin receptor gene. *Circ Res.* 1993;73:612-621.
10. Guo DF, Inagami T: The genomic organization of the rat angiotensin II receptor AT1B. *Biochem Biophys Acta.* 1994;1218:91-94.

5

CHARACTERIZATION AND REGULATION OF ANGIOTENSIN II RECEPTORS IN RAT ADIPOSE TISSUE

Angiotensin Receptors in Adipose Tissue

L. A. Cassis, M. J. Fettinger, A. L. Roe, U. R. Shenoy, and G. Howard

University of Kentucky
Division of Pharmacology and Experimental Therapeutics
College of Pharmacy
Lexington, Kentucky

SUMMARY

Characterization and regulation of angiotensin II (AII) receptor binding sites was performed in rat membrane preparations from nonadipose (liver, lung) and adipose (interscapular (ISBAT) and periaortic (PA) brown adipose tissue; epididymal (EF) and retroperitoneal (RPF) white adipose tissue). In membrane preparations from brown and white adipose sources, [^{125}I]AII saturation binding revealed a single, high affinity (Kd range of 0.3 - 0.6 nM) binding site with a modest AII receptor density (Bmax range of 17 - 120 fmol/mg protein) comparable to rat lung (130 fmol/mg protein). White adipose tissue contained a greater number of AII receptor sites than brown adipose tissue. Competition displacement studies demonstrated the AT1 receptor is the only angiotensin receptor subtype localized in adipose tissue, with the rank order for competition of [^{125}I]AII binding in all adipose tissues examined AIII > AII > losartan > angiotensin I (AI) > PD123319. The AT2 specific receptor antagonist, PD123319, was ineffective at displacing [^{125}I]AII binding in all adipose tissues examined.

Since components of the renin-angiotensin system are regulated in adipose tissue, we determined if the AII receptor is also regulated in the obese state. AII receptor binding characteristics were determined in liver, lung, ISBAT and EF membrane preparations from adult Zucker obese (fa/fa) and lean (Fa/?) rats. AII receptor density was decreased in liver from obese rats. In contrast, the affinity for [^{125}I]AII binding was not altered in tissues from obese rats. In a separate group of obese and lean rats, regulation of the AII receptor by phenobarbital (PB) was examined. Administration of PB restored AII receptor density in liver from obese rats to levels obtained in lean rats. In summary, these results demonstrate the presence of AT1 receptor sites in brown and white adipose tissue. Moreover, AII receptor density is decreased in tissues from obese rats, with restoration of receptor density by

Recent Advances in Cellular and Molecular Aspects of Angiotensin Receptors
Edited by Mohan K. Raizada et al., Plenum Press, New York, 1996

administration of PB. Future studies will determine if PB regulates the AT1 reptor at the level of gene expression.

INTRODUCTION

Angiotensin II (AII) is a physiologically active peptide with a broad range of effects, including blood pressure regulation, water and electrolyte balance and cellular growth (1-4). Based on their pharmacological and biochemical properties, several types of angiotensin receptors have been identified including AT1 (5,6), AT2 (6), AIV (7,8), A1-7 (9) and atypical binding sites (6). Molecular cloning studies have identified the genomic sequences for the AT1 (10,11) and AT2 (12,13) receptor subtypes, both of which belong to the seven-trans-membrane domain receptor superfamily but which possess minimal DNA homology (32%). These two receptors can be further distinguished by their differential sensitivity to the sulfhydryl reducing agent, dithiothreitol (14) and by their selectivity for the nonpeptide ligands losartan (AT1) and PD123319 (AT2). Extensive pharmacologic studies have indicated that the cardiovascular effects of AII are mediated through the AT1 receptor with subsequent activation of phosphoinositide/calcium pathways (15) or inhibition of adenylate cyclase activity (16). While the function of the AT2 receptor remains unclear, recent studies have demonstrated a correlation between localization of the AT2 receptor and growth and development (17,18), and a possible role of the AT2 receptor in neuronal function (19).

Previous studies in our laboratory have suggested the presence of a local renin-angiotensin system in rat adipose tissue. Evidence supporting AII formation in adipose tissue include angiotensinogen mRNA expression (20, 21), renin activity (22) and immunoreactive AII in adipose tissue extracts (23). Functional studies demonstrate AII increases catecholamine neurotransmitter release from rat brown adipose tissue (23), with potential contributions towards altered thermogenesis states such as those seen in cold-exposure (24) and genetic obesity (25). In a model white adipocyte cell line, angiotensin I and II were localized to cell extracts but independent of renin or angiotensin converting enzyme activity (26). Additionally, angiotensinogen mRNA expression was regulated by nutritional status in epididymal white adipose tissue (27). Finally, previous investigators have localized the AT1 receptor to isolated epididymal white adipocytes (28,29). Collectively, these results demonstrate that the components of the renin-angiotensin system are present in both brown and white adipose tissue.

Currently, little is known concerning tissue-specific expression or physiological and pharmacological regulation of the AT1 receptor. The purpose of the present study was 1) to characterize the angiotensin receptor subtype present in white and brown adipose tissue, 2) to determine if the AT1 receptor is regulated in response to altered lipid metabolism of the genetically obese Zucker (fa/fa) rat and, 3) to determine if administration of the barbiturate phenobarbital (PB) regulates AII receptor binding in tissues from obese and lean rats.

MATERIALS AND METHODS

Characterization of AII Receptors in Rat Adipose Tissue

Rat Adipose Membrane Preparation. Membranes were prepared from interscapular (ISBAT) and periaortic (PA) brown adipose tissue and epididymal (EF) and retroperitoneal (RPF) white adipose tissue. For characterization of AII receptor subtypes in adipose tissue, male Sprague Dawley rats (Charles River, MA ; tissues pooled from 2 rats) were killed and adipose tissues dissected, cleaned, weighed and homogenized in membrane buffer (50 mM Tris, pH 7.4; 250 mM sucrose; 10 mM EDTA). For all studies, tissues were disrupted with a Polytron homogenizer (4°C), followed by homogenization using a teflon/glass homogenizer at full speed for 30 sec (4°C). Homogenates (30 ml) were centrifuged at 1,100 g for 10 min (4°C), and the supernatant was transferred to a fresh tube. Supernatants were recentrifuged at 48,000 g in a Sorval supraspeed centrifuge (rotor F28/36) for 10 min (4°C). The supernatant was discarded and the pellet resuspended in membrane buffer and recentrifuged twice more at 48,000 g. The final membrane pellet was resuspended (200 mg wet weight/ml; 1 mg protein/ml) in buffer containing 50 mM Tris (pH 7.4), 120 mM NaCl, 1 mM $MgCl_2$.

Radioligand Binding. Initial studies determined optimum membrane protein concentration and time to equilibrium binding in adipose tissues. For saturation binding isotherms, membrane protein (50 - 100 μg for each adipose tissue) was incubated with 0.025 - 0.8 nM [^{125}I]AII (2200 Ci/mmol; New England Nuclear, Wilmington, DE) for 60 min at 26°C in a buffer (250 μl total volume) containing 120 mM NaCl, 1 mM $MgCl_2$, 0.2% fatty acid free BSA (Sigma), 50 mM Tris, 0.005% (w/v) bacitracin and 0.24 units/ml aprotinin. For competition studies, membrane protein (50 μg) was incubated with [^{125}I]AII (0.15 nM) in the presence of increasing concentrations (12 concentrations/competitor) of competitors (AI, AIII, AII, Losartan, PD123319) for 60 min at 26°C. Incubations were terminated by rapid filtration (Brandel Cell Harvestor) through GF/B glass fiber filters (Brandel) which had been presoaked in 0.2% BSA. Filters were washed three times with ice-cold buffer (50 mM Tris, pH 7.4), and particle-bound radioactivity assayed in a γ counter. Nonspecific binding was defined as radioactivity bound in the presence of 10 μM AII in the incubation medium. All samples were run in duplicate. Protein was determined according to the method of Bradford (30), with BSA as the standard.

Equilibrium binding parameters (Kd and Bmax) and binding isotherms from competition studies were obtained using the LIGAND program. The inhibitory constant (Ki) was calculated from the IC50 using the equation of Cheng and Prusoff (31). Two-tailed Student's t-tests were used to determine whether Hill slopes were significantly different from unity (one-sample tests), or to determine if AII receptor density differed between obese and lean groups.

Regulation of AII Receptor Binding in Zucker Genetically Obese (Fa/Fa) and Lean (Fa/?) Rats. For studies examining AII receptor density in Zucker genetically obese rats, AII receptor binding kinetics were determined in liver, lung, ISBAT and EF membrane preparations from adult (8 - 12 weeks of age) male Zucker obese (fa/fa; N = 4) and lean (Fa/?; N = 4) rats. Obese (body weight, 476.5 ± 91.6 g) and lean (353.8 ± 50.4 g) rats were purchased and genotyped at Charles River, MA. For examination of the effect of PB administration on AII receptor binding, 2 additional groups (obese/PB, lean/PB; N = 4/group) of obese and lean rats (8 - 12 weeks of age) were administered either PB (70 mg/kg, i.p.) or saline vehicle for 4 days. Membranes were prepared from liver, lung, ISBAT and EF from individual rats in each group.

RESULTS

Characterization of AII Binding in Nonadipose and Adipose Tissue

Binding of [^{125}I]AII to adipose membranes was saturable and of high affinity. At 26°C, binding was rapid, reaching equilibrium in approximately 30 min, and remained stable throughout 1 hr (data not shown). Linearity of the binding to increasing amounts of protein

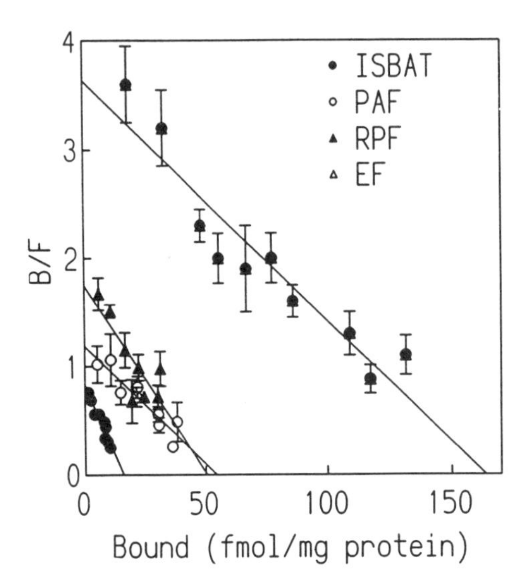

Figure 1. Scatchard analysis of AII receptor binding in adipose tissue from Sprague Dawley rats. Brown (ISBAT, PAF) and white (EF, RPF) adipose tissue membranes (N = 5/tissue) were incubated with varying concentrations (0.01 - 100 nM) of [^{125}I]AII for 60 min at 26°C. Binding was terminated by filtration. Epididymal white adipose tissue contained the greatest number of AII receptor sites (119 fmol/mg protein) with the highest affinity (Kd = 0.29 nM) of the adipose tissues examined.

was assessed by determining specific binding between 25 and 200 µg protein from each adipose tissue (data not shown). Membrane protein (50 - 100 µg) was chosen from the linear portion of protein curves for subsequent saturation binding isotherms. Nonspecific binding of [^{125}I]AII to adipose membranes was routinely 20% of specific binding, with nonspecific binding in nonadipose tissues less than 5% of specific binding.

In both brown and white adipose tissue, [^{125}I]AII interacted with a single population of binding sites (Fig. 1). Kinetic parameters derived from scatchard analysis of specific binding of [^{125}I]AII to adipose and nonadipose tissue are presented in Table 1. Epididymal white adipose tissue contained the greatest number of AII receptor sites (119 fmol/mg protein) with the highest affinity (Kd = 0.29 nM) of the adipose tissues examined and gave comparable AII receptor characteristics as demonstrated in lung (Table 1). Hill coefficients for the adipose tissues examined were not different from unity (Table 1), indicating a single class of binding sites for AII and the absence of positive or negative cooperativity.

Effect of AII Agonists and Antagonists on Specific [^{125}I]AII Binding in Rat Adipose Tissue. The ability of several angiotensin peptide analogues and nonpeptide ligands to compete for specific binding sites on white and brown adipose membranes is illustrated in Fig. 2. The rank order of potency for competition of [^{125}I]AII binding by these compounds was similar in white and brown adipose membranes: AIII ≥ AII > losartan ≥ AI > PD123319 (Table 2). The AT2 selective antagonist, PD123319, was ineffective at competing for

Table 1. Angiotensin II receptor binding characteristics

		K_d(nM)	B_{max} (f_{mol}/mg protein)	n_H
Liver	(n=7)	0.38 ± 0.15	1003.6 ± 395.5	0.94 ± 0.02
Lung	(n=4)	0.63 ± 0.05	128.4 ± 15.7	0.93 ± 0.02
Periaortic Adipose	(n=4)	0.39 ± 0.04	60.0 ± 17.3	0.85 ± 0.07
Retroperitoneal Adipose	(n=3)	0.33 ± 0.08	34.9 ± 12.6	0.96 ± 0.03
Epididymal Adipose	(n=5)	0.29 ± 0.04	119.3 ± 39.3	0.99 ± 0.01
Interscapular Adipose	(n=5)	0.47 ± 0.09	17.2 ± 3.9	0.95 ± 0.02

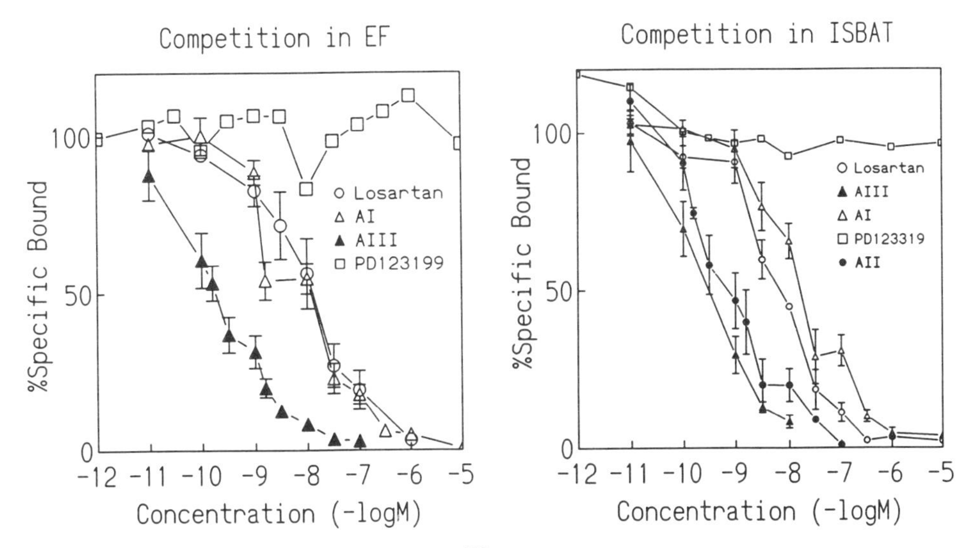

Figure 2. Competition displacement binding for [^{125}I]AII binding in white (EF) and brown (ISBAT) adipose tissue. Membranes (μg protein) were incubated with [^{125}I]AII (0.15 nM) in the presence of increasing concentrations (1 pM - 1 μM; 12 concentrations/displacer) of losartan, angiotensin I (AI), angiotensin III (AIII) or PD123319 (N = 5/displacer compound). The rank order of potency for competition of [^{125}I]AII binding by these compounds in white and brown adipose membranes was: AIII ≥ AII > Losartan ≥ AI > PD123319. The AT2 selective antagonist, PD123319, was ineffective at competing for [^{125}I]AII binding in adipose membranes.

[^{125}I]AII binding in adipose membranes. Analysis of the displacement curves demonstrated all compounds examined recognized a single class of sites (Table 2).

Alterations in [^{125}I]AII Binding in Membranes from Zucker Genetically Obese Rats Compared to Lean Controls. Saturation binding isotherms were used to quantitate [^{125}I]AII binding in liver lung, ISBAT and EF membranes in tissues from obese and lean rats (Table 3). Livers from obese rats exhibited a decrease (49%) in AII receptor sites compared to lean controls (Fig. 3). However, no significant differences were observed in AII receptor density in lungs, EF or ISBAT from obese and lean rats. The affinity of [^{125}I]AII binding was not altered in tissues from obese rats compared to lean controls (Table 3).

Table 2. Competition displacement binding in adipose tissue

	Epididymal Adipose		Interscapular Adipose	
	K_i(nM)	n_H	K_i(nM)	n_H
Angiotensin II	1.17 ± 0.4	1.15 ± 0.15	0.52 ± 0.11	0.79 ± 0.04
Angiotensin I	5.99 ± 1.6	0.85 ± 0.04	10.00 ± 2.52	0.87 ± 0.03
Angiotensin III	0.16 ± 0.03	0.86 ± 0.03	0.13 ± 0.03	0.96 ± 0.02
Losartan	4.04 ± 1.12	0.86 ± 0.06	3.51 ± 0.72	0.91 ± 0.03
PD1923199	> 10,000		> 10,000	

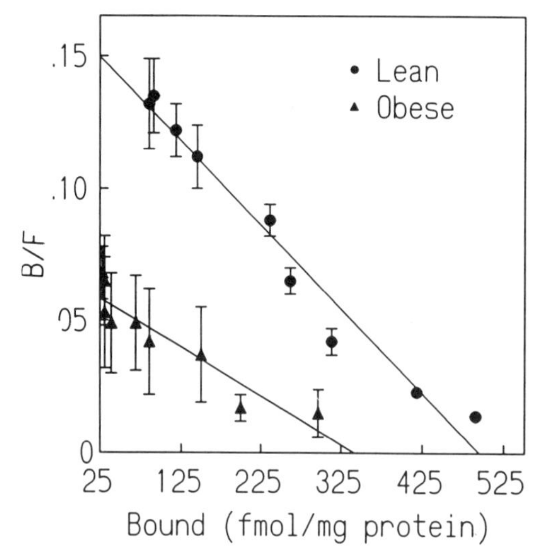

Figure 3. Scatchard analysis of AII receptor binding in liver membranes from obese and lean rats. Liver membranes (25 μg protein) from obese and lean rats (N = 4/group) were incubated with [^{125}I]AII (0.2 nM) in the presence of increasing concentrations of unlabelled AII (0.01 - 100 nM) for 60 min at 26°C. In liver membranes from both obese and lean rats, [^{125}I]AII interacted with a single population of binding sites. The affinity for [^{125}I]AII binding was not different in liver membranes from obese and lean rats; however, AII receptor density was decreased (49%) in liver membranes from obese rats compared to lean controls.

Regulation of AII Receptor Binding by Administration of PB to Obese and Lean Rats. AII receptor binding was examined 4 days following administration of PB or vehicle to obese and lean rats (Table 3). Administration of PB did not alter AII receptor affinity in tissues from either obese or lean rats. In lean rats, administration of PB resulted in a decrease in AII receptor density in EF; however, PB administration did not alter density of AII receptor binding in other tissues examined. Interestingly, administration of PB to obese rats resulted in an increase in AII receptor density in liver to levels not significantly different from lean controls.

Table 3. Angiotensin receptor binding characteristics in tissues from control- and phenobarbital-treated obese and lean rats

	Obese		Obese/Pb		Lean		Lean/Pb	
	Kd (nM)	Bmax (fmol/mg protein)	Kd (nM)	Bmax (fmol/mg protein)	Kd (nM)	Bmax (fmol/mg protein)	Kd (nM)	Bmax (fmol/mg protein)
Liver	1.20± 0.48	197.9†± 31.0	0.69± 0.11	329.9*± 37.6	0.48± 0.11	388.6± 56.6	0.63± 0.06	323.9± 49.1
Lung	0.83± 0.22	181.7± 54.5	1.52± 0.50	163.9± 44.9	1.07± 0.18	171.6± 58.6	1.02± 0.14	178.6± 27.2
ISBAT	4.97± 2.61	50.6± 16.8	5.73± 2.61	57.3± 14.8	3.60± 1.13	59.8± 14.6	4.45± 2.20	37.2± 11.4
EF	6.80± 4.20	97.4± 55.6	6.95± 5.40	185.5± 62.2	3.07± 1.68	194.9± 17.9	4.83± 2.93	111.7*± 29.1

- Values are mean ± SEM of N=4 rats/group.
- †, P>0.05 Obese control compared to lean control.
- *, P>0.05 Pb-treated compared to saline vehicle within each group (obese/lean).

DISCUSSION

Characterization of AII receptor subtypes in various tissues has provided insight into mechanisms contributing to the physiologic effects of AII. The present study demonstrates the presence of high affinity, moderate capacity AT1 receptor sites in rat brown and white adipose tissue. The density of AII receptor sites is greatest in white adipose tissue, with comparable numbers of AII receptors as seen in rat lung. Competition displacement studies demonstrate the AT1 receptor predominates in adipose tissue. In genetically obese Zucker rats, the density of AII receptor sites is decreased in liver, suggesting alterations in the AII receptor in the presence of the *fa* mutation. Moreover, administration of the barbiturate PB to obese and lean rats resulted in a restoration of AII receptor density in liver from obese rats to levels obtained in lean rats.

The present study revealed a high affinity, moderate capacity AII receptor binding site in membranes prepared from white (EF, RPF) and brown (ISF,PAF) adipose tissue. The binding was rapid, saturable, and reversible. The dissociation constant determined in saturation experiments is consistent with previously reported data obtained in several different organs of the rat, as well as with nonadipose tissues (liver (16), lung (32)) examined in the present study. The density of AII receptor sites was greatest in epididymal white adipose tissue. Interestingly, AII receptor density did not parallel the degree of tissue vascularization. For example, interscapular brown adipose, with the greatest degree of vascularization, contained the fewest number of AII receptor sites of all adipose tissues examined. AII receptor density in EF membranes in the present study was two-fold greater than that previously reported in isolated epididymal adipocytes (28,29). Thus, a portion of AII receptor binding sites identified in adipose tissue in the present study may represent AII receptors localized to nonadipocyte cell elements.

The order of potency for several AII analogs (agonists and antagonists) for competition with [^{125}I] AII binding in rat adipose membranes was AIII $\geq$ AII > losartan > AI , with Ki values ranging from 0.13 - 10 nM. This rank order of potency (Ki value) is in agreement with previously reported competition studies using these AII analogues in nonadipose tissues (33,34). The selective AT2 receptor antagonist, PD123319, did not displace specific [^{125}I]AII binding in adipose membranes, suggesting the AT2 receptor site is not localized to adipose tissue. These results are in agreement with previous studies demonstrating the AT1 receptor as the only AII receptor subtype in isolated epididymal adipocytes (28,29). Interestingly, in the present study AIII was equipotent with AII for displacement of specific [^{125}I]AII binding in adipose membranes. Previous studies have demonstrated that AIII, while capable of interacting with the AT1 receptor, binds with lower affinity than AII (35). In contrast, AIII is equipotent with AII for binding to the AT2 receptor site. Results from this study demonstrate that AIII binds with the same affinity as AII to the AT1 receptor site in adipose membranes.

Previous studies in our laboratory have demonstrated that AII increases the evoked release of norepinephrine from sympathetic nerve terminals of rat brown adipose tissue (23). Moreover, evoked release of norepinephrine was decreased in the presence of losartan, indicating endogenous AII acts at presynaptic AT1 receptors on sympathetic nerve terminals of brown adipose tissue to modulate norepinephrine release (23). In agreement with functional studies, results from the present study demonstrate the presence of high affinity, moderate capacity AT1 receptors in rat brown adipose tissue. Additionally, the AT1 receptors of white adipose tissue may be responsible for the previously reported effects of AII on prostaglandin production in isolated white rat adipocytes (36).

Previous studies in our laboratory have demonstrated alterations in AII production and function in brown adipose tissue from young and adult genetically obese (fa/fa) Zucker rats compared to lean controls (Fa/?) (25). The facilitatory effect of AII on release of norepinephrine from slices of brown adipose tissue was lost in adult obese rats compared to lean controls. In contrast, the presynaptic effects of AII were enhanced in brown adipose tissue from young obese rats compared to age-matched lean controls. Collectively, these results suggested that increased AII-mediated regulation of norepinephrine release in young obese rats resulted in downregulation of AII receptor density in adult obese rats. In the present study a decrease in AII receptor density was observed in liver from obese rats compared to lean controls. Future studies will determine if decreases in AII receptor density in Zucker obese rats is the result of agonist-induced downregulation, or in response to long-term suppression of the systemic renin-angiotensin system.

Administration of the barbiturate PB to obese rats resulted in a restoration of AII receptor density in liver to levels observed in lean rats, whereas PB administration did not alter AII receptor density in lean rats. However, administration of PB to lean rats resulted in an increase in AII receptor density in epididymal white adipose tissue. The mechanism by which PB alters gene expression is currently unknown. A recent study, however, demonstrated that administration of PB to Sprague-Dawley rats resulted in increased AP-1 binding activity in liver tissue (37). Moreover, PB treatment of hepatoma cells in culture induced AP-1-binding activity as well as AP-1-like enhancer activity (38) . Analysis of 3.2 kb of the 5′ flanking region of the AT1 receptor gene has revealed at least one putative AP-1 binding site at -387 (39,40). Although the AT1 promoter region from -558 to -331 has been shown to contain enhancer activity in cultured vascular smooth muscle cells and PC12 cells (39,40), a functional AP-1 site has not yet been defined. Future studies will determine whether PB administration regulates AT1 message levels or AP-1 binding to sequences specific for the AT1 receptor gene, and if alterations in AP-1 binding to the AT1 receptor gene occur in tissues from Zucker obese rats.

In summary, characterization of AII receptors in brown and white adipose tissue revealed high affinity, moderate capacity AII receptors of the AT1 subtype. In Zucker obese (fa/fa) rats, AII receptor density is decreased, suggesting regulation of the renin-angiotensin system in obesity. Finally, administration of the barbiturate PB restored AII receptor density in liver from obese rats to levels observed in lean rats. Future studies will determine if PB-regulation of AII receptor density is mediated through the transcription factor AP-1.

ACKNOWLEDGMENTS

The authors wish to acknowledge DuPont-Merck Pharmaceuticals for losartan, and Parke-Davis Pharmaceuticals for PD123319.

This work was supported by NIHK04HL02742 and the American Heart Association (local Kentucky affiliate) .

REFERENCES

Peach, M.J. (1977) Physiol. Rev. *57*:313-370.

Natfilan, A.J., Gilliland, G.K., Eldridge, C.S. and Draft, A.S. (1990). Mol. Cell Biol *10*:5535-5540. Owens, G.K. (1985) Circ. Res. *56*:525-536.

Schelling, Pl, Fischer, H. and Ganten, D. (1991) J. Hypertens. *9*:3-15.

Chiu, A.T., Herblin, W.F., Ardecky, R.J., McCall, D.E., Carini, D.J., Duncia, J.V., Pease, L.J., Wexler, R.R., Wong, P.C., Johnson, A.L. and Timmermans, P.B.W.M. (1989). Biochem. Biophys. Res. Commun. *165*:196-203.

immermans, P.B.M.W.M., Wong, P., Chiu, A., Herblin, W.F., Benfield, P., Carini, D.J., Lee, R.J., Wexler, R.R., Saye, J.M. and Smith, R.D. (1993) Pharm. Reviews *45*:205-251.

Swanson, G.N., Hanesworth, J.M., Sardinia, M.F., Coleman, J.K.F., Wright, J.W., Hall, K.L., Miller-Wing, A.V., Stobb, J.W., Cook, V.I., Harding, E.C. and Harding, J.W. (1992). Regul. Peptides *40*:409-419.
Harding, J.W., Cook, V.I., Miller-Wing, A.V., Hanesworth, J.M., Sardinia, M.F., Hall, K.L., Stobb, J.W., Swanson, G.N., Coleman, J.K.M. and Wright, J.W. (1992). Brain Res. *583*:340-342.
Schiavone, M.T., Khosla, M.C. and Ferrario, C.M. (1990). J. of Cardiovas. Pharmacol. *4*:S19-S24.
Murphy, T.J., Alexander, R.W., Griendling, K.K., Runge, M.S. and Bernstein, K.E. (1991). Nature(Lond.) *351*:233-236.
Sasaki, K., Yamano, Y., Bardhan, S., Iwai, N., Murray, J.J., Hasegawa, M., Matsuda, Y. and Inagmi, T. (1991). Nature (Lond.) *351*:230-232.
Mukoyama, M., Nakajima, M., Horiuchi, M., Sasamura, H., Pratt, R.E. and Dzau, V.J. (1993). J. Biol. Chem. *268*:24539-24542.
Kambayashi, Y., Bardhan, S., Takahashi, K., Tsuzuki, S., Inue, H., Hamakubo, T. and Inagami, T. (1993). J. Biol. Chem. *268*:24543-24546.
Chiu; , A.T., McCall, D.E., Nguyen, T.T., Carini, D.J., Duncia, J.V., Herblin, W.R., Wong, P.C., Wexler, R.R., Johnson, A.L. and Timmermans, P.B.M.W.M. (1989). Eur. J. Pharmacol. *170*:117-118.
Griendling, K.K., Tsuda, T., Berk, B.C. and Alexander, R.W. (1989). Am. J. Hypertens. *2*:659-665.
Crane, J.K., Campanile, C.P. and Garrison, J.C. (1982). J. Biol. Chem. *257*:4959-4965.
Grady, E.F., Sechi, L.A., Friggin, C.A., Schambelan, M. and Kalinyak, J.E. (1991). J. Clin Invest. *88*:921-933.
Viswanathan, M., Tsutsumi, K., Correa, F.M.A. and Saavedra, J.M. (1991). Biochem. Biophy. Res. Commun. *179*:1361-1367.
Sumners, C., Tang, W., Zelena, B. and Raizada, M.D. (1991). Proc. Natl. Acad. Sci. USA *88*:7567-7571.
Cassis, L.A., Lynch, K.R. and Peach, M.J. (1988). Circulation Res. *62*:1259-1262.
Cassis, L.A., Saye, J. and Peach, M.J. (1988). Hypertension *11*:591-596.
Samant, U., Painter, D. J., Howard, G., Dwoskin, L.P. and Cassis, L.A. (1993) FASEB *7*(3):A220.
Cassis, L.A. and Dwoskin, L. P.(1992). J. of Neural Transmission *34*:129-137.
Cassis, L.A. (1993). Amer. J. Physiol. *265*:E860-E865.
Cassis, L.A. (1994). Amer. J. Physiol. *264*:E453-E458.
Saye, J., Ragsdale, V., Carey, R. and Peach, J. (1989). Amer. J. Physiol. *264*:C1570-C1576.
Frederich, R.C., Kahn, B.B., Peach, M.J. and Flier, J.S. (1992). Hypertension *19*:339-344.
Crandall, D.L., Kerzlinger, H.E., Suanders, B.D., Armellino, D.C. and Kral, J.G. (1994). J. Lipid Research *35*:1378-1385.
Crandall, D.L., Herzlinger, H.E., Saunders, D., Zolotor, R.C., Feliciano, L. and Cervoni, P. (1993). Metabolism *42*:511-515.
Bradford, M. (1976). Anal. Biochem. *72*:248-254.
Cheng, Y. and Prusoff, W. (1973). Biochem. Pharmacol. *22*:3099-3108.
Entzeroth M. and Hadamovsky, S. (1991). Eur. J. Pharm. *206*:237-241.
Ernsberger, P., Zhou, J., Damon, T.H. and Douglas, J.G. (1992). Am. J. Physiol. *263*:F411-F416.
Baker, K.M., Campanile, C.P., Trachte, G.J. and Peach, M.J. (1984). Circ. Res. *54*:286-293.
Wong, P.C., Chiu, A.T., Duncia, J.V., Herblin, W.F., Smith, R.D. and Timmermans, P.B.M.W.M. (1992). Trends Endocrinol Metab *3*:211-217.
Richelsen, B. (1987). Biochem. J. *247*:389-394.
Roe, A.L., Howard, G. and Blouin, R.A. (1995). The Toxicologist *15*:57.
Pinkus, R., Bergelson, S. and Daniel, V. (1993). Biochem. J. *290*:637-640.
Takeuchi, K., Alexander, W., Nakamura, Y., Tsujino, T. and Murphy, T. (1993). Circ. Res. *73*:612-621.
Murasawa, S., Matsubara, H., Urakami, M. and Inada, M. (1993). J. Biol. Chem. *268*:26996-27003.

6

CHANGES IN ANGIOTENSIN AT_1 RECEPTOR DENSITY DURING HYPERTENSION IN FRUCTOSE-FED RATS

Shridhar N. Iyer, Michael J. Katovich, and Mohan K. Raizada[1]

Department of Pharmacodynamics and
[1] Department of Physiology
College of Pharmacy and College of Medicine
University of Florida
Gainesville, Florida 32610

ABSTRACT

Feeding carbohydrate-enriched diets to normal rats has been shown to induce insulin resistance and hyperinsulinemia associated with an elevation of blood pressure. Previously we reported that the renin-angiotensin system (RAS) is likely to be involved in the elevation of blood pressure. The purpose of this study was to determine the changes in plasma angiotensin II (AII) and AII receptor density associated with the elevation of blood pressure in fructose-treated rats. Male Sprague-Dawley rats were divided into two groups and were fed either normal rat chow or a 60% fructose-enriched diet for four weeks. Plasma insulin of fructose-treated rats was significantly elevated ($p<0.05$) by the end of first week of fructose treatment and remained elevated throughout the study. Plasma AII levels of fructose-fed rats was 3.5 fold greater than the controls at the end of second week and returned to basal levels at the end of the fourth week of dietary treatment. Blood pressure was significantly elevated in the fructose-fed rats within two weeks of fructose treatment. Elevation of blood pressure was associated with left ventricular hypertrophy. Angiotensin II type I receptor (AT1) density was determined in the left ventricle, aorta, adrenal gland and hypothalamus. There was a significant increase in AT1 receptor density in the ventricle at the end of third and fourth weeks of treatment, whereas there was a significant decrease in the receptor density in the aorta at the end of the fourth week of treatment. Receptor density in the adrenal gland and hypothalamus of fructose-fed rats was similar to their respective controls. The results of this study suggest that the RAS plays a role in the elevation of blood pressure of fructose-fed rats and also contributes to the ventricular hypertrophy observed in these rats.

Recent Advances in Cellular and Molecular Aspects of Angiotensin Receptors
Edited by Mohan K. Raizada et al., Plenum Press, New York, 1996

INTRODUCTION

Previous studies have demonstrated that ingestion of carbohydrate-enriched diets elevates blood pressure (BP) in rats and is associated with hyperinsulinemia and insulin resistance (1,2,3). Clinical studies have reported a correlation between insulin resistance and hypertension in both obese and non-obese patients (4,5). Hyperinsulinemia and/or insulin resistance have been suggested to play a role in the elevation of blood pressure. Improvement in insulin sensitivity by exercise (2) and treatment with insulin sensitizing agents such as somatostatin (3) and vanadyl sulfate (6) has shown to prevent the elevation of blood pressure in rats fed carbohydrate-enriched diets. These studies suggest that hyperinsulinemia and insulin resistance contribute to the elevation of blood pressure.

However, the precise mechanism(s) by which carbohydrate enriched diets elevate blood pressure is still not clear. Previous studies suggest that hyperinsulinemia induced by carbohydrate feeding may be activating the sympathetic system which in turn could elevate the blood pressure (7). One of the consequences of an activated sympathetic nervous system is an increase in renin release associated with an increase in plasma angiotensin II (AII), a potent endogenous vasoconstrictor, and a subsequent elevation of blood pressure (8,9). Furthermore it has been reported that insulin modulates the renin-angiotensin system (RAS) (10). Recent studies suggest that RAS is involved in the hypertension associated with fructose treatment (11,12) since blockade of AII receptors abolishes the elevation of blood pressure. In a preliminary study, we showed that Losartan, an angiotensin II type I receptor (AT_1) selective antagonist blocked the elevation of blood pressure in rats fed fructose-enriched diets. Limura et al (12) observed similar results using TCV-116, an AT_1 selective antagonist. These results strongly suggest that the RAS plays a contributory role in the elevation of blood pressure in this model of hypertension.

The purpose of this study was to further evaluate the role of RAS in fructose-induced hypertension. In this study we determined weekly plasma AII levels over a period of four weeks of fructose treatment. We also determined if changes in AII receptor density were associated with the elevation of blood pressure. We hypothesized that if the RAS is involved in the elevation or maintenance of the hypertensive state then changes in density or binding affinity of AII receptors would be observed due to increase in plasma AII.

Methods

Male Sprague Dawley rats (Charles River, Raleigh, N.C. U.S.A.) weighing 225-250 grams were used in this study. Rats were housed in individual cages in a room maintained at 25 ± 2°C with a 12 hour light/dark cycle. All rats were placed on standard rat chow for a week. Animal's were then divided into two groups. One of these groups served as the control group and thus continued to be maintained on standard rat chow whereas rats from the remaining group were fed a 60% fructose diet (Teklad, Madison, WI, U.S.A.) for four weeks. The percentage of sodium in the two diets are approximately the same.

At the end of each week of fructose treatment, 16 rats (8 from each group) were utilized for the receptor binding studies. Indirect blood pressures was determined by tail plethysmography at the end of each week from rats that were subsequently used for receptor binding studies. Readings were taken one minute apart and to be accepted, BP measurements had to be stable for at least three consecutive readings. Previously we have shown that indirect blood pressure correlates with blood pressure determined directly by carotid cannulation (11). Indirect blood pressures and the receptor binding was performed in an additional 16 rats during weeks 2, 3 and 4 of the dietary fructose treatment.

Plasma AII and Insulin Determinations. Rats were fasted overnight and trunk blood was collected in chilled vacutainer tubes containing EDTA, captopril (9.2 mM) and 1,10-O-phenanthroline (2.5 mM) for determination of plasma AII. Blood samples were kept on ice and centrifuged within 15-20 minutes at 3000 g for 10 minutes at 4°C and plasma separated and immediately frozen on dry ice and stored at -85°C. Plasma AII levels were determined by radioimmunoassay (Alpco, Windham, NH). For determination of serum insulin, blood samples were collected in separate glass tubes and serum separated and stored at -20°C. Insulin levels were subsequently determined by radioimmunoassay using rat insulin standard (Linco Research Inc, St. Charles, MO).

Radioligand Binding Studies. Receptor binding studies was performed in left ventricle, adrenal gland, aorta and hypothalamus. On the day of the experiment, rats were killed by decapitation and the heart, aorta, adrenal gland and brain were removed rapidly and immediately placed in cold phosphate-buffered saline (pH 7.4). The left ventricle was separated from the atria and right ventricle and minced into small pieces. Ventricular membranes were prepared by a slight modification of the method previously used by Baker et al (13). A homogenate of the left ventricle was prepared in 0.25 mol/L sucrose and 25 mmol/L Tris at pH 7.5, containing 0.5 mmol/L phenylmethylsulfonyl fluoride (PMSF), 15 mg/L bacitracin, 2 μg/ml leupeptin, 5 μg/ml trasylol and 0.2% bovine serum albumin (BSA) using a Brinkman polytron at half-maximal speed for 30 seconds (twice). The homogenate was sedimented at 10,000g for 10 minutes and the supernatant centrifuged at 45,000g for 30 minutes. The pellet was resuspended in 0.6 mol/L potassium chloride and 5 mmol/L histidine (pH 7.0) containing 0.5 mmol/L ethylenediamine tetracetic acid (EDTA), 15 mg/L bacitracin, 0.5 mmol/L PMSF, 2 μg/ml leupeptin, 5 μg/ml trasylol and 0.2% BSA to solubilize actin and myosin filaments and resedimented at 45,000g for 30 minutes. The pellet was resuspended in the assay buffer containing 25 mmol/L Tris pH 7.5, 10 mmol/L Magnesium chloride ($MgCl_2$), 0.5 mmol/L PMSF, 15 mg/L bacitracin, 2 μg/ml leupeptin, 5 μg/ml trasylol, 1 μg/ml pepstatin and 0.2% BSA. The membranes were used immediately for radioligand binding studies.

The homogenate for the hypothalamus was prepared by the method of Baksi et al (14) with slight modifications. The hypothalamus-thalamus-septum (HTS) region of the brain was dissected as previously published by Speth et al (15). The HTS was then homogenized in the assay buffer containing 150 mmol/L sodium chloride, 10 mmol/L $MgCl_2$, 50 mmol/L sodium phosphate, 5 mmol/L ethylene glycol tetracetic acid (EGTA), 15 mg/L bacitracin, 0.5 mmol/L PMSF, 2 μg/ml leupeptin, 5 μg/ml trasylol, 1 μg/ml pepstatin and 0.2% BSA (pH 7.2). The homogenate was spun at 3,000g for 15 minutes. The supernatant was separated and centrifuged at 45,000g for 35 minutes. After the final centrifugation, the supernatant was discarded and the pellet resuspended in the assay buffer.

The adrenal glands were separated and dissected from all adherent fat and a homogenate prepared by the method of Chiu et al (16). The adrenal glands were placed in ice-cold sucrose buffer containing 0.2M sucrose, 1mM EDTA, 10mM Tris (pH 7.5), 15 mg/L bacitracin, 0.5 mmol/L PMSF, 2 μg/ml leupeptin, 5 μg/ml trasylol, 1 μg/ml pepstatin and 0.2% BSA and homogenized using a polytron. The homogenate was spun at 3,000g for 15 min. The pellet was discarded and supernatant centrifuged at 45,000g for 35 minutes. The final pellet was resuspended in the assay buffer containing 50 mM Tris (pH 7.5), 5mM $MgCl_2$, 15 mg/L bacitracin, 0.5 mmol/L PMSF, 2 μg/ml leupeptin, 5 μg/ml trasylol, 1 μg/ml pepstatin and 0.2% BSA. All centrifugations were carried out at 4°C. Protein concentration was determined by the Bradford method using an aliquot from all final suspensions.

AII receptor binding assays were done in microfuge tubes in a total volume of 0.3 ml made up as follows: 100 μl of freshly prepared homogenate containing 150 μg protein, 50 μl of ^{125}I-Sar-AII, 100 μl of varying concentrations of losartan (10^{-10}M to 10^{-7}M) and 50 μl of assay buffer. All incubations were carried out at 25°C for 60 minutes. ^{125}I-Sar-AII was added in concentrations ranging from 0.15-0.3 nM. Non-specific binding determined in the presence of 10 μM losartan was subtracted from total binding. In all cases the reaction was stopped by addition of 1 ml of the assay buffer and centrifuged for 3 minutes at 4°C. The supernatant was aspirated and the pellet washed with 1 ml of the assay buffer. The tip of the microfuge tube was cut and radioactivity counted in a gamma counter. All determinations were done in triplicate.

The density of AII receptors (B_{max}) and the dissociation constant (K_d) were calculated from the specific binding data with the computer program LIGAND. Hill coefficients were calculated from the same program. All data were analyzed by analysis of variance and an appropriate post-hoc test was used for multiple comparison. Data are expressed an mean ± s.e.m.

Potassium chloride, histidine, EDTA, bacitracin, PMSF, trasylol, sodium chloride, magnesium chloride, sodium phosphate, EGTA, tris and sucrose were obtained from Sigma Chemical Company (St. Louis, MO). Pepstatin and leupeptin were purchased from Bachem California (Torrance, CA).

RESULTS

Throughout the study growth patterns were similar for the fructose-fed and control groups. The body weight of the control and fructose-treated rats at the end of the fourth week of treatment were 458 ± 11 and 425 ± 12 respectively. Table 1 summarizes the weekly changes in serum insulin and plasma AII levels associated with fructose treatment. Serum insulin levels were significantly elevated in the fructose-fed animals at the end of first week of treatment and remained elevated throughout the study. At the end of the first week of fructose treatment, there was no significant change in plasma AII levels in the fructose-fed rats. However, by the end of second week of dietary treatment, plasma AII levels in fructose-fed rats were elevated 3.5 fold compared to the controls ($p<0.01$). At the end of the third and fourth weeks of fructose treatment, plasma AII levels still remained significantly elevated compared to week 1, but were lower than that at the end of second week. However, at the end of the fourth week of fructose treatment, there was no significant difference in plasma AII levels between fructose-treated rats and their respective controls. Throughout the four week study period there was no significant difference in plasma AII or serum insulin levels of the control group.

Table 1. Weekly changes in plasma angiotensin II (AII) and serum insulin levels associated with dietary fructose treatment

	Plasma AII (nM)		Serum Insulin (μU/ml)	
	Control	Fructose	Control	Fructose
Week 1	0.044 ± 0.009	0.041 ± 0.01	29.6 ± 8	123.6 ± 42*
Week 2	0.058 ± 0.007	0.203 ± 0.06*	29.3 ± 4	115.2 ± 25*
Week 3	0.056 ± 0.009	0.088 ± 0.01*	42.2 ± 10	99.7 ± 25*
Week 4	0.068 ± 0.013	0.085 ± 0.02	33.4 ± 6	104.8 ± 19*

Data are expressed as mean ± SEM; n = 8/group
* $p < 0.01$ compared to their respective controls

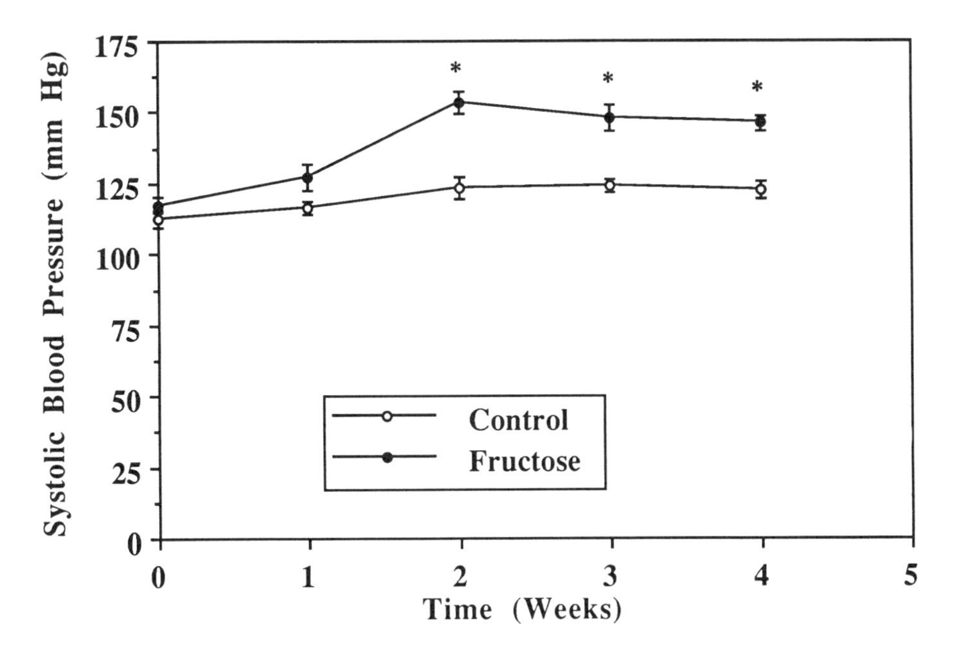

Figure 1. Systolic blood pressure of control and fructose-treated rats. Results are expressed as mean ± S.E.M. n=8/group. *p<0.05 compared to control group and fructose-treated rats at the end of week 1.

Figure 1 shows the effect of fructose treatment on the systolic blood pressure. A significant time and treatment interaction was observed (p<0.01). Subsequent analysis revealed significant time and treatment main effects. Blood pressures were elevated in the fructose-fed rats by the end of the second week of treatment and remained elevated throughout the study compared to the control group. There was no significant change in blood pressure during the four week period of the control group.

Table 2 represents the changes in left ventricular weight in the two groups throughout the four week period. There was no significant change in the LV weight at the end of the first week of fructose treatment. However, by the end of the second week of treatment cardiac hypertrophy was observed in fructose-treated rats. The cardiac hypertrophy in the fructose-fed rats persisted throughout the remainder of the study whereas left ventricular weight was unchanged in the controls during the four weeks of observation (table 2).

Ligand analysis of the displacement of ^{125}I-Sar-AII by losartan demonstrated a one-site model suggesting the presence of a single class of receptors. Table 3 shows the

Table 2. Changes in weight of the left ventricle (mg/100 gm body weight) associated with fructose treatment

Group	Week 1	Week 2	Week 3	Week 4
Control	207 ± 5	196 ± 4	183 ± 5	185 ± 8
Fructose	222 ± 9	214 ± 5*	208 ± 4*	211 ± 6*

Data are expressed as mean ± SEM;c n = 8/group
* indicates p < 0.05 compared to their respective controls.

Table 3. Changes in AT_1 receptor density (Bmax) and binding affinity (K_d) in the control and fructose-treated rats

Tissue	Group	Time of Binding Assay	B_{max}(fmol/mg protein	k_d (nM)
Ventricle	Control		46 ± 4	9.7 ± 0.9
	Fructose	Week 1	41 ± 8	7.6 ± 1.0
		Week 2	27 ± 6	10.4 ± 1.7
		Week 3	98 ± 9*	10.8 ± 5.5
		Week 4	103 ± 8*	14.6 ± 2.9
Aorta	Control		33 ± 4	10.8 ± 1.5
	Fructose	Week 1	39 ± 2	9.3 ± 0.9
		Week 2	47 ± 4	7.8 ± 1.9
		Week 3	30 ± 8	8.9 ± 0.6
		Week 4	13 ± 1**	7.5 ± 1.2
Adrenal Gland	Control		485 ± 27	7.7 ± 0.9
	Fructose	Week 1	418 ± 7	6.8 ± 1.1
		Week 2	474 ± 11	8.9 ± 0.5
		Week 3	464 ± 40	8.7 ± 3.9
		Week 4	556 ± 40	6.8 ± 0.3
Hypothalamus	Control		29 ± 5	9.1 ± 0.9
	Fructose	Week 1	31 ± 2	6.4 ± 1.7
		Week 2	33 ± 1	8.7 ± 0.4
		Week 3	21 ± 1	6.3 ± 1.2
		Week 4	25 ± 6	8.8 ± 1.7

Data are expressed as mean ± SEM.
* $p < 0.01$ compared to respective control, week 1 and week 2 fructose-treated rats.
** $p < 0.05$ compared to respective control, week 1 and week 2 fructose-treated rats.

weekly changes in B_{max} and K_d values in the four tissues studied in both the control and fructose-treated rats. Throughout the four week study period, the density of AT_1 receptors in the ventricle, aorta, adrenal gland and hypothalamus of the control group were not different (data not shown). Hence the data for the control groups over the four week period were combined for the respective tissues. No change in receptor number was observed in the ventricles of the fructose-fed rats at the end of the first and second weeks of treatment. However, fructose treatment was associated with an significant increase ($p<0.05$) in receptor number at the end of the third and fourth weeks of treatment and was significantly greater than the receptor number of the control group. No significant change in binding affinity was observed between the two groups of rats.

The density of AT_1 binding sites in the adrenal gland and hypothalamus are shown in table 3. There was no significant change in receptor number or binding affinity between the two groups in either of these tissues. In the aorta there was a significant decrease in receptor number in the fructose-treated rats ($p<0.05$) at the end of the fourth week of treatment with no changes in receptor affinity. In all studies the Hill coefficients ranged from

0.98 ± 0.1 to 1.1 ± 0.08 suggesting that cooperativity was not involved in the binding of the radioligand.

DISCUSSION

Results of this study demonstrate that fructose feeding is associated with an increase in blood pressure and an increase in plasma insulin. Furthermore, dietary fructose treatment was associated with an increase in circulating AII levels by the end of the second week of treatment which tended to normalize by the fourth week of treatment. The elevation of blood pressure in fructose-treated rats was associated with ventricular hypertrophy and an increase in AII receptor density in the ventricle. Collectively the results of this study suggests that with chronic fructose treatment the RAS is activated and plays a role in the development of hypertension.

Despite the fact that carbohydrate-induced hypertension in experimental animals has been studied for several years the mechanism by which blood pressure is elevated is still not clearly understood. Recent studies suggest that the RAS contributes to the development of hypertension (11,12); however, the precise mechanism by which the RAS is activated is not known. Previous studies (17,18) suggest that when rats are fed carbohydrate-enriched diet the sympathetic nervous system is activated which could directly contribute to elevation of blood pressure. One of the consequences of an activated sympathetic nervous system is an increase in renin release (8) which in turn would be associated with an increase in plasma AII and blood pressure as observed in this study. In a previous study Kobayashi et al (19) also showed an increase in plasma AII in fructose-treated rats at the end of twelve days of fructose feeding. However, Hwang et al (17) reported that plasma renin activity (PRA) was decreased in fructose-fed rats as compared to controls after four weeks of dietary fructose treatment. The latter authors concluded that the RAS was not involved in the elevation of blood pressure in fructose-treated rats. Furthermore, in the study of Hwang et al (17), PRA activity was measured after the hypertension was fully developed. Thus it is possible that the elevated blood pressure could have suppressed the PRA. Our results confirm those of Kobayashi et al (19) in that plasma AII levels were elevated at 14 days. Had we monitored plasma AII levels only at the end of the fourth week of fructose treatment, our results would have supported the data of Hwang et al (17) and our conclusion would have been similar suggesting that the RAS was not involved in the hypertension associated with fructose feeding in rats.

The results of the present study are the first to demonstrate in a time dependent manner that the RAS may be a significant contributor to the development of hypertension in this model. The fact that plasma AII levels is first elevated at which time BP is also elevated suggests that the RAS has a profound effect in the early stages of the development of hypertension in fructose-fed rats. With continuous fructose feeding plasma AII levels gradually decline and by the end of the fourth week of treatment the concentration of plasma AII were not significantly different than age-matched controls although levels were still two fold elevated compared to the fructose-fed animals at the end of week 1. However, despite this fall in plasma AII levels from the four fold increased observed after two weeks of dietary fructose treatment, blood pressure remained elevated in fructose-treated rats compared to the controls. The reason for the lack of a sustained elevation of AII during dietary treatment may be due to a decrease in renal sympathetic nerve activity possibly by a reflex mechanism due to elevated blood pressure. This in turn would be associated with a decrease in plasma renin activity (PRA) and thus a decrease in AII levels. This hypothesis is consistent with the previous observation of Hwang et al (17) who reported a decrease in PRA at the end of four weeks of fructose treatment. The fact that blood pressure remained elevated despite a reduction in the four fold increase of plasma AII levels suggests that the RAS may not be as critical in the maintenance of the hypertensive state although local RAS activation cannot be dissociated. It is also possible that less of an increase in plasma AII is necessary to maintain on already elevated blood pressure. Thus the blockade of the elevation of blood pressure with AII antagonists as reported previously (11,12) and the elevation of plasma AII as observed in this study together strongly suggest the involvement of the RAS in the development of hypertension in fructose-fed rats. Although it cannot be ruled out that the maintenance of the hypertensive state could also be due to altered vasorelaxant activities. Further studies will be required to assess the vasorelaxant activity of fructose-treated rats.

Associated with an increase in BP in fructose-treated rats was an increase in weight of the left ventricle. This manifestation is additional evidence to suggest that the fructose-treated rats were hypertensive. Cardiac hypertrophy has been previously demonstrated in experimental models of hypertension (20). Although the mechanism of ventricular hypertrophy in fructose-treated rats is not certain, results of previous studies suggest that AII may be involved in the process of cardiac hypertrophy (21). Recent studies suggest the RAS is involved in the development of cardiac hypertrophy by induction of growth factors (21). In this study we observed significant elevation in the weight of the left ventricle of fructose-treated rats and a significant increase in AII receptor density compared to the controls at the end of the third and fourth weeks of treatment. The reason for an increase in receptor density at the end of third and fourth weeks of treatment could be a compensatory response to the decrease in plasma AII following the initial increase. Alternately it is possible that the tissue RAS may be downregulated during the third and fourth week of treatment and could thereby lead to an upregulation of receptors. We have not assessed tissue RAS in this study and further studies will be required to address this aspect. Ventricular hypertrophy in fructose-treated rats observed in this study confirms the earlier results of Kobayashi et al (19) who also demonstrated left ventricular hypertrophy in fructose-fed rats. Furthermore, Kobayashi et al (19) showed that TCV-116, an AT_1 antagonist completely prevented the cardiac hypertrophy in fructose-fed rats. These results strongly suggest that AII mediates the ventricular hypertrophy via AT_1 receptors. However, it is possible that the cardiac hypertrophy may be secondary to the elevation of blood pressure.

In this study we identified a single class of high-affinity angiotensin II receptors in the all the tissues in both control and fructose-treated rats. To our knowledge this is the first study to report a time course of AII receptor density changes in fructose-treated rats. Unlike most peptide-receptor systems in which an increase in concentration of the peptide results in downregulation or a loss of membrane receptors, we did not see a significant fall in AII receptors in the aorta at the second week. Although B_{max} in the aorta of fructose-treated rats appeared elevated over that observed in the controls, the increase was not significant compared to the controls despite a 3.5 fold increase in plasma AII levels in fructose-treated rats at the end of the second week of treatment. It is possible that the plasma AII levels achieved during the second week of the study may not have been high enough to down-regulate the receptor density. Previous studies have demonstrated that changes in AII receptor density is prevented when elevation in plasma AII levels are associated with an increase in plasma levels of aldosterone (22). Theoretically this could be achieved by activation of the RAS which in turn would result in increased secretion of aldosterone and either upregulation or maintainance of the AII vascular smooth muscle receptor number by mechanisms that are still not clear. We have not determined plasma aldosterone levels in this study. Similar results have been observed in other models of hypertension (23). For instance, in the two-kidney, one-clip Goldblatt hypertensive rats, an experimental model of hypertension, there is an increase in vascular AII receptors despite an increase in PRA (23). Alternately it is possible that the expression of AII receptors may be increased due to potassium depletion (24,25) which was not assessed in this study. Furthermore, in a recent study we have observed the in vitro response to AII in the aorta of fructose-treated rats was much greater than the control at the end of second week of treatment whereas at the end of fourth week of fructose treatment the in vitro response of the aortic vessel to AII was less that of the control (manuscript in preparation). These observations are consistent with the results of the present study in that the AII receptor number tended to increase in the aorta of fructose-treated rats at the end of the second week although they did not attain statistical significance. This observation of an enhanced vascular reactivity to circulating AII may be the mechanism by which fructose-treated rats develop hypertension.

Since no change in receptor number were observed in the hypothalamus or adrenal gland these results suggest that no disturbance in angiotensinergic mechanism in these tissues. Although the expression of AT_1 receptor gene normally depends on the circulating AII, it is possible that in the adrenal gland and hypothalamus the local RAS may regulate the AII receptor density. The fact that no change in AII receptor density was observed in the hypothalamus does not preclude that the central RAS may not be involved in the development or maintenance of hypertension in the fructose model of hypertension since the RAS within the brain is known to modulate blood pressure in other experimental models of hypertension (26).

In conclusion, the results of this study demonstrate that chronic fructose treatment is associated with an initial increase in plasma AII at which time there is an elevation of blood pressure. These results suggest that the RAS contributes, at least in part to the development of hypertension in fructose-fed rats. Furthermore, the increase in blood pressure is associated with hyperinsulinemia, left ventricular hypertrophy and an increase in AII receptor density in the left ventricle after hypertension was fully developed. Any one or all of these factors may be involved in maintaining the elevated blood pressure in this model of hypertension.

ACKNOWLEDGMENTS

We would like to thank Drs. Ronald Smith of DuPont-Merck Pharmaceutical Co and Robert C. Speth of Peptide Radioiodination Center, Washington State University (Pullman, WA) for providing Losartan and ^{125}I-Sar-AII respectively. We also thank Ms. Kim Hanley, Ms. Gena Strubbe, Johanna Studer, Aubrey Johnson for their technical assistance and Ms. Ana Morgan for preparing this manuscript. This work was supported in part by a grant to Dr. Michael J. Katovich from the American Heart Association-Florida Affiliate.

REFERENCES

1. Hwang, I.S. Ho, H., Hoffman, B.B. and Reaven, G.M. (1987). Hypertension *10*:512-516.
2. Reaven, G.M., Ho, H. and Hoffman, B.B. (1988). Hypertension *12*:129-132.
3. Reaven, G.M., Ho, H. and Hoffman, B.B. (1989). Hypertension *14*:117-120.
4. Bonoza, E., Zavaroni, I., Alpi O., Pezzarossa, A., Bruschi, F., Guena, L., Coscelli, C. and Butturini, U. (1987) *30*:719-723.
5. Haffner, S.M., Fong, D., Hazada, H.P., Pugh, J.A. and Patterson, J.K. (1988). Metabolism 338-345.
6. Bhanot, S., McNeill, J. and Bryer-Ash, M. (1994) Hypertension *23*:308-312.
7. Rowe, J.W., Young, J.B. Minaker, R.J., Stevens, A.L., Pallotta, J. and Landsberg, L. (1981) Diabetes *30*:219-225.
8. Scammell, J.G. and Fregly, M.J. (1981) Proc. Soc. Exp. Biol. Med. *167*:117-121.
9. Fregly, M.J., Shectman, O., Van Bergen, P., Reeber, C. and Papanek, P.E. (1991) Pharmacol. Biocem. Behav. *38*:837-842.
10. Trovati, M., Massucco, P., Anfossi, G., Cavalot, F., Mularoni, E., Mattiello, L., Rocca, G. and Emanuelli, G. (1989) Metabolism *38*:501-503.

11. Limura, O., Shimamoto, K., Matsuda, K., Masuda, A., Takizawa, H., Higashiura, K. and Nakagawa, M. (1994) Am J. Hypertens *7*:12A.
12. Iyer, S.N. and Katovich, M.J. (1994) Life Sciences *55*:PL 139-144.
13. Baker, K.M., Campanile, C.P., Trachte, G.J. and Peach, M.J. Circ Res. (1984) *54*:286-293.
14. Baksi, S.N., Abhold, R.H. and Speth, R.C. (1989) J. Hypertens *7*:423-427.
15. Speth, R.C. Singh, R., Smeby, R.R., Ferrario, C. and Husain, A. (1984) Neuroendocrinology *38*:387-392.
16. Chiu, A.T., Duncia, J.V., McCall, D.E., Wong, P.C., Perice, W.A., Thoolen, M.J.M.C., Carini, D.J., Johnson, A.L. and Timmermans, P.B.M.W.M. (1989). J. Pharmacol Exp. Ther. *250*:867-875.
17. Hwang, I.S., Hwang, W.C., Wu, J.N., Shian, L.R. and Reaven, G.M. (1989) J. Hypertens *2*:424-427.
18. Young, J. and Landsberg, L. (1982) J. Chron Dis. *35*:879-886.
19. Kobayashi, R., Nagano, M., Nakamura, P., Hiyaki, J., Fujioka, Y., Ikagami, H., Mikanni, H., Kawaguchi, N., Onishi, S. and Ogihara, T. (1993) Hypertension *21*:1051-1055.
20. Korner, P.I. (1982) Cli Sci. *63*:5S-62S.
21. Villarreal, F.J. and Dillman, W.H. (1992) Am J. Physiol. *362*:H1861-H1866.
22. Schiffrin, E.L. Franks, D.J. and Gutkowska, J. (1985) Can J. Physiol Pharmacol *64*:1522-1527.
23. Schiffrin, E.L. Thome, F.S. and Genest, J. (1983) Hypertension *5*(Suppl V): V-16-V-21.
24. Paller, M.S., Douglas, J.G. and Linas, S.L. (1984) J. Clin. Invest. *73*:79-86.
25. Linas, S.L., Marzee-Calvert, R. and Ullian, M.E. (1990) Am. J. Physiol. *258*:C849-C854.
26. Harrap, S.B. (1991) Am J Hypertens. *4*:2125-2165.

7

CARDIAC EFFECTS OF AII

AT_{1A} Receptor Signaling, Desensitization, and Internalization

W. G. Thomas, T. J. Thekkumkara, and K. M. Baker

Weis Center for Research
Geisinger Clinic
North Academy Avenue
Danville, Pennsylvania 17822

SUMMARY

Angiotensin II receptors present in cardiomyocytes, nonmyocytes (predominantly fibroblasts), nerve terminals, and the heart vasculature mediate the multiple actions of angiotensin II (AII) in the heart, including modulation of normal and pathophysiological cardiac growth. Although the cellular processes that couple AII receptors (principally the AT1 subtype) to effector responses are not completely understood, recent studies have identified an array of signal transduction pathways activated by AII in cardiac cells. These include: the stimulation of phospholipase C which results in the activation of protein kinase C and the release of calcium from intracellular stores; an enhancement of phosphaditic acid formation; the coupling to soluble tyrosine kinase phosphorylation events; the initiation of the mitogen activated protein kinase (MAPK) cascade; and the induction of the STAT (Signal Transducers and Activators of Transcription) signaling pathway. It is tempting to speculate that these latter responses, which have been previously associated with growth factor signaling pathways, are involved in AII-induced cardiac growth. Interestingly, some of these novel pathways are apparently not under the same strict control imposed upon the more classical signaling pathways. Thus, while AII-induced calcium transients are rapidly (within minutes) desensitized following exposure to AII, the MAP kinase pathway is not, and activation of the STAT pathway requires hours of agonist exposure for maximal induction. These observations support an emerging picture in which the downstream signal transduction pathways of AII receptors are initiated and terminated with a distinct temporal arrangement. This organization allows appropriate rapid responses (e.g. vascular contraction) to transient AII exposure, some of which are rapidly ter-

Recent Advances in Cellular and Molecular Aspects of Angiotensin Receptors
Edited by Mohan K. Raizada et al., Plenum Press, New York, 1996

minated, perhaps for protective reasons, and others not. In contrast, additional responses (e.g. growth) probably require prolonged exposure to agonist.

INTRODUCTION

Comprehensive reviews are available concerning the multiple actions of AII on the heart, the contribution of systemic and locally produced AII, via a putative cardiac renin-angiotensin system, to cardiac function, and the signal transduction pathways activated by AII in cardiac cells [1-4]. It is not our intention to reiterate the contents of these reviews, but rather to provide an overview of the actions of AII on the heart in order to orientate the reader and to form a basis for understanding the central position occupied by AII receptors in the control of heart function. This article focuses on some recently described novel signaling pathways activated by AII and addresses the role of the cytoplasmic tail of the AT_{1A} receptor in controlling desensitization and internalization.

Cardiac Actions of AII

Angiotensin II (AII) promotes multiple short-term and long-term cardiac effects by (1) acting directly through AII receptors present on cardiac cells, (2) indirectly through the release of other factors and (3) possibly via crosstalk with the intracellular signaling pathways of various cardiac hormones. AII indirectly affects cardiac function through interaction with cardiovascular regulatory sites in the brain, stimulation of aldosterone production by the adrenal gland, and by modulation of sympathetic tone [2]. Directly, AII induces positive chronotropic and inotropic effects on cardiac muscle, as well as the alteration of cardiac metabolism and the vasoconstriction of coronary blood vessels. In addition, many studies implicate AII in the modulation of normal (neonatal period) and pathological (pressure and volume overload and post-myocardial infarction) cardiac growth [2,5,6].

These diverse cardiac actions may occur in response to either systemically-derived AII or to AII generated locally by an intra-cardiac renin-angiotensin system (RAS)[2,7]. Cardiac tissue contains the precursor components (angiotensinogen, renin and angiotensin converting enzyme) and the end-effectors (principally AII and AT_1 receptors) of the RAS [2,5,7-10]. The presence of these components and their regulation during development and in cardiac hypertrophy and following myocardial infarction, provides support for a functional cardiac RAS [5]. Sadoshima et al. [11] have recently demonstrated that the hypertrophic growth of cultured neonatal rat cardiomyocytes, in response to static stretch, is blocked by an AT_1 receptor antagonist, providing the most compelling evidence yet that locally produced AII is important for biological responses in cardiac cells. Moreover, the development and maintenance of left ventricular hypertrophy, in the pressure-overload rat model, is prevented by ACE inhibition or blockage of the AT_1 receptor with losartan [12-14]. Thus, both *in vitro* and *in vivo* evidence strongly support a functional role for locally generated AII.

Cardiac AII Receptors

AII receptors have been detected throughout the rat heart, with high densities over cardiac nerves and lower levels associated with the atria, ventricles and vasculature [15]. AII receptors are upregulated following myocardial infarction [16,17] and hypertrophy [18], but downregulated with endstage heart failure [19]. The major subtypes of AII receptors (AT_{1A}, AT_{1B} and AT_2) have all been identified in cardiac tissue [17], but the contribution of, and interplay between, each receptor subtype in the control of AII-induced cardiac function is unclear. Interestingly, most cardiac effects of AII can be blocked by specific AT_1 receptor antagonists. While these antagonists do not discern between the highly homologous AT_{1A} and AT_{1B} receptors, recent studies using selective DNA probes have provided evidence that these two subtypes are differentially regulated and subserve different functions. In rats, AII treatment increases adrenal AT_{1B}, but decreases AT_{1A} mRNA in the vasculature [20]. Conversely, AT_{1A}, but not AT_{1B}, mRNA is elevated in the hearts of rats following myocardial infarction [17], and AT_{1A}, but not AT_{1B}, mRNA is detected in the hearts of embryonic and neonatal rats [21]. These data, apart from underscoring the differential expression of AT_{1A} and AT_{1B} receptors, also infer a key role for the AT_{1A} subtype in normal and abnormal heart growth.

The role of AT_2 receptors is less clear, but the presence of high levels in fetal tissue [22] implies a role in fetal growth and development. Although some controversy exists as to whether the AT_2 receptor couples to heterotrimeric G-proteins, the cellular signaling pathways evoked by AT_2 involve a G-protein mediated stimulation of a protein tyrosine phosphatase [23]. Given the association of tyrosine phosphorylation events and growth pathways, the inhibition of tyrosine phosphorylation following AT_2 receptor activation, as recently demonstrated by Nahmias et al. [24], suggests anti-growth actions for AT_2. Indeed, AT_2 receptors have been implicated in the apoptosis (programmed cell death) of rat ovarian granulosa cells [25] and in the inhibition of coronary endothelial cell proliferation [26]. In addition, the hypertrophic effects of AII on neonatal cardiomyocytes are enhanced by treatment with an AT_2 receptor antagonist [Booz and Baker, unpublished data]. These observations raise the interesting possibility that cross-talk and antagonism occurs between AT_1 and AT_2 receptors. Interestingly, the AT_2 receptors increase in parallel with AT_{1A} in the infarcted rat heart model [17] and in cardiac hypertrophy [18], perhaps in an attempt to compensate for the deleterious growth effects of AT_1 receptors.

AII Growth Effects on Cultured Cardiac Cells

While animal studies have proven useful in establishing a key role for AII in heart function, these do not delineate direct versus indirect effects of AII or dismiss possible "non-specific" effects of some treatments such as ACE inhibition (e.g. decreased bradykinin degradation and reduced cardiac afterload). Therefore, cell culture models have been utilized to address the questions of direct growth effects of AII and to unravel the myriad of signaling pathways and cellular responses activated by AII.

Using cultured chick cardiomyocytes, our group first described the presence of functional AII receptors in myocytes [27] and the AII-receptor mediated hypertrophy of these cells in response to AII [28,29]. AII-stimulated hypertrophy has also been demonstrated in cultured rat neonatal myocytes, where AII induces immediate-early genes (*c-fos*, *c-jun*, *jun B*, *Egr-1* and *c-myc*), genes for growth factors (TGF-b) and the hypertrophic marker genes, a-actin and atrial natriuretic factor [30]. Other cardiac cells are also targets for AII. Cultured neonatal cardiac fibroblasts express abundant, high affinity G-protein coupled AT_1 receptors [31] that couple AII stimulation to proliferative growth responses [32,33]. These responses of cardiomyocytes and fibroblasts were blocked by AT_1, but not AT_2 receptor antagonists [30,31]. AII also induces the production of collagen and the deposition of extracellular matrix by cardiac fibroblasts, consistent with the observation that myocardial fibrosis, which often exists in diseased hearts, can be prevented by ACE inhibition [34]. Thus, AII induced changes in cardiac growth and function presumably reflect a combination of AT_1 receptor-mediated cardiomyocyte hypertrophy as well as proliferation and deposition of extracellular matrix by fibroblasts.

AII SIGNAL TRANSDUCTION

AT_1 receptors mediate most of the physiological actions of AII and, as discussed above, this subtype is predominant in the control of AII-induced cardiac functions. The coupling of these receptors to phosphatidylinositide hydrolysis, via a G-protein mediated stimulation of phospholipase C(PLC), is well documented in tissues that express AT_1 receptors, including cardiac myocytes and fibroblasts [3]. Activation of PLC leads to the formation of diacylglycerol (DAG) and phosphatidylinositol 4,5-bisphosphate (IP_3) which transiently activate protein kinase C (PKC) and the release of calcium from intracellular stores, respectively. While these second messengers play a role in mediating AII-induced *c-fos* gene expression in cardiac myocytes and fibroblasts [35], recent evidence, however, suggests strongly that the cellular processes initiated by increasing intracellular calcium and PKC are insufficient for AII mediated growth [3,34,36]. Thus, attention has now been focused on the observation that AT_1 receptors, couple to multiple novel second messenger pathways, which have traditionally been associated with signaling by growth factor receptors.

In cardiac fibroblasts [32] and in CHO-K1 cells transfected with the AT_{1A} receptor [37], AII promotes a rapid and robust activation of mitogen activated protein kinase (MAPK). MAPK is activated by phosphorylation and is recognized as a key regulator of cell proliferation and differentiation because it represents the convergence point for numerous intracellular signals, initiated at the cell membrane, and the relay of these signals into the nucleus. MAPK phosphorylates and regulates a variety of transcription factors (e.g. ribosomal protein S6-kinase), protooncogene products (e.g. c-myc), other enzymes (e.g. phospholipase A2) and receptors (e.g. epidermal growth factor receptor). While the mechanism through which AT_1 receptor couples to MAPK is unclear, the activation can be attenuated by inhibition of G-protein function with pertussis toxin treatment [32]. We recently showed that AII stimulation of cultured neonatal rat fibroblasts (NRF) rapidly induced the formation of phosphatidic acid and that phosphatidic acid treatment was capable of inducing MAPK activity [36]. Interestingly, phosphatidic acid, which is formed through the combined activation of phospholipase D, PKC and DAG kinase, has been proposed as a participant in growth factor induced mitogenesis [38]. In addition to MAPK, we have shown in NRF that AII induces the rapid tyrosine phosphorylation of members of the SHC family of proteins and the focal adhesion kinase FAK^{125}, as well as other unidentified proteins [39]. Leduc and Meloche [40] recently demonstrated the AII induced tyrosine phosphorylation of Paxillin, another focal adhesion-associated protein. Activation of tyrosine phosphorylation events has also been demonstrated for other G-protein coupled receptors linked to stimulation of cell growth, such as those for thrombin, bombesin, vasopressin and endothelin. Moreover, for these receptors, blockade of tyrosine phosphorylation abolished growth responses [discussed in 40]. Whether AII growth effects are mediated through tyrosine phosphorylation events, the mechanism by which G-protein coupled receptors activate these pathways and the identity of the induced substrates and kinases represent important areas for future investigation.

Of recent interest, is the demonstration that many growth factors and cytokines activate a common direct signal transduction pathway that involves tyrosine phosphorylation of the cytoplasmic STAT proteins (STAT, Signal Transducers and Activators of Transcription), and the association of these STAT proteins into complexes which then translocate to the nucleus and activate gene transcription [41]. These processes are initiated by tyrosine phosphorylation of reisdues within the cytoplasmic regions of the growth factor and cytokine receptors [42]. We have shown that AII stimulates the tyrosine phosphorylation and nuclear translocation of STAT1 (stat91) [43] and STAT3 (stat92) [Bhat and Baker, unpublished]. Using electrophoretic mobility shift assays, we have also shown that AII, directly through the AT_{1A} receptor, induces the formation of a particular complex of STAT proteins termed SIF (*sis* inducing factor) which binds the DNA sequence, SIE (sis inducing element) present in the promoter region of many genes. This observation represents the first evidence for coupling of a G-protein linked receptor to this direct STAT pathway, and suggests that AII may use this pathway to activate growth factor-like gene programs. Interestingly, in contrast to growth factors and cytokines which activate this pathway in minutes, activation by AII is maximal only after 1-2h of agonist exposure. This delayed activation is in keeping with the observation that exposure to AII for extended periods is required for growth effects [31]. It also implies that the SIF-mediated AII gene induction, which is temporally displaced from that of growth factors, mobilizes a unique assembly of genes. Such a delay may also be necessary to prevent inappropriate cell growth in response to the transient surges of AII essential for the minute to minute regulation of blood pressure and salt and water homeostasis. The mechanism responsible for this delay, the question of whether SIF activation by AII is a G-protein coupled phenomena, and the role of tyrosine residues within the cytoplasmic regions of the AT_{1A} receptor in coupling to STAT pathways are presently under investigation.

CONTROL OF AT_{1A} RECEPTOR FUNCTION

Because plasma membrane AII receptors recognize extracellular AII and initiate the intracellular signaling pathways described above, they provide a critical control point for cellular responses to AII. Given the central importance of AT_1 receptors to cardiac AII function, it is important to understand the processes, occurring at the level of the receptor, which are responsible for initiation, maintenance and termination of AT_1 receptor signaling

Receptor Desensitization

Desensitization is a term used to describe the rapid loss of receptor responsiveness subsequent to an initial exposure of receptor to agonist. Rapid desensitization of AII mediated responses *in vivo*, in preparations of isolated blood vessels and in cell culture are well documented [44]. Using CHO-K1 cells stably expressing the AT_{1A} receptor, we have shown that AII-induced intracellular calcium transients are rapidly desensitized in response to agonist, and that, upon removal of agonist, up to 1 h is required for full resensitization [45]. For the prototypical G-protein coupled receptor, the b_2-adrenergic receptor, the process of desensitization has been extensively studied and appears to result from a functional uncoupling of the agonist stimulated receptor from its associated G-protein. this uncoupling coincides with phosphorylation of serine and threonine residues, located within the cytoplasmic regions of the receptor, by general cellular kinases such as PKC and PKA or by specific G-protein coupled receptor kinases [46]. In contrast, little is known regarding the processes that control AII mediated desensitization, and only two preliminary reports [47, 48] have described phosphorylation of AT_1 receptors. Paxton et al. [47] reported a constitutive phosphorylation of the receptor on serine and tyrosine residues that was unchanged upon AII stimulation, an observation that is difficult to reconcile given the dynamic agonist-driven phosphorylation of many G-protein coupled receptors. The carboxy-terminal region, which contains numerous serine, threonine and tyrosine residues, was not a substrate for PKA, PKC or MAPK, but it was an *in vitro* substrate for the Src family of kinases. In the other study, [48], a rapid agonist-induced phosphorylation on serine and tyrosine residues was demonstrated, which was unresponsive to activation of PKC and calcium mobilization, but enhanced ~2-fold by PKA activation. Issues not addressed in these somewhat conflicting reports, were the identification of the specific residues phosphorylated in response to agonist and an association, if any, between receptor phosphorylation and desensitization. Controversy also exists over the role of PKC in AII receptor desensitization [44]. While Dzau et al. [44] argue in favor of a role for PKC in AT_1 receptor desensitization, we have observed no effect of PKC activation or inhibition on AII induced desensitization of calcium signaling in CHO-K1 cells transfected with the AT_{1A} receptor [45] or in primary cultures of rat neonatal cardiac fibroblasts (Thomas, Booz, Baker and Thekkumkara, manuscript in preparation). Moreover, in agreement with the phosphorylation studies described above, we find that a peptide corresponding to the carboxy-terminal residues 325-351 of the AT_{1A} receptor, encompassing three putative PKC sites, is a poor substrate for PKC. Clearly, these issues require clarification, as does the possible phosphorylation and desensitization of the AT_{1A} receptor by specific receptor kinases.

In order to more directly access the role of the serine/threonine-rich carboxy-terminus of the AT_{1A} receptor in desensitization, we stably expressed in CHO-K1 cells a mutated form of the AT_{1A} receptor, truncated after Leucine314 to remove 45 amino acids including 11 serine and 2 threonine residues [37]. These truncated receptors displayed high affinity for AII, coupled to G-protein(s) and responded to AII stimulation with a transient increase in intracellular calcium and MAPK activity. We hypothesized that cells expressing these truncated receptors would be unable to terminate AII mediated signaling or would be repetitively responsive to multiple AII challenges. Interestingly, we observed normal initiation and termination of calcium transients in response to AII and refractoriness to additional AII stimulation suggestive of receptor desensitization. While we can not dismiss the possibility that downstream mechanisms are responsible, these results suggest that phospho-

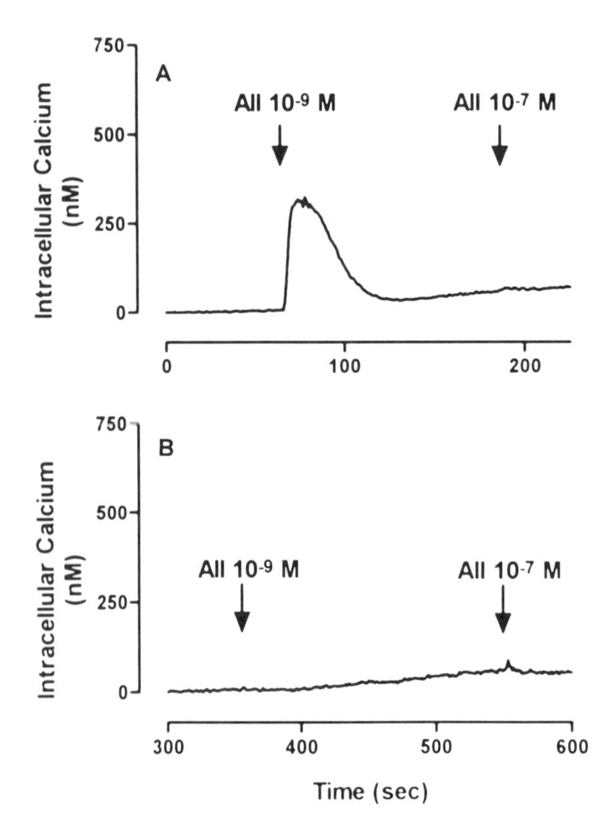

Figure 1. Prolonged desensitization of AII induced calcium responses in CHO-K1 cells expressing the AT_{1A} receptor (T3CHO/AT_{1A}). We have previously shown that T3CHO/AT_{1A} cells respond to a maximal stimulation of AII (100-1000nM) with a transient increase in intracellular calcium of around 1000nM [45]. In A, T3CHO/AT_{1A} cells were loaded with Fura-2AM as previously described [45] and stimulated with an initial submaximal dose of 1nM AII and a second maximal dose of 100nM AII, 100sec later, as indicated. In B, after an initial dose of AII (1nM for 2min), the cells were washed to remove AII and then restimulated either 6min later with 1nM AII or at 9min with 100nM AII. These results demonstrate that stimulation of intracellular calcium transients by a submaximal dose of AII completely prevents subsequent stimulation by similar or maximal doses up to 10min after the initial exposure.

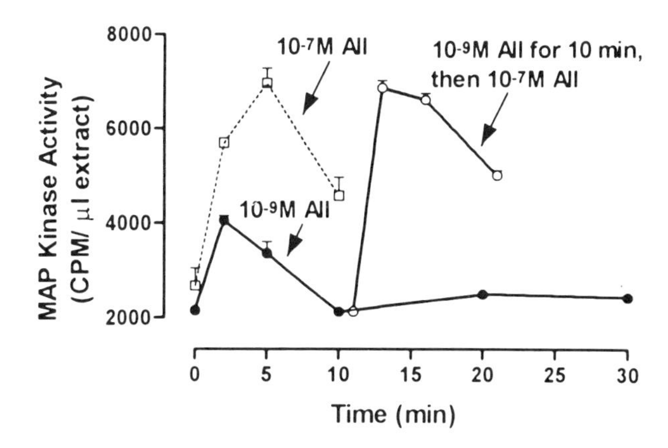

Figure 2. MAPK activation is not desensitized in response to AII. Serum-starved T3CHO/AT_{1A} cells were stimulated, for varying time periods, with either 1nM AII (filled circles) or 100nM AII (open squares), or 1nM AII for 10min followed by a second dose of 100nM (open circles), as indicated. Preparation of cytoplasmic extracts and MAPK assays were as previously described [37]. As was observed for AII induced calcium transients, 1nM AII produced a submaximal stimulation of MAPK activity compared to 100nM AII. However, in contrast, 10 min after an initial dose of 1nM AII, 100 nM AII was able stimulate a MAPK response equivalent in magnitude to that observed to a single 100nM dose of AII. Thus, using a similar experiment paradigm, AII induces desensitization of calcium, but not MAPK, responses.

rylation of the cytoplasmic tail is not a prerequisite for desensitization and that agonist occupied receptor may spontaneously transform to an inactive form and uncouple from signaling pathways. Similar conclusions were drawn by Boulay et al. [49] in their study of short-term desensitization of AII receptors in bovine adrenal glomerulosa cells, and by Robertson et al. [50] who proposed a two-state receptor model to explain their observation that losartan, a specific non-peptide AT_1 receptor antagonist, was capable of reversing AII mediated receptor desensitization.

The desensitization process is even more perplexing when other AII stimulated signaling pathways are examined. While the IP_3 mediated calcium mobilization is rapidly desensitized by submaximal (nanomolar) doses of AII [45; Fig. 1], the activation of the MAPK cascade by AII, under similar conditions is not [Fig. 2]. This observation raises the very interesting possibility that receptor signaling can be differentially desensitized. Such a possibility is further illustrated by our observation that stimulation of the STAT pathway by AII in cardiac fibroblasts, and in CHO-K1 cells stably expressing the AT_{1A} receptor, is delayed (in comparison to rapid activation of calcium and MAPK signaling) and maximal at 2-3 h [43]. This activation of the STAT pathway can be prevented or diminished by competition with EXP3174, an AT_1 receptor antagonist, at times up to 1 h after the initiation of AII treatment (Bhat and Baker, unpublished data), suggesting that prolonged receptor occupancy is required for induction. The mechanisms utilized by the AT_{1A} receptor to segregate these divergent pathways and separate the temporal initiation and termination of responses are unclear, but may include processes distinct from coupling and uncoupling from G-proteins. Such a possibility is supported by recent observations in some non-G-protein coupled receptors. First, activation of signal transduction can occur, in the absence of ligand occupancy, if receptor dimerization or aggregation is induced [51]. Second, in an elegant study on integrin receptors [52], a clear distinction was demonstrated between responses (tyrosine phosphorylation signaling and cytoskeletal organization) and the type of transmembrane signal (ligand occupancy, receptor aggregation, or a combination of both). Although

AII receptors reportedly dimerize and aggregate, these processes are poorly characterized. Interestingly, receptor aggregation precedes internalization and there is some evidence that internalized receptors retain the capacity to couple to signaling pathways [53]. Hence, in order to gain a better understanding of the molecular processes that control receptor aggregation and internalization, we have performed mutations on the AT_{1A} receptor and investigated the effect on agonist-stimulated endocytosis.

Receptor Internalization

Binding of AII to its receptor in primary cell cultures, isolated from a variety of tissues, initiates rapid endocytosis, and cloned AT_1 (AT_{1A} and AT_{1B})[54,55], but not AT_2[55], receptors expressed in CHO-K1 and COS-7 cells, rapidly internalize in response to AII exposure. AT_1 receptor internalization occurs independently of G-protein coupling [56], and is stimulated effectively by both peptide agonists and antagonists [54], but not by the non-peptide antagonist, DuP753 [54]. Although the mechanism for AT_1 receptor endocytosis has not been established, it probably proceeds via clathrin-coated pits, based on a susceptibility to high concentrations of sucrose [54]. Upon agonist stimulation, receptors presumably undergo an allosteric change that exposes cytoplasmic domains to adaptor proteins which bind the receptors and mediate the selective recruitment of receptors into clathrin-coated pits [57]. The exact mechanism by which the adaptor proteins couple to transmembrane receptors is not yet resolved, but it is generally accepted that internalization motifs are present within receptor cytoplasmic domains (commonly the cytoplasmic tail) and that these motifs are recognized and bound by the adaptors. A variety of such consensus motifs for receptor endocytosis exist among receptors that utilize the clathrin-coated pit pathway [58], including those of the G-protein coupled receptor superfamily. Typically, motifs include: (1) crucial tyrosine residues (e.g. NPXY, NPXXY or YXX-hydro, where X is any amino acid and hydro is a large bulky hydrophobic amino acid); (2) stretches of hydrophobic or serine/threonine residues; and (3) dileucine motifs. Sometimes multiple motifs are present within a single receptor.

Regions of the AT_{1A} receptor involved in AII mediated endocytosis have recently been identified. Deletion of the N-terminal segment of the third intracellular loop inhibited internalization [56], but extensive point mutations in this region did not [59]. Thus, the contribution of amino acids in this area to endocytosis is unclear. We recently described the stable expression in CHO-K1 cells of a truncated AT_{1A} receptor [37; detailed above], devoid of the carboxy-terminal 45 amino acids. This truncated receptor displayed severely impaired agonist-induced endocytosis, implicating the carboxy-terminal tail in receptor internalization. This region of the AT_{1A} receptor has a number of particularly attractive candidates for the internalization motif: multiple tyrosine residues (Y302(in a NPXXY motif), Y312, Y319 and Y339); clusters of serine and threonine residues; an AKS motif similar to the DAKSS motif for the yeast a-factor receptor [60];and a dileucine motif at L316L317. In a follow-up study [Thomas, Baker, Thekkumkara, manuscript in preparation], we have utilized various truncations and deletions of the carboxy-terminus of the AT_{1A} receptor, as well as point mutations to identify the critical amino acids required for internalization. Truncation of the AT_{1A} receptor after K333 resulted in a receptor that was inhibited in its capacity to undergo endocytosis, but not to the extent of truncation after L314. Moreover, a deletion mutant (315-329), demonstrated reduced capacity for internalization, similar to the K333 truncated receptor. These observations raised the interestingly possibility that two separate motifs are present; one distal to K333 and one located in the region 315-329. Very recently, Hunyday et al. [55] identified the contribution of a serine335-leucine336-threonine337 motif, distal to K333, to AT_{1A} receptor internalization. However, the deficient internalization observed for our deletion mutant (315-329), where this SLT motif is maintained, suggests that this collection of residues, although critical, are not sufficient for AT_{1A} receptor endocytosis. Thus, we have focused our attention to more proximal regions of the carboxy-terminus. We have mutated Y302, Y312, Y319, and K325 to alanine and L316 to phenylalanine, and stably expressed these mutated receptors in CHO-K1 cells. Mutants Y302A, Y312A and K325A showed internalization profiles indistinguishable from the wild type receptor. Interestingly, two of the mutants (L316F and Y319A), which fall within the 315-329 region, displayed significantly reduced internalization rates. Substitution of L316 is the most effective single-point mutant yet described in inhibiting AT_{1A} receptor endocytosis, implying a key role for this residue and region in the internalization process. Just how, and if, L316 and Y319 combine with each other to form a site utilized for internalization, and the interaction of this region with the more distally positioned sites [55] and other regions of the receptor implicated in the endocytotic pathway [56] remains to be determined.

FUTURE PERSPECTIVES

Our present focus is to understand the signaling, desensitization and internalization processes that occur immediately following agonist stimulation of AT_{1A} receptors. With the recent observation that AII activates previously identified growth factor and cytokine signal transduction pathways, the question of control and separation arises. At present, no information is available regarding the mechanism(s) by which the AT_{1A} receptor couples to tyrosine phosphorylation and activation of the STAT pathway, or whether these processes are mediated through heterotrimeric G-proteins. Possible crosstalk between G-protein coupled receptors, such as AT_{1A}, and the receptors for growth factor and cytokines also merits investigation. Equally important will be the elucidation of possible crosstalk and antagonism among different classes (AT_{1A}, AT_{1B}, AT_2, etc.) of AII receptors, as well as establishing a role for secondary receptor processes, such as aggregation and internalization, in the cellular responses to AII. We and others have identified residues within the AT_{1A} receptor important for internalization. What regions of the receptor interact with the endocytotic machinery? What are the agonist induced allosteric switches that differentiate signaling and internalization? Is internalization simply a process for clearing the membrane of ligand bound receptors, or do internalized receptors retain the capacity for signaling? Finally, is all internalized AII degraded or does it perhaps serve to activate putative cytoplasmic and nuclear receptors?

ACKNOWLEDGMENTS

The authors thank Dr. H.E. Morgan for critically reading this manuscript and appreciate many productive discussions regarding this work with members of the Baker Laboratory. Studies described in this article were supported by the Geisinger Clinic Foundation, NIH grants (HL44883 and HL44379 to K.M.B.), Amercian Heart Association grants (AHA94013470 to T.J.T. and AHA91003020 to K.M.B.) and a grant from the Pennsylvania Affiliate of the American Heart Association to K.M.B. K.M.B is an Established Investigator of the American Heart Association. W.G.T. is the recipient of a CJ Martin Fellowship from the National Health and Medical Research Council of Australia.

REFERENCES

1. Morgan, H.E., and Baker, K.M. (1991). Circulation *83*:13-25.
2. Baker, K.M., Booz, G.W., and Dostal, D.E. (1992). Annu. Rev. Physiol. *54*:227-241.

3. Booz, G.W., Dostal, D.E., and Baker, K.M. (1994), in The Cardiac Renin Angiotensin System (eds. K. Lindpainter, D. Ganten), Futura, NY, pp 101-123.
4. Rogers, T.B., and Lokuta, A.J. (1994). Trends Cardiovasc. Med. *4*:100-116.
5. Dostal, D.E., and Baker, K.M. (1993). Trends Cardiovasc. Med. *3*:67-74.
6. Dostal, D.E., Baker, K.M., and Peach, M.J. (1991), in Horizons in Endocrinology (eds. M. Maggi, V. Greenen), Raven Press, NY, *76*:265-272.
7. Dostal, D.E., Booz, G.W., and Baker, K.M. (1994), in The Cardiac Renin Angiotensin System (eds. K. Lindpainter, D. Ganten), Futura, NY, pp 1-20.
8. Paul, M., Wagner, J., and Dzau, V.J. (1993). J. Clin. Invest. *91*:2058-2064.
9. Dostal, D.E., Rothblum, K.C., Chernin, M.I., Copper, G.R., and Baker, K.M. (1992). Am. J. Physiol. *263*:C838-863.
10. Dostal, D.E., Rothblum, K.C., Conrad, K.M., Cooper, G.R., Baker, K.M. (1992). Am. J. Physiol. *263*:C851-C863.
11. Sadoshima, J., Xu, Y., Slayter, H.S., and Izumo, S. (1993). Cell *75*:977-984.
12. Baker, K.M., Chernin, M.I., Wixson, S.K., and Aceto, J.F. (1990). Am. J. Physiol. *259*: H324-332.
13. Everett, A.D., Tufro-McReddie, A., Fisher, A., and Gomez, R.A. (1994). Hypertension 23:587-592.
14. Bruckschlegel, G., Holmer, S.R., Jandeleit, K., Grimm, D., Muders, F., Kromer, E.P., Riegger, G.A.J., and Schunkert, H. (1995). Hypertension *25*:250-259.
15. Zhou, J., Allen, A.M., Yamada, H., Sun, Y., and Mendelsohn, F.A.O. (1994), in The Cardiac Renin Angiotensin System (eds. K. Lindpainter, D. Ganten), Futura, NY, pp 63-88.
16. Meggs, L.G., Coupet, J., Huang, H., Li, W.C.P., Capasso, J.M., Homcy, C.J., and Anversa, P. (1993). Circ. Res. *72*:1149-1162.
17. Nio, Y., Matsubara, H., Murasawa, S., Kanasaki, M., and Inada, M. (1995). J. Clin. Invest. *95*:46-54.
18. Suzuki, J., Matsubara, H., Urakami, M., and Inada, M. (1993). Circ. Res. *73*:439-447.
19. Regitz-Zagrosek, V., Friedel, N., Heymann, A., Bauer, P., Neuss, M., Rolfs, A., Steffen, C., Hildebrandt, A., Hetzer, R., and Fleck, E. (1995). Circulation *91*:1461-1471.
20. Inagami, T., Iwai, N., Sasaki, K., Guo, D.F., Furata, H., Yamano, Y., Bardhan, S., Chaki, S., Makito, N., and Badr, K. (1993). Drug Res. *43*:226-228.
21. Shanmugan, S., Corvol, P., and Gasc, J. (1994). Am. J. Physiol. *267*:E828-836.
22. Grady, E.F., Sechi, L.A., Griffin, C.A., Schambelan, M., and Kalinyak, J.E. (1991). J. Clin. Invest. *88*:921-933.
23. Buisson, B., Laflamme, L., Bottari, S.P., De Gasparo, M., Gallo-Payet, N., and Payet, M.D. (1995). J. Biol. Chem. *270*:1670-1674.
24. Nahmias, C., Cazaubon, S.M., Briend-Sutren, M.M., Lazard, D., Villgeios, P., and Strosberg, A.D. (1995). Biochem. J. *306*:87-92.
25. Tanaka, M., Ohnishi, J., Ozawa, Y., Sugimoto, M., Usuki, S., Naruse, M., Murakami, K., and Miyazaki, H. (1995). Biochem. Biophys. Res. Commun. *207*:593-598.
26. Stoll, M., Steckelings, M., Paul, M., Bottari, S.P., Metzger, R., and Unger, T. (1995). J. Clin. Invest. *95*:651-657.
27. Baker, K.M., Singer, H.A., and Aceto, J.F. (1989). J. Pharmacol. Exp. Ther. *251*:578-585.
28. Aceto, J.F., and Baker, K.M. (1990). Am. J. Physiol. *258*:H806-813.
29. Baker, K.M., and Aceto, J.F. (1990). Am. J. Physiol. *259*:H610-618.
30. Sadoshima, J., and Izumo, S. (1993). Circ. Res. *73*:413-423.
31. Schorb, W., Booz, G.W., Dostal, D.E., Chang, K.C., and Baker, K.M. (1993). Circ. Res. *72*:1245-1254.
32. Schorb, W., Conrad, K.C., Singer, H.A., Dostal, D.E., and Baker, K.M. J. Mol. Cell. Cardiol. (in press).
33. Booz, G.W., Dostal, D.E., Singer, H.A., and Baker, K.M. (1994). Am. J. Physiol. *267*:C1308-1318.
34. Booz, G.W., and Baker, K.M. Cardiovasc. Res. (in press).
35. Sadoshima, J, and Izumo, S. (1993). Circ. Res. *73*:424-438.
36. Booz, G.W., Taher, M.M., Baker, K.M., and Singer, H.A. (1994). Mol. Cell. Biochem. *141*:135-143.
37. Thomas, W.G., Thekkumkara, T.J., Motel, T.J., and Baker, K.M. (1995). J. Biol. Chem. *270*:207-213.
38. Inui, H., Kondo, T., Konishi, F., Kitami, Y., and Inagami, T. (1994). Biochem. Biophys. Res. Commun. *205*:1338-1344.
39. Schorb, W., Peeler, T.C., Madigan, N.N., Conrad, K.M., and Baker, K.M. (1994). J. Biol. Chem. *269*:19626-19632.
40. Leduc, I., and Meloche, S. (1995). J. Biol. Chem. *270*:4401-4404.
41. Darnell, J.E., Kerr, I.M., and Stark, G.R. (1994). Science *264*:1415-1421.
42. Stahl, N., Farruggella, T.J., Boulton, T.G., Zhong, Z., Darnell, J.E., and Yancopoulos, G.D. (1995). Science *267*:1349-1353.

43. Bhat, G.J., Thekkumkara, T.J., Thomas, W.G., Conrad, K.C., and Baker, K.M. (1994). J. Biol. Chem. *269*:31443-31449.
44. Sasamura, H., Dzau, V.J., and Pratt, R.E. (1994). Kidney Int. *46*:1499-1501.
45. Thekkumkara, T.J., Du, J., Dostal, D.E., Motel, T.J., Thomas, W.G., and Baker, K.M. Mol. Cell. Biochem. (in press).
46. Premont, R.T., Inglese, J., Lefkowitz, R.J. (1995). FASEB J. *9*:175-182.
47. Paxton, W.G., Marrero, M.B., Klein, J.D., Delafontaine, P., Berk, B.C., and Bernstein, K.E. (1994). Biochem. Biophys. Res. Commun. *200*:260-267.
48. Kai, H., Griendling, K.K., Lassegue, B., Ollerenshaw, J.D., Runge, M.S., and Alexander, R.W. (1994). Hypertension *24*:523-527.
49. Boulay, G., Chretien, L., Richard, D.E., and Guillemette, G. (1994). Endocrinology *135*:2130-2136.
50. Roberston, M.J., Dougall, I.G., Harper, D., McKechnie, K.C.W., and Leff, P. (1994). Trends Pharmacol. Sci. *15*:364-369.
51. Rui, H., Lebrun, J.J., Kirken, R.A., Kelly, P.A., and Farrar, W.L. (1994). Endocrinology 135:1299-1306.
52. Miyamoto, S., Akiyama, S.K., and Yamada, K.M. (1995). Science *267*:883-885.
53. Kapas, S., Hinson, J.P., Puddefoot, J.R., Ho, M.M., and Vinson, G.P. (1994). Biochem. Biophys. Res. Commun. *204*:1292-1298.
54. Conchon, S., Monnot, C., Teutsch, B., Corvol, P., and Clauser, E. (1994). FEBS Lett. *349*:365-370.
55. Hunyday, L., Bor, M., Balla, T., and Catt, K.J. (1994). J.Biol. Chem. *50*:31378-31382.
56. Hunyday, L., Baukal, A.J., Balla, T., and Catt, K.J. (1994). J. Biol. Chem. *40*:24798-24804.
57. Pearse, B.M.F., and Robinson, M.S. (1990). Annu. Rev. Cell. Biol. *6*:151-171.
58. Trowbridge, I.S., Collawn, J.F., and Hopkins, C.R. (1993). Annu. Rev. Cell. Biol. *9*:129-161.
59. Chaki, S., Guo, D., Yamano, Y., Ohyama, K., Tani, M., Mizukoshi, M., Shirai, H., and Inagami, T. (1994). Kidney Int. *46*:1492-1495.
60. Rohrer, J., Benedetti, H., Zanolari, B., and Riezman, H. (1993). Mol. Biol. Cell *4*:511-521.

8

AT_1-RECEPTORS AND CELLULAR ACTIONS OF ANGIOTENSIN II IN NEURONAL CULTURES OF STROKE PRONE-SPONTANEOUSLY HYPERTENSIVE RAT BRAIN

Mohan K. Raizada, Di Lu, Hong Yang, and Kan Yu

Department of Physiology
University of Florida
Gainesville, Florida

ABSTRACT

AT_1-receptors, its mRNA and cellular actions of angiotensin II (Ang II) have been compared between neuronal cultures of Wistar Kyoto (WKY) and stroke-prone spontaneously hypertensive (SP-SH) rat brains. Bmax for AT_1-receptor binding is 2-fold higher and is associated with a parallel increase in the levels of AT_1-receptor mRNA in SP-SH rat brain neurons compared with WKY rat brain neurons. Ang II causes stimulation of both c-fos and norepinephrine transporter (NET) mRNAs in both strains of neurons and this stimulation is also 2-3-fold higher in SP-SH rat brain neurons compared with WKY rat brain neurons. In contrast, Ang II stimulation of PAI-1 mRNA in SP-SH neurons is only 50% that of in WKY rat brain neurons suggesting that SP-SH neurons express a decrease in AT_1-receptor coupling with PAI-1 response. These observations demonstrate that SP-SH neurons express AT_1-receptor-functions similar to those described for SHR neurons.

INTRODUCTION

It is well established that renin-angiotensin system is intrinsic to the brain and plays a key role in the central control of blood pressure (BP). Activation by Ang II of neuronal Ang II receptors of AT_1 subtype, localized in cardioregulatory-relevant areas of the brain, is the first cellular event leading to hypertensive action of Ang II which include changes in catecholaminergic system, stimulation of sympathetic pathways, release of vasopressin and modulation of baroreceptor reflexes (1-3). Further support for the involvement of brain Ang II system in the control of BP is derived from experiments with various animal models of

Recent Advances in Cellular and Molecular Aspects of Angiotensin Receptors
Edited by Mohan K. Raizada et al., Plenum Press, New York, 1996

hypertension including the spontaneously hypertensive (SH) rat (4-6) and renin-transgenic rat (7-8). These studies have shown that intervention in the expression of brain Ang II system either by pharmacological or by genetic means lowers BP (9-13). Thus, elucidation of the cellular and molecular mechanism of Ang II in the brain could be of great significance in designing new and improved approaches to control Ang II-dependent hypertension.

Our research group has been involved in understanding of the cellular actions of Ang II in the brains of WKY and SH rats with the use of an in vitro neuronal cell culture model. These studies have established that hyperactivity of the brain Ang II system in the SH rat is, in part, a result of an increased AT1-receptor gene expression with a parallel increase in the responsiveness of SH rat brain neurons to Ang II (4, 14-16). These are significant observations because they, for the first time, show that increased AT1-receptor activity is genetically-linked and is not a result of high BP in SH rat. The objective of this study was to determine if the above hypothesis would also apply for other genetic models of hypertension. Thus, AT1-receptor gene expression and AT1-receptor-mediated actions of Ang II on c-fos, NET and PAI-1 genes were compared in neuronal cultures of WKY and SP-SH rat brains. Our observations show that, as in SH rat neurons, AT1-receptors and its cellular functions are increased in SP-SH rat neurons.

MATERIALS

One-day-old WKY and SP-SH rats were obtained from our breeding colony. SP-SH adult rats of F2 prodigy were supplied by Dr. Ralph Dawson, Department of Pharmacodynamics, College of Pharmacy, University of Florida, Gainesville, FL, which originated from Dr. Carl Hansen's laboratory, at the Genetic Research Center, NIH, Bethesda, MD. Mean arterial blood pressure (BP) of adult breeders of WKY strain was 100 æ 10 mm Hg whereas BP of SP-SH was 225 æ 20 mm Hg. Dulbecco's Modified Eagles Medium (DMEM), plasma derived horse serum (PDHS) and 1x crystalline trypsin were from Central Biomedia (Irwin, MO). a-[32P]-dCTP (3000 Ci/mmol) was from DuPont/NEN (Boston, MA). PCR kit containing Taq DNA Polymerase was purchased from Perkin Elmer Cetus (Norwalk, CT). RNase inhibitor, MMLV reverse transcriptase, dNTP, and other reagents for PCR were purchased from Boehringer Mannheim Biochemicals (Indianapolis, IN). Dynal-beads and other reagents for poly(A+) RNA isolation were from Dynal, Inc. (Lake Success, NY). Losartan potassium (formally DUP753) was a gift from Dr. A. T. Chiu of DuPont/Merck (Wilmington, DE). All other biochemicals were from Fisher Scientific (Pittsburgh, PA) and were of molecular biology grade. Primers for NET, c-fos, PAI-1 and b-actin and sense and antisense oligonucleotides for c-fos were synthesized in the DNA synthesis facility of the ICBR, University of Florida, Gainesville. The sequences of these primers are listed as follows:

NET:	sense:	5′CCGCATCCATGCTTCTGGCGCGGATGAA3′
	antisense:	5′GGGCAGGCTCAGATGGCCAGCCAGTGTT3′
PAI-1:	sense:	5′GCTCCAGGATGCAGATGTCT3′
	antisense:	5′GCTCTCGTTCACCTCGATCT3′
β-actin:	sense:	5′GAGAAGATGACCCAGATCATGT3′
	antisense:	5′ACTCCATGCCCAGGAAGGAAG3′
c-fos:	sense:	5′AGGAGGGAGCTGACAGATA3′
	antisense:	5′CCTGGCTCACATGCTACTA3′

METHODS

Preparation of Neuronal Cultures from WKY and SP-SH Rat Brains

Hypothalamus-brainstem areas of one-day-old WKY and SP-SH rat brains were dissected and brain cells were dissociated by trypsin. The hypothalamic block contained the paraventricular nucleus, the supraoptic, anterior, lateral, posterior, dorsomedial and ventromedial nuclei whereas the brainstem block contained medulla oblongata and pons. Trypsin-dissociated brain cells were plated in poly-L-lysine precoated tissue culture 35-mm dishes ($3x10^6$ cells/dish) in DMEM containing 10% PDHS and neuronal cultures established essentially as described previously (14, 17). The cultures were allowed to grow for 15 days prior to their use in the experiments. Previous studies have shown that neuronal cultures prepared in this fashion were comparable from both WKY and SH rat strains by many biochemical, physiological and immunohistochemical criteria (4, 15).

RT-PCR of C-Fos, NET, PAI-1 and β-Actin

Neuronal cells were lysed by the addition of 0.5 ml GTC solution (4 M guanidium isothiocyanate, 0.01% B-mercaptoethanol, 25 mM NaOAc, and 0.5% sarcosyl) at room temperature for 10 min. Lysis solution was moved to sterile RNase-free tubes and mixed with equal volume of H_2O-saturated phenol and incubated on ice for 15 min. This was followed by addition of 1/10 volume of chloroform. The total RNA was precipitated by equal volume of isopropanol from the top aqueous phase. The resulting pellet was resuspended in Dynal-beads (dT) 25 binding buffer and poly(A^+) RNA was isolated with the use of Dynal-beads exactly as described in the protocol provided by the company. RT-PCR for NET, c-fos, PAI-1 and β-actin was run essentially as described previously (18).

After PCR, 5 μl of samples containing [^{32}P]-labelled PCR products were mixed with 5 μl 2x gel loading buffer (4% Ficoll 400, 20 mM EDTA, pH 8.0, 0.2% SDS, 0.05% bromophenol blue and 0.05% xylene cyanol) and applied to a 5% acrylamide gel (29:1 acrylamide/bis-acrylamide ratio) in a mini-cell prepared in 1x TBE buffer (89 mM Tris base, 89 mM boric acid and 2 mM EDTA, pH 8.0). Gel was run for 45 min at 200 V in a Bio-Rad Mini-Gel System (Bio-Rad Laboratories, Hercules, CA). Gel was decasted and wrapped in a plastic bag and exposed to an x-ray film overnight at -70°C. X-ray film was developed and bands representing PCR products imaged on floppy disk by UVP Gel Documentation System and transferred to a PC computer for analysis (with the SW5000 Software Analysis).

@Table =

Table 1: AT_1-receptors in neuronal culture of WKY and SP-SH rat brain (n=3)

Strain	Kd nM	Bmax (fmol/mg protein)	AT_1 receptor mRNA (% of β-actin mRNA)
WKY	7.3 ± 0.24	90.5 ± 10.1	13.6 ± 2.1
SP-SH	7.5 ± 0.36	208.4 ± 15.1	57.2 ± 4.3

Kd and Bmax values were calculated from the binding data essentially as described previously. AT_1 receptor mRNA was quantitated as described in Methods. Data are mean ± SE (n=3). * A representative autoradiogram showing the density of AT_1-R mRNA

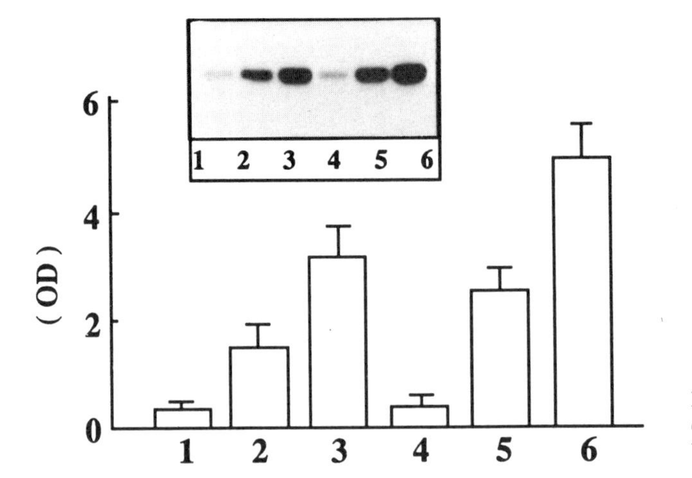

Figure 1. Ang II stimulation of c-fos nRNA in neuronal cultures of WKY and SP-SH rat brains.

Densitometric quantification of DNA bands was normalized by β-actin cDNA (14, 18, 19). The above described RT-PCR technique has been successfully used to quantitate mRNAs for NET and c-fos in neuronal cultures (18, 20).

AT_1-Receptor Binding. Kd and Bmax for AT_1-receptors in neuronal cultures of WKY and SP-SH rat brains was measured essentially as described previously (17, 21).

RESULTS

Table 1 shows the comparison of Kd and Bmax for AT_1-receptor and AT_1-receptor mRNA levels in neuronal cultures of WKY and SP-SH rat brains. Binding of [^{125}I]-sar^1-Ile^8-Ang II to specific AT_1-receptors was 2-fold higher in SP-SH rat brain neuronal cultures compared with WKY normotensive control. The increase was a result of an increase in the Bmax (208.4 ± 15.1 fmol/mg protein in SP-SH vs. 90.5 ± 10.1 fmol/mg protein in WKY)

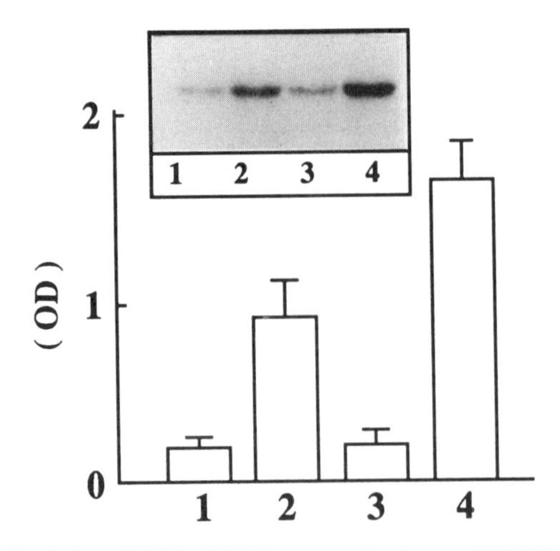

Figure 2. Ang II stimulation of NET mRNA in neuronal cultures of WKY and SP-SH rat brains.

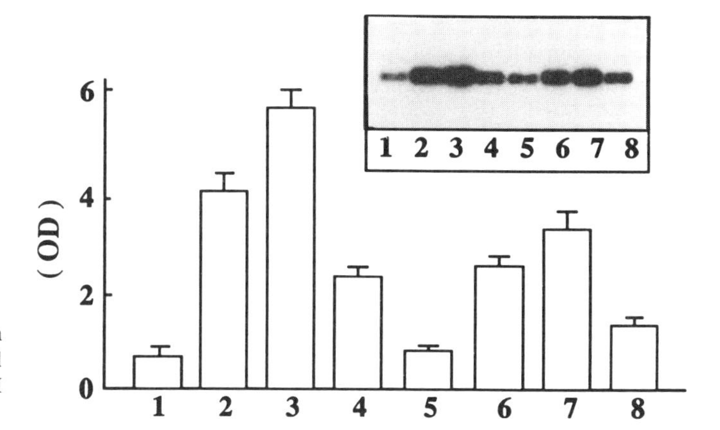

Figure 3. Ang II stimulation of PAI-1 mRNA in neuronal cultures of WKY and SP-SH rat brains.

without any change in the Kd. Parallel with this increase in AT_1-receptor numbers was a 4-fold increase in the mRNA for AT_1-receptors in SP-SH rat brain neurons.

Next, we studied the cellular actions of Ang II mediated by its interaction with AT_1-receptors in an attempt to determine if increased AT_1-receptors in SP-SH are physiologically functional.

Figure 1 shows the effect of Ang II on c-fos mRNA levels in WKY and SP-SH rat brain neuronal cultures. Ang II caused significant stimulation of c-fos mRNA levels in neuronal culture of both strains of rats, and this stimulation was 2-fold higher in SP-SH when compared with WKY rat brain neurons. The effect of Ang II on chronic stimulation of NET system was determined by quantitation of NET mRNA levels since our previous studies have shown that chronic stimulation of NE uptake by Ang II is associated with NET gene expression and involves c-fos gene (22). Figure 2 shows that 100 nM Ang II caused a 4-fold stimulation of NET mRNA in WKY rat brain neurons. This stimulation was 7-fold in SP-SH rat brain neurons under identical conditions. Thus, Ang II stimulated NET mRNA was ~2-fold higher in SP-SH neurons compared with WKY rat brain neurons and was consistent with increase in AT_1-receptors in the neurons of this strain of rat.

In previous studies, we have shown that Ang II stimulates PAI-1 gene expression in neurons of normotensive rat brain and that the stimulation is significantly attenuated in neurons from SH rat brain (23). The object of our next experiment was to determine if neuronal cultures from SP-SH rat brains responds to Ang II in a similar fashion in SH rat brain neurons. Figure 3 shows that 100 nM Ang II caused a 8-fold stimulation of PAI-1 mRNA in WKY and only 4-fold stimulation in SP-SH neurons. Thus, SP-SH neurons were only 50% as responsive to Ang II in PAI-1 mRNA stimulation compared with WKY rat brain neurons.

DISCUSSION

This study demonstrates that neuronal cultures of SP-SH rat brain express higher numbers of AT_1-receptors compared with neurons of WKY rat brain. This increase is associated with a parallel increase in Ang II stimulation of c-fos and NET mRNAs and a decrease in coupling of AT_1-receptor with stimulation of PAI-1 mRNA in SP-SH neurons.

Previous studies have shown that AT_1-receptor gene expression and neuromodulatory actions of Ang II are increased in neuronal cultures of SH rat brain (14-15). These observations were consistent with our hypothesis that hyperactivity of the brain Ang II system observed in the adult SH rat is a result of an increase in AT_1-receptor expression and AT_1-receptor-mediated cellular actions. Both of these properties are genetically regulated in SH rat brain since they are expressed in neurons *in vitro* from one-day-old animals (4, 14-15). Observations of this study, which was conducted in SP-SH model of hypertension, clearly support this hypothesis. Bmax for AT_1-receptor was 2-fold higher in SP-SH neuronal cultures compared with normotensive control. In addition, mRNA levels for AT_1-receptors was also 4-fold higher which is consistent with the conclusion that AT_1-receptor gene transcription and translation is increased in these models of genetic hypertension (14, 19). Ang II stimulation of c-fos and NET mRNAs was studied to determine if increased AT_1-receptors are physiologically functional. Ang II stimulated both these genes in a manner proportional with the increase in AT_1-receptors in SP-SH neurons and was consistent with the data reported for SH rat brain neurons (22) suggesting that c-fos may act as a third messenger in AT_1-receptor mediated signal transduction pathway. Finally, the effect of Ang II on PAI-1 gene expression was compared in WKY and SP-SH neurons to confirm the similarities between the neurons of SH and SP-SH strains. AT_1-receptor mediated PAI-1 mRNA stimulation is significantly uncoupled in SP-SH neurons. These observations, taken together, support our conclusion that changes in AT_1-receptors and Ang II actions are significantly higher in neuronal cultures of both SH and SP-SH rat brains compared with normotensive WKY rat brain neuronal cultures.

Increase in the brain Ang II activity has been implicated in development and maintenance of hypertensive state in many other animal models of hypertension, in addition to the SH rat. For example, blood pressure response to centrally injected Ang II is significantly heightened and is associated with increases in Ang II receptors in hypothalamic-thalamus-septum (HTS) and subfornical organ in DOCA-salt model (24-25). This increase in the brain Ang II receptors is preserved in neuronal cultures from the HTS areas of DOCA-salt rat indicating a possible effect on the receptor gene expression rather than a hormonal or humeral change induced by this pathological state (24). Similarly, brain Ang II levels are higher in two kidney, one clip and anephric rat models of hypertension (26) suggesting that the brain renin-angiotensin related-gene expression is critical during the development of hypertension in these models as well. Finally, hyperactive brain Ang II system is proposed in transgenic rat model of hypertension expressing mouse Ren-2 gene (8, 11, 27). These observations, collectively considered, point out that the brain Ang II system, in part, contributes to the development and establishment of hypertensive state in these rat models. Data presented here support this notion at the cellular level in at least two models of hypertension. Thus, it is reasonable to suggest that intervention in the expression of brain Ang II system at a genetic level would be of highly significant therapeutic value in the control of Ang II-dependent hypertension. In fact, such an approach has been used both in vivo and in vitro (13, 18, 28). Similar experiments with other models should be conducted to further support the central role of brain Ang II in hypertension.

ACKNOWLEDGMENTS

We wish to thank Dr. Ralph Dawson, University of Florida, for providing us SP-SH rats. The research was supported by the NIH grant #33610. Di Lu and Kan Yu are supported by American Heart Association, Florida Affiliate.

Neuronal cultures of WKY (1-3) and SP-SH (4-6) strains of rats were incubated without (1,4) or with 100 nM Ang II (2, 3, 5, 6) for 10 min (2, 5) or 30 min (3, 6). c-fos mRNA was determined as described in the “Methods”. Data are mean ± SE (n=3). Inset shows a representative autoradiogram.

Neuronal cultures of WKY (1-2) and SP-SH (3-4) strains of rats were incubated without (1, 3) or with 100 nM Ang II (2, 4) for 4 hr at 37°C. NET mRNA was measured (22). Data are mean ± SE (n=3). Inset shows a representative autoradiogram.

WKY (1-4) and SP-SH (5-8) neuronal cultures were incubated without (1, 5) or with 100 nM Ang II (2-4, 6-8) for 2 hr (2, 6), 4 hr (3, 7) and 24 hr (4, 8). PAI-1 mRNA was determined as described previously (28). Data are mean æ SE (n=3). Inset shows a representative autoradiogram.

REFERENCES

1. Saavedra, J.M., 1992, Brain and pituitary angiotensin, *Endocrine Rev.* 13:329-380.
2. Timmermans, P.B.M.W.M., Wong, P.C., Chiu, A.T., Herblin, W.F., Benfield, P., Carini, D.J., Lee, R.J., Wexler, R.R., Saye, J.A.M., and Smith, R.D., 1993, Angiotensin II receptors and angiotensin II receptor antagonists, *Pharmacol. Rev.* 45:205-251.
3. Wright, J.W., and Harding, J.W., 1994, Brain angiotensin receptor subtypes in the control of physiological and behavioral responses, *Neurosci. Biobehavioral Rev.* 18:21-53.
4. Raizada, M.K., Lu, D., and Sumners, C., 1994, AT_1-receptors and angiotensin actions in the brain and neuronal cultures of normotensive and hypertensive rats, In: Current concepts: Tissue renin-angiotensin system as local regulators in Reproductive and Endocrine Organs, (Plenum Press, New York), pp. 331-348.
5. Trippodo, N.C., and Frolich, E.D., 1981, Similarities of genetic (spontaneous) hypertension: Man and rat, *Circ. Res.* 48:309-319.
6. Johnson, M.I., Ely, D.L., and Turner, M.E., 1992, Genetic divergence between WKY rat and SH rat, *Hypertension* 19:425-427.
7. Bader, M., Zhao, Y., Sander, M., Lee, M.A., Bachmann, J., Böhn, M., Djavidani, B., Peters, J., Mullins, J.J., and Ganten, D., 1992, Role of tissue renin in the pathophysiology of hypertension in TGR (mREN2)27 rats, *Hypertension* 19:681-686.
8. Senanayake, P.D., Moriguchi, A., Kumagai, H., Ganten, D., Ferrario, C.M., and Brosnihan, K.B., 1994, Increased expression of angiotensin peptide in the brain of transgenic hypertensive rats, *Peptides* 15:919-926.
9. Phillips, M.I., Mann, J.F.E., Haebara, H., Hoffman, W.E., Dietz, R., Schelling, P., and Ganten, D., 1977, Lowering of hypertension by central saralasin in the absence of plasma renin, *Nature* 270:445-447.
10. McDonald, W., Wickre, C., Aumann, S., Ban, D., and Meffitt, B., 1980, The sustained antihypertensive effect of chronic ICV infusion of angiotensin antagonist in SH rat, *Endocrinol.* 107:1305-1308.
11. Hutchinson, J.S., Mendelsohn, F.A.O., and Doyle, A.E., 1980, Blood pressure responses of conscious normotensive and spontaneously hypertensive rats to IVT and peripheral administration of captopril, *Hypertension* 2:546-550.
12. Okuno, T., Nagahama, S., Lindheimer, M.D., and Oparil, S., 1983, Attenuation of the development of spontaneously hypertension in rats of chronic central administration of captopril, *Hypertension* 5:653-663.
13. Gyurko, R., Weilbo, D., and Phillips, M.I., 1993, Antisense inhibition of AT_1-receptor mRNA and angiotensinogen mRNA in the brain of spontaneously hypertensive rat reduces hypertension of neurogenic origin, *Regul. Pept.* 49:167-174.
14. Raizada, M.K., Lu, D., Tang, W., Kurian, P., and Sumners, C., 1993, Increased angiotensin II type 1 receptor gene expression in neuronal cultures from spontaneously hypertensive rats, *Endocrinol.* 132:1715-1722.
15. Sumners, C., and Raizada, M.K., 1993, In: Cellular and molecular biology of renin-angiotensin system, (CRC Press, Boca Raton, FL), pp. 379-411.
16. Sumners, C., Raizada, M.K., Kang, J., Lu, D., and Posner, P., 1994, Receptor- mediated effects of angiotensin II on neurons, *Frontiers in Neuroendocrinol.* 15:203-230.
17. Sumners, C., Tang, W., Zelezna, B., and Raizada, M.K., 1991, Angiotensin II receptor subtypes are coupled with distinct signal transduction mechanisms in neurons and astroglia from rat brain, *Proc. Natl. Acad. Sci. USA* SS7567-7571.
18. Lu, D., and Raizada, M.K., 1995, Delivery of angiotensin II type 1 receptor antisense inhibits angiotensin action in neurons from hypertensive rat brain, *Proc. Natl. Acad. Sci. USA* 92:2914-2918.
19. Raizada, M.K., Sumners, C., and Lu, D., 1993, Angiotensin II type 1 receptor mRNA levels in the brains of normotensive and spontaneously hypertensive rats, *J. Neurochem.* 60:1949-1952.

20. Wang, A.M., Doyle, M.V., and Mark, D.F., 1989, Quantitation of mRNA by the polymerase chain reaction, *Proc. Natl. Acad. Sci. USA* 86:9719-9721.
21. Zelezna, B., Rydzewski, B., Lu, D., Olson, J.A., Shiverick, K.T., Tang, W., Sumners, C., and Raizada, M.K., 1992, Angiotensin II induction of plasminogen activator inhibitor-1 gene expression in astroglial cells of normotensive and spontaneously hypertensive rat brain, *Mol. Endocrinol.* 6:2009-2017.
22. Raizada, M.K., and Lu, D., 1995, Angiotensin II (AII) stimulation of norepinephrine transporter (NET) gene expression in neuronal cultures in normotensive and spontaneously hypertensive (SH) rat brain (abstr), *FASEB J.* part I:A90.
23. Yu, K., and Raizada, M.K., 1995, Angiotensin II (AII) stimulation of plasminogen activator inhibitor-1 (PAI-1) gene expression in neuronal cultures (NC) of normotensive (WKY) and spontaneously hypertensive (SH) rat brains (abstr), *FASEB J.* part I:A90.
24. Wilson, K.M., Sumners, C., Hathaway, S., and Fregly, M.J., 1986, Mineralocorticoids modulate central angiotensin II receptors in rats, *Brain Res.* 382:87-96.
25. Wilson, S.K., Lynch, D.R., and Ladenson, P.W., 1989, Angiotensin II and atrial natriuretic factor-binding sites in various tissues in hypertension: comparative receptor localization and changes in different hypertension models in the rat, *Endocrinol.* 124(6):2799-2808.
26. Morishita, R., Higaki, J., Okuniski, H., Nakamura, F., Nagano, M., Mikami, H., Ishii, K., Miyazaki, M., and Ogihara, T., 1993, Role of tissue renin angiotensin system in two-kidney, one-clip hypertensive rats, *Am. J. Physiol.* 264:F510-F514.
27. Moriguchi, A., Brosnihan, K.B., Kumegai, H., Ganten, D., and Ferrario, C.M., 1994, Mechanisms of hypertension in transgenic rats expressing the mouse Ren-2 gene, *Am. J. Physiol.* 266(4Pt2):R1273-1279.
28. Lu, D., Yu, K., and Raizada, M.K., 1995, Retrovirus-mediated transfer of AT_1-receptor antisense sequence decreases AT_1-receptors and angiotensin II action in astroglial and neuronal cell in primary culture from the brain, *Proc. Natl. Acad. Sci. USA* 92:1162-1166.

9

ANTISENSE OLIGONUCLEOTIDES FOR *IN VIVO* STUDIES OF ANGIOTENSIN RECEPTORS

M. Ian Phillips, Philipp Ambühl, and Robert Gyurko

Department of Physiology
College of Medicine, Box 100274
University of Florida
Gainesville, Florida 32610-0274

INTRODUCTION

Synthetic antisense (AS) oligodeoxynucleotides inhibit genetic expression by sequence-specific hybridization to mRNA that renders the mRNA inactive for translation. We have been using AS oligos to lower blood pressure by inhibiting angiotensin receptors and angiotensinogen in freely moving, whole animals. This recent application of antisense technology to *in vivo* studies, opens a new way of approaching physiological problems with the precision of molecular biology. The possibility of blocking specific gene expression by AS inhibition without multiple, non-specific side effects has potential for therapeutic uses in many diseases. Antisense inhibition is an extremely attractive pharmacological and investigative approach since it offers base-to-base specificity to the target protein and versatility appropriate to the complexity of the genetic code. However, there are a number of issues to be considered before using antisense in any experimental or clinical setting. These include (1) selection of target sequence, (2) the mechanism of cellular uptake, (3) stability of antisense oligos in cells and body fluids, (4) possible intracellular sites of action and (5) effectiveness, in terms of specificity, and duration of action.

An antisense oligodeoxynucleotide is a short (15-22 bases long) synthetic DNA molecule, complementary sequence to the specific messenger RNA (mRNA) sequence. Thus the sequence will bind to the mRNA in a specific manner and inhibit the translation of the targeted mRNA. This will inhibit the expression of the mRNA to produce the specific protein (Fig. 1). Thus in principle, the concept is very simple.

Selection of the Target Sequence

Though it might seem easy, designing an antisense oligonucleotide that works is not a straightforward matter. The ambiguity in antisense oligonucleotide design stems from

Recent Advances in Cellular and Molecular Aspects of Angiotensin Receptors
Edited by Mohan K. Raizada et al., Plenum Press, New York, 1996

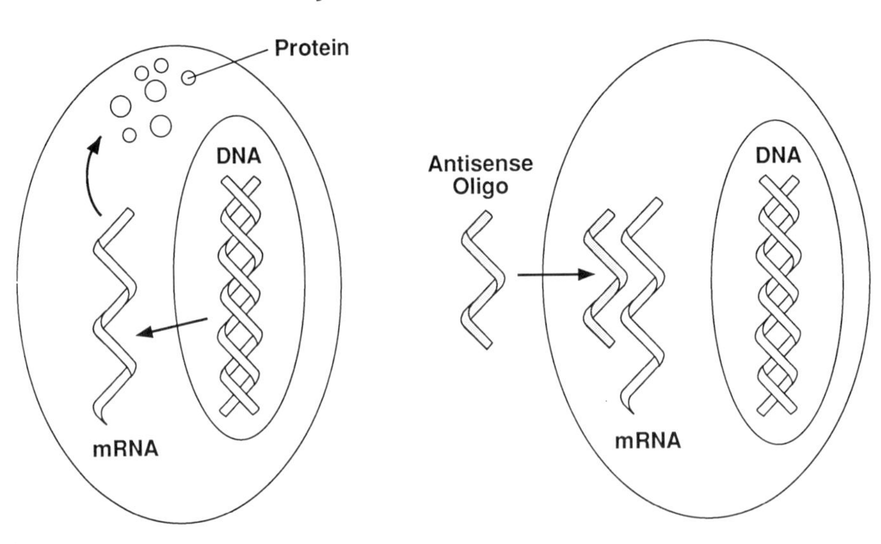

Figure 1. The principle of antisense oligodeoxynucleotide inhibition. Left: DNA in the nucleus encodes the message for the production of protein via RNA. The RNA is transported out of the nucleus and into the ribosomes of the cytoplasm, and the readout from the mRNA results in the production ofa specific protein. Right: When a strand of DNA is synthesized in the reverse direction of normal DNA, it will bind to the mRNA in the cytoplasm or in the nucleus and prevent it from being translated through the ribosomes into protein. Because of this antisense-directed oligonucleotide, there is antisense inhibition of the mRNA. To direct the antisense oligo to a specific site, the antisense must be synthesized with complementary base pairs to the mRNA. Generally, antisense oligos are small (15 to 22 bases).

incomplete knowledge about the mechanism of action of such molecules, as well as from conflicting experimental results.

The concept of antisense inhibition assumes that antisense molecules bind to the mRNA of the target protein in the cytoplasm, and prevent either ribosomal assembly or read-through. Most oligonucleotides therefore are targeted to the initiation codon AUG, or part of the coding region downstream from it. When designing such an antisense oligonucleotide, one has to consider two antagonistic factors: the affinity of oligonucleotide to its target sequence, which is dependent on the number and composition of complementary bases, and the availability of the target sequence, which is dependent on the folding of the mRNA molecule (1).

There have been attempts to correlate oligonucleotide efficiency and these two factors: affinity and availability. There is a general agreement that increasing affinity of antisense oligonucleotides positively influences effectiveness, but results are conflicting on the importance of the secondary structure of the target molecule. Several reports suggest that antisense oligonucleotides targeted to different regions of the RNA have unequal efficiencies (2,3). These differences may be related to the predicted secondary structure of the target mRNA (4,5,6), but other experimenters did not find significant differences in antisense efficiency between oligonucleotides targeting different segments of the RNA (1,7,8). It would seem logical that the folding of the mRNA influences target sequence availability, but one also has to consider that the stretches of the RNA double helices which are responsible for the secondary structure of the mRNA incorporate a weaker G-U base pairing besides A-U and G-C and are generally short and rarely perfect. Therefore an ODN which has strong base

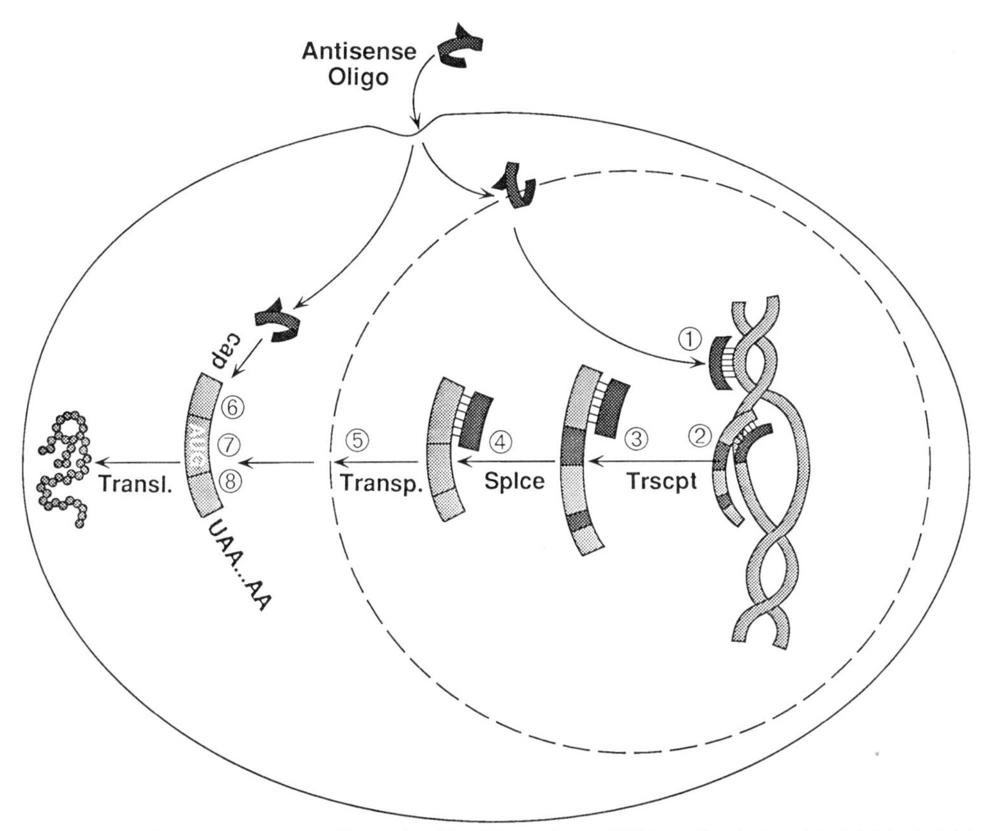

Figure 2. The design of antisense oligonucleotides. The antisense DNA can be designed to inhibit at eight different sites or more in the cell. 1) The antisense oligo may form a triple helix with the DNA in the nucleus. 2) The antisense may bind the nascent RNA and prevent transcription. 3) The antisense binds to the exon and prevents transcription. 4) The antisense binds to the exon-intron junction and prevents splicing. 5) The antisense binding to the mRNA prevents transport from the nucleus. 6) The antisense binds to the 5′-CAP, or 7) binds to the AUG initiation codon, or 8) binds to the 3′-untranslated region of the mRNA, all of which prevent translation.

pairing with 100% complementarity will form the most thermodynamically favorable structure with its target RNA (9). If this is the case, the antisense DNA will displace the RNA strand to hybridize with the targeted strand.

Besides inhibition of translation, other possible antisense mechanisms of action have been proposed. Based on studies of cellular uptake of labelled oligonucleotides, the picture emerged that most antisense oligonucleotides quickly migrate to the cell nucleus, suggesting an intranuclear site of action (10). This suggests a more direct action on gene expression: antisense molecules might inhibit pre-RNA splicing, transport of mRNA from the nucleus to the cytoplasm, or bind to the DNA inhibiting transcription. Accordingly, effective antisense oligonucleotides were designed targeting exon-intron splicing sites (11) or the major groove of the DNA (12).

The experimental evidence accumulated in the past few years of antisense research suggest that a large number of antisense oligonucleotides targeted to different regions of a variety of mRNAs were successful in downregulating their target proteins, although the number of failed attempts can not be gauged. It is, however, hard to recognize a general pattern by which one can design an effective antisense oligonucleotide to a particular target.

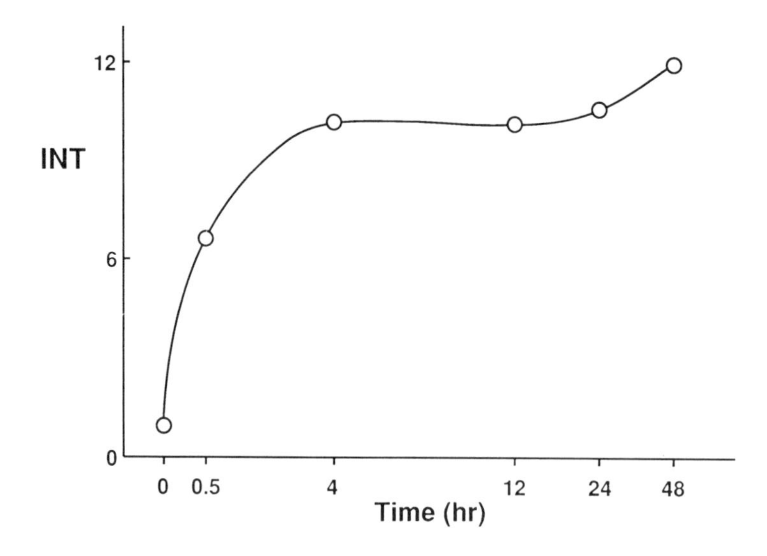

Figure 3. Uptake of fluorescent-labelled oligo in NG108 cells. Cells were incubated with FITC-labelled antisense oligonucleotide (15). The intensity (INT) of fluorescence was measured using confocal microscopy/spectrophotometry. There was rapid uptake of the oligos in the first 30 min which reached an asymptote at about 4 hrs. Oligos remained in the cells for more than 48 hrs. The FITC labelling was found in the nucleus region within 4 hrs after incubation. Further study of the lysate with thin-layer chromatography showed that the FITC-labelled oligo was maintained intact at 48 hrs. There was no difference in uptake between 15mer- and 18mer-length oligos.

The three regions that are repeatedly suggested to be better targets are the 5′ cap region, the AUG translation initiation codon and the 3′ untranslated region of the mRNA (1,13,14). Since most mRNAs have an AUG initiator codon site, targeting 12-15 of the neighboring bases should produce inhibitory oligos. Unfortunately, many mRNAs in vertebrates have the Kozak consensus sequence (9) of 10-12 bases around the AUG start codon and there is a danger of creating nonspecific oligos by taking this approach. In any case, all design oligos must be checked on GenBank for existing sequences to avoid homology with other mRNAs.

A possible reason why there is no general rule of thumb for antisense oligonucleotide design is that different mechanisms of action are present in different experimental situations (6). If this is the case, the safest way for finding the optimal target sequence is through trying a series of oligonucleotides. Common rules for antisense action may exist, but factors like differential cellular uptake and discharge of different oligonucleotides or intracellular feedback mechanisms on mRNA or protein production make them difficult to identify. Further systematic investigation is needed to elucidate the molecular mechanisms of antisense inhibition to be able to design antisense molecules rationally. For our studies of angiotensin II actions we have used the following sequences targeted to the coding region of the mRNA by testing their effectiveness in cell culture and then in vivo. We currently use an 18mer for Angiotensinogen: 5′-CCG-TGG-GAG-TCA-TCA-CGG-3′ targeted to the -5 to +13 bases of Ao mRNA encompassing the AUG translation initiation codon; and for AT_1 receptor: 5′-TAA-CTG-TGC-CTG-CCA-3′. [Note: In Gyurko et al., 1993, there was an error in the published sequence.] The AT_1 antisense oligo is targeted to the +63 to +88 bases of AT_1 mRNA.

Stability of Oligonucleotides

Oligonucleotides in their natural form as phosphodiesters (PDE) are subject to rapid degradation in the blood, intracellular fluid or cerebrospinal fluid by exo- and endonucleases. The half-life of PDE oligonucleotides is in the range of minutes in blood and tissue culture media (15). Wahlestedt detected evidence of active PDE oligonucleotides 24 hours after injection into the cerebral ventricles (16). Data from our lab also indicated that PDE oligos have effects up to 24 hours after injection into the brain ventricles (17).

Several chemical modifications have been proposed to prolong the half-life of oligonucleotides in biological fluids while retaining their activity and specificity. The majority of these modifications effect the phosphodiester bond between nucleotides: the backbone-modifications. Although the phosphodiester DNA backbone has poor nuclease resistance and does not produce a very stable antisense oligonucleotide, modifying the ODN can lead to stable, long-lasting analogues. In our laboratory we use phosphorothioated backbone and have found stability of the antisense ODN after injection into the brain for 3-7 days. In cell culture, NG108-15 cells (18) and adrenal bovine cells (19) took up phosphorothioated antisense and sense oligos rapidly and retained them intact (Fig. 3). This was studied using FITC-labelled ODNs and testing the stability of the ODNs on thin layer chromatography after extraction from the cells up to 4 days after incubation.

Phosphorothioate modification is where one of the oxygen atoms in the phosphodiester bond between nucleotides is replaced with a sulfur atom. These phosphorothioate oligonucleotides have greater stability in biological fluids than normal oligos. The half-life of a 15-mer phosphorothioate oligo is 9 hours in human serum (15) up to 48 hours in tissue culture media (18) with 10% fetal bovine serum and 19 hours in cerebrospinal fluid (15).

Cellular uptake is generally faster for shorter oligonucleotides than for longer ones. Decreasing the temperature prevented the oligonucleotide uptake almost completely, indicative of an active uptake mechanism. Any sequence or size of unlabeled oligonucleotide competed with the labelled oligonucleotides for uptake, even DNA as small as a single nucleotide or as big as yeast DNA. Free nucleotides, however, were not competitive, showing that the phosphate group is essential for this uptake mechanism. An 80 kDa oligonucleotide binding protein was isolated using oligo (dT) cellulose beads and this protein has given rise of the theory of the receptor mediated uptake of oligonucleotides(20). An efflux mechanism has also been described indicating temperature-dependent secretion of the oligonucleotides from the cells to the cytoplasm(21). Other groups describe fast translocation of the polynucleotides from the cytoplasm to the nucleus which was found to be independent of ATP pool or temperature suggesting diffusion through nuclear pores (22,23). Uptake rate has been shown between different cell lines (24). It is generally believed that most types of oligonucleotides enter the cell by absorptive or fluid phase endocytosis, but the existence of an oligonucleotide receptor that effects this transport has yet to be proven by purification of the protein.

Mechanism of Action

The action of antisense ODNs to prevent the expression of mRNA may or may not lead to reduction of mRNA. One theory is that AS-ODNs, by hybridizing to the mRNA, stimulate the actions of RNase H which degrades the RNA strand of an RNA-DNA duplex. In cell free systems and injected oocytes RNase-H has been shown to digest the RNA part of any RNA-DNA hybrid. This action cuts the target mRNA into two pieces which can be detected by Northern blotting. The antisense oligo is not digested by RNase-H, so it can hybridize to another mRNA. This way the action of the antisense oligo becomes catalytic instead of stoichiometric.

In experiments using AS-ODN targeted to angiotensinogen mRNA, there was a significant reduction of angiotensinogen but no change in mRNA, as measured by Northern blotting. However, Wahlestedt et al (25) also reported a lack of change in mRNA after antisense oligos significantly reduced NMDA receptors. It is possible that the Northern blot is simply not sensitive enough to detect small changes in the mRNA and RTPCR would be more sensitive. Although tumor cells have an RNase H mechanism to explain AS effects, the RNase H concept has not been accepted as the mechanism of AS inhibition in nervous system studies. There are alternative mechanisms: antisense DNA can hybridize to its target mRNA or pre-mRNA in the nucleus forming a partially double-stranded structure which would inhibit its transport from the nucleus to the cytoplasm thus preventing translation. Antisense oligonucleotides targeted to intron-exon junction sites can prevent proper slicing and consequently the maturation of the transcript.

There is also a theoretical possibility of the antisense oligonucleotides' binding to the DNA by triple helix formation or by binding to the locally opened looped created by RNA polymerase (26). These appear to be possible only in special cases, e.g. triple helix formation by antisense oligonucleotide binding in the major groove of the DNA can only happen if the target sequence contains only purines on one strand and pirimidines on the other (26).

Specificity

Antisense inhibition can be considered pharmacologically a drug-receptor interaction where the oligonucleotide is the drug and the target sequence is the receptor. In order for binding to occur between the two, a minimum level of affinity is required which is provided by hydrogen bonding between the Watson-Crick base pairs and base stacking in the double helix that is formed. In order to achieve pharmacological activity, a minimum number of 15 bases in necessary to provide the minimum level of affinity (27).

The specificity of the antisense oligonucleotide-target sequence interaction is provided by the Watson-Crick base pairing. Considering that roughly 5% of the human genome contains coding regions, an oligonucleotide of 11-15 nucleotides long is specific enough statistically to be complementary to a single sequence. Increasing the length of the antisense should result higher level of specificity, but there is an upper limit of about 30 nucleotides above which oligonucleotides are not taken up effectively by cells. Instead of increasing length, it is advisable to test the antisense sequence against genetic databases for possible cross-reaction with other mRNAs. In our experiments on uptake in NG108-15 cells, there was no difference between 15 and 18 mer oligos. Both were taken up equally and rapidly within 30 mins of incubation and reached a saturable level after 4 hours (19).

Sequence-specificity can be ascertained experimentally by using appropriate controls. These are discussed later but include oligonucleotides such as sense ODN, which is identical to the target sequence; scrambled ODN, which has the same base composition as the antisense, but in a scrambled order, and mismatch, which is similar to the antisense but contain 3-4 mismatches while maintaining the same base composition. Naturally the latter is the most stringent control even although thermodynamic considerations tell us that a single base mismatch results in a 500-fold decrease in affinity.

Effectiveness

Antisense oligonucleotides can inhibit protein synthesis in cultured cell in μM doses. This inhibition however never results in a total knockout of the target protein, but rather in a partial decrease. Increasing the dose generally does not result in complete inhibition, and with very high doses nonspecific effects might take place. The nucleotides used to build

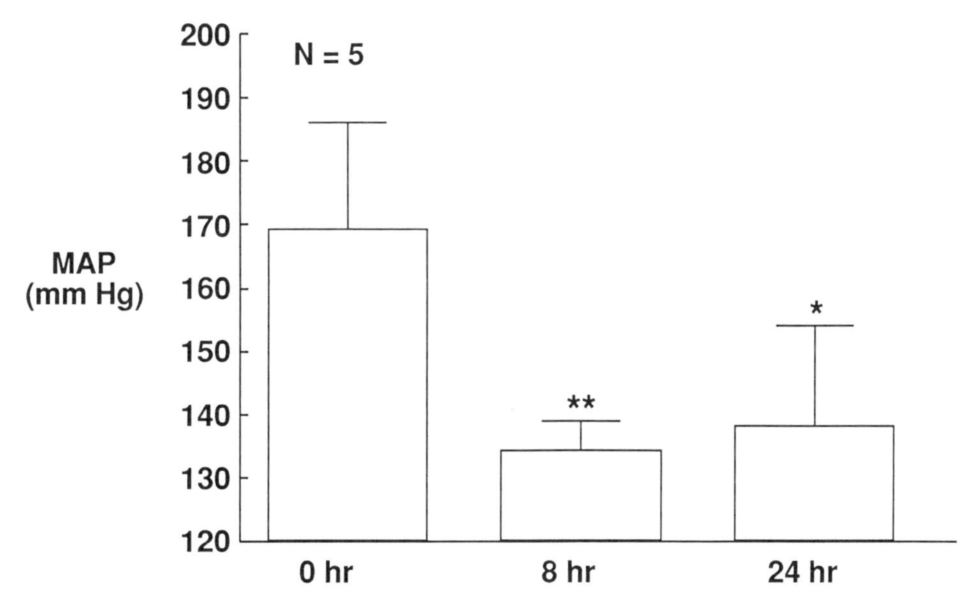

Figure 4. The effect of antisense oligonucleotide to angiotensinogen mRNA with a phosphorothioated backbone, delivered intracerebroventricularly (i.c.v.) in spontaneously hypertensive rats (SHR) at three different time periods. The rats an average of 169 mmHg blood pressure before injection. By 8 hrs post-injection there was a significant fall in blood pressure of 35 mmHg. This fall in blood pressure was observed 24 hrs after the single injection of antisense and recovery occurred slowly over the next 3 to 4 days. No effect on heart rate was observed. Scrambled and sense control oligos did not produce a decrease in blood pressure (from Wielbo et al, *Hypertension*, 1995).

phosphodiester oligonucleotides are chemically identical to the building block of the naturally occurring DNA, therefore even quite high doses should be tolerated in living animals without toxicity. Reports of toxicity in vivo may reflect improper procedures, such as high pressure injection causing mechanical damage and unsterile conditions inducing inflammation.

ANGIOTENSIN ANTISENSE OLIGONUCLEOTIDES IN THE BRAIN

Antisense Inhibits Hypertension

We injected AS-ODNs into the brains of rats to produce significant changes in physiological responses. Initially we injected phosphodiester oligos to angiotensinogen mRNA in hypertensive rats and found a lowering of blood pressure but the effect lasted for only a few hours (18,28). Next we used phosphorothioated oligos, injected into the lateral ventricles of spontaneously hypertensive rats (SHR), to test the hypothesis that the brain of SHR has an overactive renin-angiotensin system which contributes to the hypertension. We showed that phosphorothioated oligos targeted to the angiotensinogen mRNA (29) and to the AT_1 receptor mRNA significantly reduced blood pressure in these rats (Fig. 4 and 5) and the effects lasted for several hours, even days. These reductions in blood pressure did not occur in the normotensive (WKY) rats using the same dose of AS-oligo. The protein products

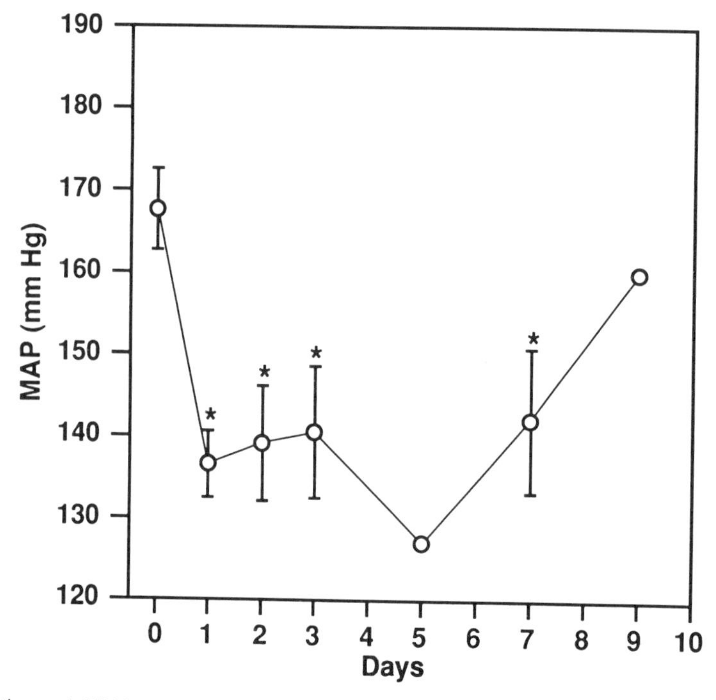

Figure 5. Antisense inhibition of blood pressure by an antisense oligonucleotide directed to AT_1 receptor mRNA, injected i.c.v. in SHR. Prior to injection rats had mean blood pressures of 167 mmHg. Twenty-four hrs following the injection, blood pressure was reduced by 35 mmHg (n = 11). The rats were catheterized and blood pressure was recorded through implanted catheters. There was some attrition due to blocked catheters, but on days 2, 3 and 7 there were still significant decreases of blood pressure. By day 9, blood pressure had recovered to the pre-injection levels. Receptor binding in these rats was reduced by ~30% in the anterior hypothalamic region (data from Gyurko, Tran and Phillips, unpublished).

of the targeted genes, namely angiotensinogen levels and AT_1 receptor levels, were significantly decreased by the AS-ODN but not by control scrambled ODNs or sense ODNs.

AT_1 Receptor Antisense Effects

Angiotensin injected into the brain elicits thirst, a blood pressure increase and vasopressin release (30). Sakai et al (31) and Meng et al (32), using an antisense to angiotensin AT_1 receptors, demonstrated that AS-ODN would inhibit the effects of angiotensin injected into the brain ventricles. Drinking was blocked when antisense to AT_1 receptors (Fig. 6) was injected. There were differences between the studies. Meng et al used a high dose of 50 µg, compared to Sakai et al who used a dose of 200 ng but there was no indication of toxicity with the 50 µg dose. Wielbo et al (29) showed that the 50 µg dose did not leak out of the brain to reduce blood pressure since an intracarotid injection of 50 µg AS-ODN had no effect on blood pressure. Despite the difference in dose, in both the Meng et al study and the Sakai et al study, the decrease in response was 50 to 60%. This seems characteristic of antisense inhibition that it does not achieve 100% inhibition. There are many potential reasons for this. One is that the route of injection (lateral ventricles in this case) is not optimal

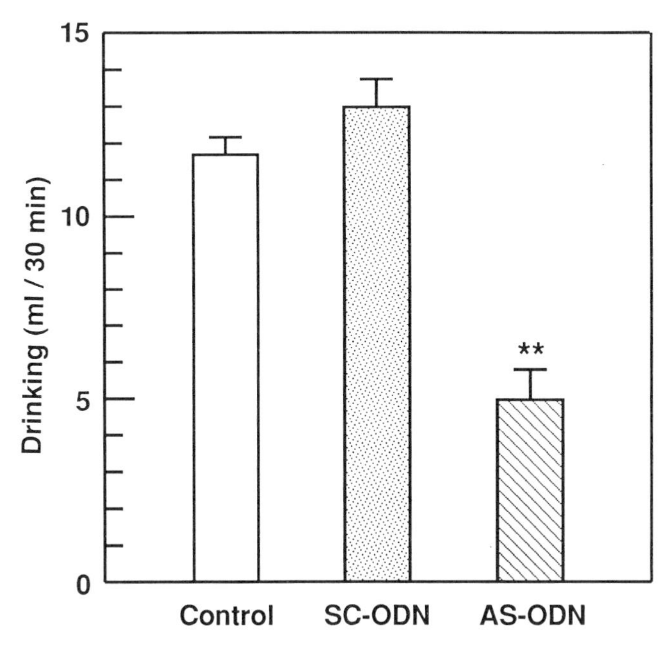

Figure 6. Effect of antisense oligonucleotide directed to AT_1 receptor mRNA on the drinking response to Angiotensin II i.c.v. Rats with implanted catheters in the lateral ventricle were tested for the drinking response (in 30 min) to 50 ng AngII i.c.v. Control rats received saline instead of antisense. SC-ODN received scrambled oligos and AS-ODN received the antisense oligo. The data are expressed as mean ± SEM ($n = 5$; $p < 0.01$) (Data from Meng et al., *Regul. Pept.*, 1994, reprinted with permission.

to reach all cells involved in the responses. As noted above, oligos are avidly taken up by cells. Therefore with central injections, the oligos will become concentrated in cells close to the site of injection. Both studies are in agreement that brain stem sites showed no change in binding, suggesting that the rostrally injected oligos did not reach the caudal brain stem in sufficient concentration to have an effect.

Additional studies on angiotensin receptor binding after administration of antisense to the AT_1 receptor also showed only a moderate decrease in binding (33). In a first series of experiments, three daily injection of 50 μg (3x50 μg) antisense into the third ventricle of normotensive Sprague-Dawley rats decreased binding in hypothalamic tissue blocks by 25%. Autoradiography showed decreases ranging from 15 to 30% in hypothalamic nuclei. The biggest decreases were seen in the suprachiasmatic (SCh) and the paraventricular (PVN) hypothalamic nuclei. Decreases were also seen in the anteroventricular region of the third ventricle, which is known to have involvement in the actions of Angiotensin II (30). In these experiments a lower dose of 3x2 μg antisense had bigger effects than the higher dose of 3x50 μg. No changes in binding were seen in nuclei located in the brainstem or in nuclei, which contains exclusively AT_2 angiotensin receptors. In a second autoradiographic study, SH rats received single injections of AS oligonucleotides into the lateral ventricle. In these experiments also, modest decreases of about 15% in hypothalamic nuclei, including PVN, SCh, supraoptic nucleus and median preoptic nucleus, were seen. There are several possible explanations for the limited effectiveness of antisense inhibition on the decrease of angiotensin receptor number in these studies. As mentioned earlier, a restricted distribution, limited

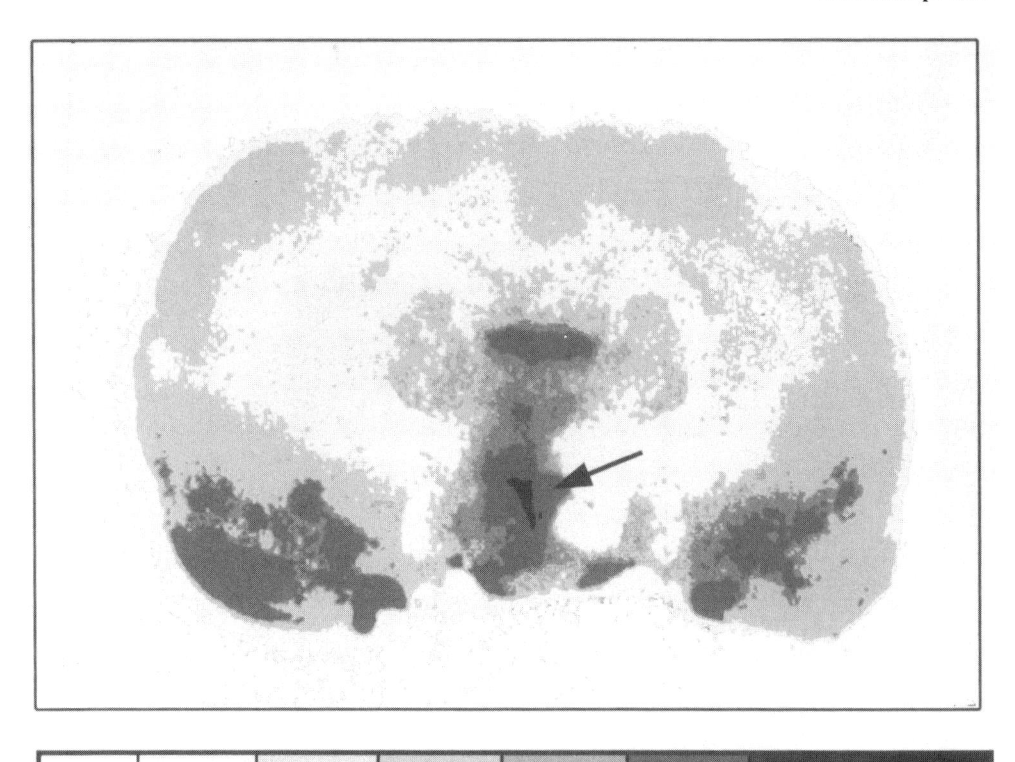

Figure 7. Autoradiography of rat hypothalamus after unilateral injection (25 μg) of antisense oligonucleotide targeted to AT_1 receptor mRNA. The injection on the right side, aimed at the paraventricular nucleus, has produced a decrease in the number of receptors compared to the non-injected side. (Arrows) The area of anterior hypothalamus close to the site of injection also shows a decrease in AT_1R binding. The antisense was designed to inhibit AT_{1A} receptors. The binding was to ^{125}I-Sar^1,Ile^8-Angiotensin II in Sprague-Dawley rats. Gradient bar indicates the code for radioactivity from low (white) to high (black).

by rapid uptake into cells may play a role. Some of the angiotensin receptors in areas to which the oligonucleotides have access may be presynaptic receptors, synthesized in distant sites, not reached by the oligonucleotides, and transported by axonal transport. This also would limit the effectiveness of the antisense inhibition.

There is heterogeneity of the AT_1 receptor. The rat AT_1 receptor (although not the human) has two subtypes, the AT_{1A} and the AT_{1B} (34). They share 91% identity but are encoded by different genes. The antisense oligonucleotides that were used in these binding studies were targeted to the AT_{1A} receptor mRNA. However, the brain has been reported to contain both AT_{1A} and AT_{1B} receptors (34,35). Notably the hypothalamus has a mixture of both types, although the AT_{1A} type seems more abundant (34,35). The presence of AT_{1B} receptors may to some extent explain the relatively small decrease in AT_1 receptor binding. Finally, antisense inhibition may lead to indirect regulatory effects that limit its effectiveness. For example, it may upregulate receptor production, increase recycling or reduce receptor metabolism, which would offset a fall in receptor number. The explanation for the modest decreases may also lie in the limitations of autoradiographic techniques in distinguishing between active and non-active AT_1 receptors.

Another way to inject is with osmotic mini-pumps. This has been used to inhibit the dopamine D-2 receptor and NMDA receptor (25). These injections made into the lateral ventricle of the rat brain produce significant reductions in receptor concentration. However, the total doses used were very large and a single much lower dose can be effective. More accuracy may be achieved by directly injecting the antisense into specific nuclei. This has been done, for example, in the paraventricular nucleus of the hypothalamus (Fig. 7), where one side shows a reduced intensity of AT_1 receptor binding after injection with AS-ODN, compared to the uninjected side.

Antisense in Peripheral Tissues

While antisense has been successful for direct injections into the brain, much less has been done with injections of antisense into peripheral tissues. Direct injection has been made into the carotid artery wall after balloon catheterization to test antisense for the prevention of restenosis or growth of the vascular smooth muscle at the site of the injury (36). In our laboratory we have injected AT_1R antisense into the adrenal glands and shown a decrease in AT_1 receptors in the zona glomerulosa. However, the direct injections only target a small region and therefore physiological changes are harder to measure. Reluctance to try i.v. injections of antisense ODNs is partly because of the loss of specificity with regard to particular tissues. This can be overcome by designing site specific carriers for the antisense oligo. Kaneda et al (37) have used liposome carriers enveloped in a viral coat of the sendai virus. This concept has a potential to allow delivery to specific tissues. By attaching antibodies or antigen to the viral coat, it should be possible to direct the carriers to a specific tissue. For example, glycogen on the coat would be selectively taken up from blood in the liver. The stability of oligos in the plasma and peripheral tissues appears to be quite good and preliminary results with oligos administered i.v. indicate a long half-life.

Appropriate Controls

It is so imperative that appropriate controls be used that the editors of the specialists journal in antisense technology, *"Antisense Research and Development"*, have stipulated that they will not consider papers which do not have at least two of the following controls (38). The most logical control is sense ODN i.e. an ODN with a normal sequence of nucleotides found in the targeted mRNA. Since the sequence would not have base pairs to bind with it should have no effect. A second control is to use a scrambled sequence i.e. a sequence that is made up by mixing the position of the bases in the antisense ODN. The third control is mismatch. In this control 2 or 3 of the nucleotides in the antisense sequence are purposely mismatched. This should render the ODN incapable of binding to the targeted mRNA. A fourth control is more appropriate for purely cellular studies, and that is to use a mismatch cell i.e. a cell that is known to have a mutated mRNA which the antisense should not bind to, and therefore should produce no effect on the protein. In each case, during the design of the oligo, the sequences should be checked on an appropriate DNA data base to ensure that one has not inadvertently targeted more than one naturally occurring sequence. In the studies mentioned before, of injecting antisense ODN to angiotensinogen or AT_1R into SHR rats (28,29), we used both sense, scrambled and the vehicle control. Only the antisense produced a decrease in blood pressure (Fig. 4).

Viral Vectors for Antisense DNA

For long-lasting effects of antisense, the antisense sequence will be integrated into the genome so that cells can constantly express an antisense RNA. This is being done with

DNA in the antisense sequence being delivered with viral vectors. There are several viral vectors being tested, including Herpes virus, which requires a helper virus to activate it, adeno-virus which is fairly short-acting and adeno-associated virus which is potentially therapeutically useful because it has no known pathological effect.

Antisense expression vectors produce antisense RNA, which can hybridize with the target mRNA and prevent its translation by either promoting its degradation, preventing its transport from the nucleus to the cytoplasm or interfering with ribosomal function.

Adeno-associated virus (AAV) -derived vectors have been gaining attention because of their safety and efficiency (39). AAV-derived expression vectors have been successfully used to inhibit alpha-globin expression in an erythroleukemia cell line where a 1.0-kb fragment of the alpha-globin gene was inserted in the antisense direction under either thymidine kinase, sv40 or the alpha-globin promoter. Ninety-one percent inhibition in alpha-globin mRNA production was achieved with the alpha-globin's own promoter, whereas the other two promoters proved to be less efficient (40).

A much shorter antisense sequence (64 bp) targeting the human immunodeficiency virus (HIV) long terminal repeat was expressed with an AAV vector using the sv40 promoter in 293 cells. Greater than 99% reduction in infectious HIV-1 production and the absence of cellular toxicity was observed for over a period of 20 days (41).

For a chronic disease such as hypertension, where genetic influences can be identified (42), the concept of genomic antisense has an appeal. Those mutated genes which are producing hypertension through overexpression can be neutralized by antisense. Angiotensinogen has been one gene implicated in hypertension (42). It would be desirable in this case to reduce overproduction of angiotensinogen down to levels where it is physiologically necessary. Perhaps the future will bring a benign vector to insert an antisense DNA into the genome that will control the overexpressing mutant genes. The goal is to have efficacy without side effects and a delivery system that is required so infrequently that there will be no problems of patient compliance.

ACKNOWLEDGMENTS

This work was supported by NIH grant HL-27334 and a Grant-in-Aid from the American Heart Association (Florida Affiliate). Philipp Ambühl and Robert Gyurko are recipients of the AHA (Florida Affiliate) postdoctoral and predoctoral fellowships, respectively.

REFERENCES

1. Stull, R.A., Taylor, L.A. and Szoka, F.C., Predicting antisense oligonucleotide inhibitory efficacy: a computational approach using histograms and thermodynamic indices, *Nucl.Acids Res.*, 20(13) (1992) 3501-3508.
2. Cowsert, L.M., Fox, M.C., Zon, G. and Mirabelli, C.K., in vitro evaluation of phosphorothioate oligonucleotides targeted to the E2 mRNA of papillomavirus: potential treatment for genital warts, *Antimicrob. Agents Chemother.* 37(2) (1993) 171-177.
3. Wakita, T. and Wands, J.R., Specific inhibition of hepatitis C virus expression by antisense oligonucleotides. In vitro model for selection of target sequence, *J.Biol.Chem.*, 269(19) (1994) 14205-14210.
4. Lima, W.F., Monia, B.P., Ecker, D.J. and Freier, S.M., Implication of RNA structure on antisense oligonucleotide hybridization kinetics, *Biochemistry*, 31(48) (1992) 12055-12061.
5. Rittner, K and Sczakiel, G., Identification and analysis of antisense RNA target regions of the human immunodeficiency virus type 1, *Nucl.Acid Res.*, 19(7) (1991) 1421-1426.

6. Jaroszevski, J.W., Syi, J.L., Ghosh, M., Ghosh, K. and Cohen, J.S., Targeting of antisense DNA: comparison of activity of anti-rabbit beta-globin oligodeoxyribonucleoside phosphorothioates with computer predictions of mRNA folding, *Antisense Res.Dev.*, 3(4) (1993) 339-348.
7. Fakler, B., Herlitze, S., Amthor, B., Zenner, H.P. and Ruppersberg, J.P., Short antisense oligonucleotide-mediated inhibition is strongly dependent on oligo length and concentration but almost independent of location of the target sequence, *J.Biol.Chem.*, 269(23) (1994) 16187-16194.
8. Puskas, L.G. and Bottka, S., Prokaryotic test system for evaluation of oligonucleotide-affected antisense inhibition, *Anal.Biochem.*, 222(2) (1994) 305-309.
9. Singer, M. and Berg, P. In: Genes and Genomes, University Science Books, 54-59.
10. Iversen, P.L., Zhu, S., Meyer, A. and Zon, G., Cellular uptake and subcdellular distribution of phosphorothioate oligonucleotides into cultured cells, *Antisense Res.Dev.*, 2(3) (1992) 211-222.
11. Colige, A., Sokolov, B.P., Nugent, P., Baserge, R. and Prockop, D.J., Use of an antisense oligonucleotide to inhibit expression of a mutated human procollagen gene (COL1A1) in transfected mouse 3T3 cells, *Biochemistry*, 32(1) (1993) 7-11.
12. Helene, C., The anti-gene strategy: control of gene expression by triplex-forming-oligonucleotides, *Anticancer Drug Res.*, 6(6) (1991) 569-584.
13. Bennett, C.F., Condon, T.P., Grimm, S., Chan, H. and Chiang, M.Y., Inhibition of endothelial cell adhesion molecule expression with antisense oligonucleotides, *J.Immunol.*, 152(7) (1994) 3530-3540.
14. Bacon, T.A. and Wickstrom, E., Walking along human c-myc mRNA with antisense oligonucleotides: maximum efficacy at the 5′ cap region, *Oncogene Res.*, 6(1) (1991) 13-19.
15. Campbell, J.M, Bacon, T.A. and Wickstrom, E., Oligodeoxynucleotide phosphorothioate stability in subcellular extracts, culture media, ser and cerebrospinal fluid, *J. Biochem. Biophys. Methods* 20(3) (1990) 259-267.
16. Wahlestedt, C., Antisense oligonucleotide strategies in neuropharmacology, *TiPS*, 15 (1994) 42-46.
17. Gyurko, R., Wielbo, D., Phillips, M.I., Antisense inhibition of AT_1 receptor mRNA and angiotensinogen mRNA in the brain of spontaneously hypertensive rats reduces hypertension of neurogenic origin, *Regul. Pept.* 49(2) (1993) 167-74.
18. Gyurko, R., Li, B., Sumners, C. and Phillips, M.I., Antisense inhibition of angiotensin type-1 receptor expression in NG108-15 cells, *Soc. for Neurosci.* (1994) 220.7
19. Li, B. and Phillips, M.I., Cellular uptake of oligonucleotides in bovine adrenal cells (in press) (1995).
20. Yu, C., Brussaard, A.B., Yang, X., Listerud, M. and Role, L.W., Uptake of antisense oligonucleotides and functional block of acetylcholine receptor subunit gene expression in primary embryonic neurons, *Dev-Genet* 14(4) (1993) 296-304.
21. Marti, G., Egan, W. Noguichi, P. Zon, G., Matsukura, M. and Broder, S., Oligodeoxyribonucleotide phosphorothioate fluxes and localization in hematopoietic cells, *Antisense Res. Rev.* 2(1) (1992) 27-39.
22. Clarenc, J.P., LeBleu, B. and Leonetti, J.P., Characterization of the nuclear binding sites of oligodeoxyribonucleotides and their analogs, *J. Biol. Chem.* 268(8) (1993) 5600-5604.
23. Sixou, S., Szoka, F.C., Jr., Green, G.A., Giusti, B., Zong, G. and Chin, D.J., Intracellular oligonucleotide hybridization detected by fluorescence resonance energy transfer (FRET), *Nuclei Acids Res.* 22(4) (1994) 662-668.
24. Temsamani, J., Kubert, M., Tang, J., Padmapriya, A., Agrawal, S., Cellular uptake of oligodeoxynucleotide phosphorothioates and their analogs, *Antisense Res. Dev.* 4(1) (1994) 35-42.
25. Wahlestedt, C., Golanov, E., Yamamoto, S., Yee, F., Ericson, H., Yoo, H., Inturrisi, C.E., Reis, D.J., Antisense oligonucleotides to NMDA-R1 receptor channel protect cortical neurons from excitotoxicity and reduce focal ischemic infarctions, *Nature* 363(6426) (1993) 260-263.
26. Rougee, M., Faucon, B., Mergny, J.L., Barcelo, F., Giovannangeli, C., Garestier, T., Helene, C., Kinetics and thermodynamics of triple-helix formation: effects of ionic strength and mismatches, *Biochemistry* 31 (1992) 9269-9278.
27. Crooke, R.M., *In vitro* toxicology and pharmacokinetics of antisense oligonucleotides, *Anticancer Drug Des.* 6(6) (1991) 609-646.
28. Phillips, M.I., Wielbo, D. and Gyurko, R., Antisense inhibition of hypertension: A new strategy for renin-angiotensin candidate genes, *Kidney Intl.* 46 (1994) 1554-1556.
29. Wielbo, D., Sernia, C., Gyurko, R. and Phillips, M.I., Antisense inhibition of hypertension in the spontaneously hypertensive rat, *Hypertension* 25 (1994) 314-319.
30. Phillips, M.I., Functions of angiotensin II in the central nervous system, *Ann. Rev. Physiol.* 49 (1985) 413-415.
31. Sakai, R.R., He, P.F., Yang, X.D., Ma., L.Y., Guo, Y.F., Reilly, J.J., Moga,C.N. and Fluharty, S.J., Intracerebroventricular administration of AT_1 receptor antisense oligonucleotide inhibits the behavioral actions of angiotensin II, *J. Neurochem.* 62 (1994) 2053-2056.

32. Meng, H-B., Wielbo, D., Gyurko, R. and Phillips, Mi.I. AT_1 receptor mRNA antisense oligonucleotide inhibits central angiotensin induced thirst and vasopressin, *Regul. Pept.* 54 (1994) 543-551.
33. Ambühl, P., Gyurko, R. and Phillips, M.I., A decrease in angiotensin receptor binding in rat brain nuclei by antisense oligonucleotides to the angiotensin AT_1 receptor (in press) (1995).
34. Kakar, S.S., Riel, K.K. and Neill, J.D., Differential expression of angiotensin II receptor subtype mRNAs (AT-1A and AT-1B) in the brain, *Biochem. Biophys. Res. Commun.*, 185 (1992) 688-692.
35. Lenkei, Z., Corvol, P. and Llorens-Cortes, C., The angiotensin receptor subtype AT_{1A} predominates in rat forebrain areas involved in blood pressure, body fluid homeostasis and neuroendocrine control, *Mol. Brain Res.*, 30 (1995) 53-60.
36. Morishita, R., Gibbons, G.H., Kaneda, Y., Ogihara, T. and Dzau, V.J., Pharmacokinetics of antisense oligodeoxyribonucleotides (cyclin B1 and CDC 2 kinase) in the vessel wall *in vivo*: enhanced therapeutic utility for restenosis by HVJ-liposome delivery, *Gene* 149(1) (1994) 13-19.
37. Kaneda, Y., Iwai, K. and Uchida, T., Increased expression of DNA cointroduced with nuclear protein in adult rat liver, *Science* 243 (1989) 375-378.
38. Stein, C.A. and Krieg, A.M., Problems in interpretation of data derived from *in vitro* and *in vivo* use of antisense oligodeoxynucleotides (Editorial), *Antisense Res. Dev.* 4 (1994) 67-69.
39. Muzyczka, N., Use of adeno-associated virus as a general transduction vector for mammalian cells, *Curr. Top Microbiol. Immunol.* 158 (1992) 97-129.
40. Ponnazhagan, S., Nallari, M.L. Srivastava, A., Suppression of human alpha-globin gene expression mediated by the recombinant adeno-associated virus 2-based antisense vectors, *J. Exp. Med.* 179(2) (1994) 733-738.
41. Chatterjee, S., Johnson, P.R., Wong, K.K., Jr., Dual-target inhibition of HIV-1 *in vitro* by means of an adeno-associated virus antisense vector, *Science* 258 (1992) 1485-1488.
42. Caulfield, M., Lavender, P., Farrell, M., Munroe, P., Lawson, M., Turner, P., Clark, A., Linkage of angiotensinogen gene to essential hypertension, *New Eng. J. Med.* 330 (1994) 1629-1633.

10

INTERACTIONS OF ANGIOTENSIN II WITH CENTRAL DOPAMINE

T. A. Jenkins, A. M. Allen, S. Y. Chai, D. P. MacGregor,[1] G. Paxinos, and F. A. O. Mendelsohn

University of Melbourne
Department of Medicine, Austin Hospital
Heidelberg, Victoria, 3084, Australia
[1] School of Psychology
University of New South Wales
Kensington, New South Wales, 2033, Australia

INTRODUCTION

Circulating angiotensin II (Ang II) maintains fluid balance and blood pressure by actions on the renal, endocrine, cardiovascular and peripheral autonomic nervous systems. Ang II also acts in the brain to augment its peripheral actions by stimulating salt appetite and drinking, elevating blood pressure, and by modulating the release of pituitary hormones. Blood borne Ang II can access the brain via the circumventricular organs which have a permeable blood brain barrier. The central actions of Ang II in other brain regions beyond the blood brain barrier are probably mediated by Ang II synthesized within the brain (1,2).

The precursors and enzymes necessary for formation of Ang II have been localized in brain tissue, along with subtypes of angiotensin receptors, AT_1 and AT_2 (1,2). There is a striking conservation of Ang II AT_1 receptor distributions between mammalian species, with Ang II receptors previously identified in the central nervous systems of rat (3), rabbit (4), dog (5), sheep (6), monkey (7) and human (8-10) using *in vitro* autoradiography. In the rat brain, Ang II receptors are distributed in areas that mediate control of blood pressure and body fluid balance, including the paraventricular hypothalamic nucleus (Pa), subfornical organ (SFO), the organum vasculosum of the lamina terminalis (OVLT), median preoptic nucleus (MnPO), area postrema (AP), and nucleus of the solitary tract (Sol). These areas have been shown to contain mainly the AT_1 receptor subtype. Other areas include the amygdala, piriform cortex, nucleus of the lateral olfactory tract. The septum and cerebellum contain both AT_1 and AT_2 sites. Ang II AT_2 receptors predominate in many nuclei of the thalamus, locus coeruleus (LC), superior colliculus, and the inferior olive (IO) (11).

In the human brain, high densities of AT_1 receptors are also found in the hypothalamus, circumventricular organs and medulla oblongata (8,10). In addition, Ang II AT_1 receptors occur in the striatum and the substantia nigra pars compacta of the human where

Recent Advances in Cellular and Molecular Aspects of Angiotensin Receptors
Edited by Mohan K. Raizada et al., Plenum Press, New York, 1996

binding is localized over pigmented dopaminergic neurones (12). In the hind brain, in the caudal and rostral ventrolateral medulla (CVLM and RVLM), AT_1 receptors are found in abundance (9). All Ang II sites in the human brain appear to be AT_1, except for the cerebellum, which has both AT_1 and AT_2 subtypes (13).

ANGIOTENSIN II AND CATECHOLAMINES

Anatomical Association

There is strong anatomical, pharmacological and physiological evidence for an interaction between brain renin-angiotensin and catecholamine systems. Ang II receptors and Ang II immunoreactivity occur in many areas rich in catecholamine cell bodies and terminals, such as the Sol, LC, MnPO, PaP and RVLM and CVLM. These areas are involved in autonomic and cardiovascular control, and fluid balance (3,14-16).

Cardiovascular Control

Central administration of Ang II increases in blood pressure through activation of the sympathetic nervous system and release of vasopressin (17,18,19). Central Ang II infusions at pressor concentrations increase noradrenaline (NA) concentration in cerebrospinal fluid in rabbits (20), and rhesus monkey (21), suggesting that central catecholamine neurons may be activated. After central administration of the catecholamine selective toxin, 6-hydroxy dopamine (6-OHDA), central Ang II induced blood pressure increase is diminished (22-26). This pretreatment also prevents development of one-kidney hypertension, and abolishes the associated increase in water consumption (23). Under physiological conditions these responses may be produced by Ang II generated by a brain renin-angiotensin system (RAS) as intracerebroventricular (icv) administration of renin also increases mean arterial pressure through activation of sympathetic nerve discharge (27).

In several brain regions, Ang II interacts directly with catecholamine-containing neurons to regulate blood pressure. A high density of AT_1 receptors occurs in Sol, in the region of the C2/A2 catecholamine cell groups. At least some of the AT_1 receptors are located presynaptically on terminals of vagal afferent neurons (28). Microinjections of Ang II into Sol elicits complex effects on blood pressure with low doses being depressor, and high doses displaying biphasic depressor/pressor responses. Microinjections of Ang II into Sol also inhibit baroreflex sensitivity (17,29). In contrast, microinjections of α_2 agonists increase baroreflex sensitivity (30). Autoradiographic studies reveal an antagonistic angiotensin AT_1/α_2-adrenoceptor interaction, suggesting that AT_1 receptors may reduce transduction and therefore the α_2-mediated vasodepressor response (33).

In the RVLM, site of the C1 cell group, microinjections of Ang II increases blood pressure through activation of sympathetic nerve discharge (18,31). In slices of rat medulla Ang II excites a subpopulation of RVLM neurones which have properties characteristic of C1 adrenaline neurones (32).

Microinjections of Ang II into the CVLM of the rabbit increase secretion of vasopressin (34), presumably through activation of the A1 noradrenergic neurones.

Drinking

Ang II induced drinking is also influenced by central catecholamines. Intracerebroventricular administration of 6-OHDA reduces central Ang II induced drinking (23,35). More specifically, noradrenergic lesions of the periventricular tissue of the anteroventral

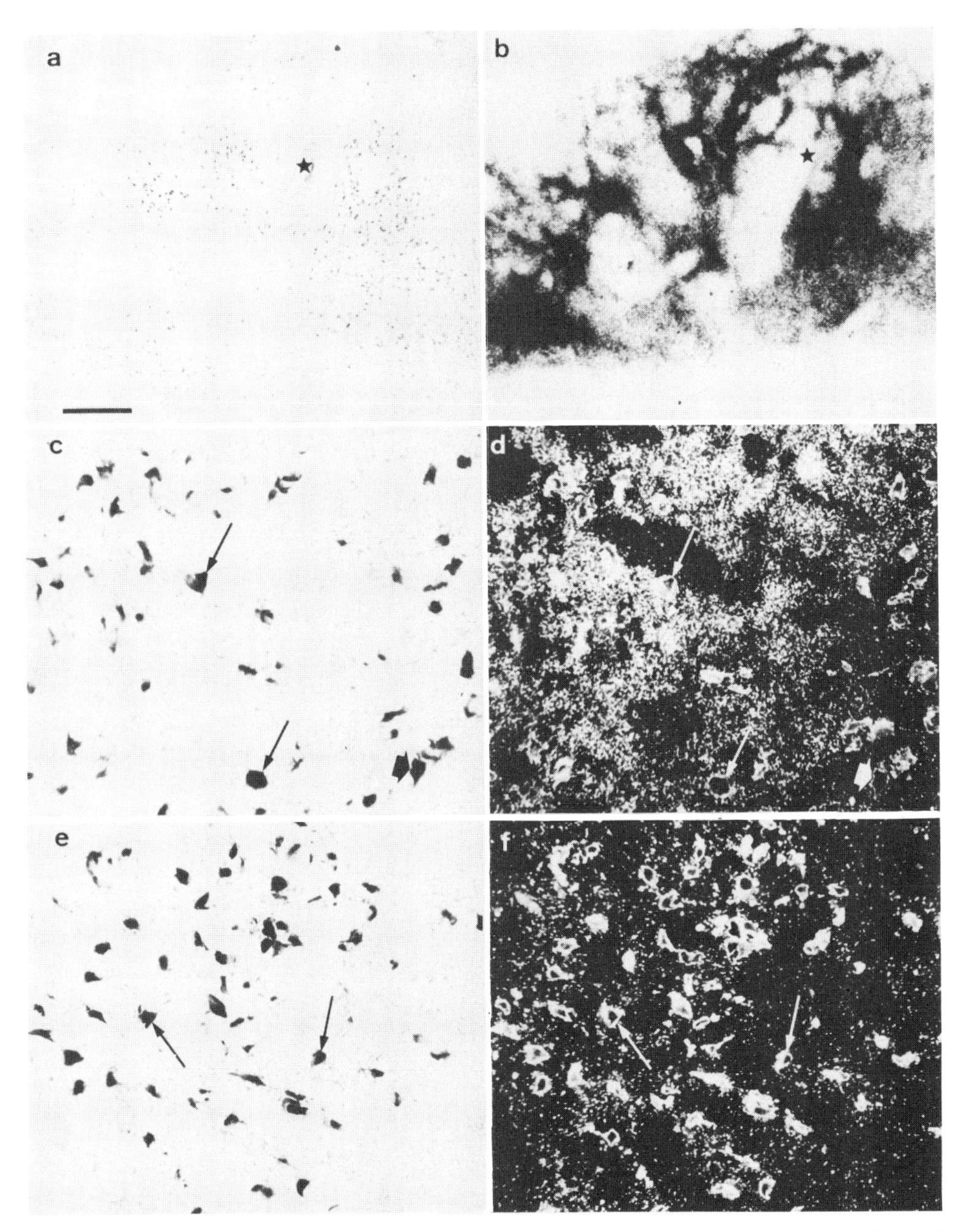

Figure 1. ^{125}I-[Sar^1, Ile^8]Ang II binding in the substantia nigra pars compacta. a: Low power photomicrograph of a thionin-stained section of the substantia nigra. b: Low power photomicrograph of an autoradiograph from a 20μM section adjacent to that shown in a. Black depicts Ang II receptor binding. The star in a and b is over the same blood vessel in each section. Scale bar = 750μm.

region of the third ventricle (AV3V) region, including MnPO and OVLT, (36) block Ang II thirst, suggesting that noradrenergic innervation of the ventral lamina terminalis region is necessary for Ang II thirst (24,37). Central coinfusion of NA with Ang II restores the drinking response to Ang II after these lesions (38), as does transplantation of foetal noradrenergic cell suspension into the OVLT/ventralMnPO region (39).

Intracerebroventricular administration of both α and β adrenoceptor antagonists, prazosin, phentolamine and propanolol, inhibit Ang II induced drinking (32,40,41). Infusion of phentolamine into the anterior hypothalamic-preoptic area also decreases thirst (31). Interestingly, α_2 agonists clonidine and phenylephrine (40,42,43) also block Ang II-stimulated drinking, with this antidipsogenic effect blocked by the antagonist yohimbine (44). Subcutaneous administration of the AT_1 receptor antagonist Losartan inhibits peripherally administered Ang II induced drinking. This effect can be blocked by subcutaneous preadministration of β-agonist isoproterenol (45). Thus it appears that Ang II induced thirst involves an adrenergic component, but the mechanism of action is unclear.

Pituitary Hormones

In some cases the regulation of pituitary hormone release by Ang II may involve interactions with catecholamine neurons.

Vasopressin release from the magnocellular neurones of the Pa and supra optic nucleus is stimulated by catecholamines acting through α_1 receptors (46).

In vivo experiments in the hypothalamus reveal Ang II increases release of NA which would probably modify release of vasopressin (47).

Behaviour

An interaction with central catecholamines may be involved in the behavioural actions of Ang II, such as stimulation of locomotor activity and stereotypy, and modification of learning (48-51). The increased rate of learning of conditioned avoidance stimuli by icv Ang II is abolished by treatment with the α_1 and α_2 antagonists prazosin and yohimbine, (52), or phentolamine (53). Yohimbine has also been reported to reverse the improvement of recall by Ang II (52). It has also been reported that in acute stress, endogenous Ang II facilitates the release of noradrenergic transmission, via the AT_1 receptor (54).

Na Release

In Vitro. In cultured neurons from rat brain Ang II increased neuronal and media catecholamine levels (55), which was blocked by prazosin (56). In medullary slices, Ang II enhances NA efflux (57). Ang II also increases electrical (58), or K^+ (57,59) stimulated NA release from hypothalamic slices. This effect is blocked by low sodium diet (58,59). Slices from the pineal gland, the richest source of renin in the brain (60,61), release electrically stimulated NA after Ang II addition (62).

The Ang II effect on NA release *in vitro* is not always excititatory: after prolonged Ang II stimulation in rat hypothalamic-brainstem neuronal cultures, NA uptake appears to be inhibited. It was suggested that this inhibitory action of Ang II represents a stimulation of NA release, with subsequent auto inhibition of NA release (63,64). In the spontaneously hypertensive rat (SHR), Ang II inhibited the K^+-evoked release of $[^3H]$NA from hypothalamic synaptosomes (65).

Noradrenaline elicits significant decreases in $[^{125}I]$Ang II specific binding to cultured neurones from normal rats, by an effect of specific α_1-cell surface receptors (64,66). In cultured neurones from SHR, NA has no effect on in $[^{125}I]$Ang II binding (64,66). Synaptosomes prepared from neonatal or adult rat hypothalamus and brainstem exhibited increased Ang II receptor binding after treatment with NA (67).

In Vivo. Central Ang II administration increased NA uptake, turnover and release in specific brain regions. Release of NA in the LC, hypothalamus, A1 region, and raphe magnus nucleus (68), and increased monoamine oxidase A activity (69) have been reported after icv injection of Ang II. The paraventricular nucleus (70) and supraoptic nucleus (71), nuclei responsible for vasopressin synthesis and secretion, exhibit increased NA release after Ang II infusion.

Studies in the anterior hypothalamus using *in vivo* microdialysis reveal that Ang II increases K^+ stimulated NA release (72).

Conclusion

Thus Ang II increases sympathetic nerve discharge, activating central catecholamine neurones, and therefore modulating central catecholamine receptors.

ANGIOTENSIN II AND DOPAMINE

Anatomical Association

The anatomical association of Ang II receptors with catecholamine-containing areas extends to specific dopaminergic pathways in the mammalian brain. High densities of Ang II receptors are found within the arcuate hypothalamic nuclei (3), the site of the tubero-infundibular dopamine (DA) neurones (73). In the striatum, [125I]Ang II binding shows functional Ang II receptors exist on membranes of intrinsic neurones in the rat neostriatum (74), while in the human brain, our autoradiographic studies reveal Ang II receptors associated with pigmented dopamine containing cell bodies in the substantia nigra, and in the striatum (9) (Figure 1).

Lightfield (c,e) and darkfield (d,f) photomicrographs showing total (c,d) and non-specific (e,f) binding of ^{125}I-[Sar^1, Ile^8]Ang II to the substantia nigra pars compacta, visualized by dipping emulsion autoradiography. In darkfield the white grains show areas of radioligand binding. The arrows show the position of the same cell in dark- and lightfield in the paired photomicrographs (i.e. c,d and e,f). In c and d the thick arrow points to a pigmented neuron which does not show Ang II receptor binding. Scale bar in a = 165μm in c-f. (reprinted with permission from Allen et al, J. Comp. Neurol. 312:291-298 (ref no. 9))

The Ang II receptor binding is markedly reduced in both the striatum and substantia nigra of brains from patients dying with Parkinson's Disease, a condition characterized by degeneration of this ascending nigrostriatal dopaminergic system (12).

On the basis of these anatomical results we proposed that Ang II receptors are produced in the substantia nigra and transported to the striatum where they are located presynaptically in the caudate and putamen (12).

Cardiovascular Control

Haloperidol and propanolol inhibit the pressor response to central Ang II (22) and central 6-OHDA administration also diminishes the Ang II induced blood pressure increase (22-26). Metoclopramide, a DA antagonist, enhances Ang II stimulated increase in plasma vasopressin levels (75).

Drinking

Ang II actions on drinking may also involve DA. First, intracerebroventricular injection of DA alone causes a dipsogenesis (35). Furthermore, intracranial and peripheral haloperidol (35,76), and metoclopramide (75), inhibit Ang II-induced drinking, whereas chronic haloperidol (77) treatment stimulates Ang II induced drinking, perhaps due to an increase in the number of DA receptors (78).

Pituitary Hormones

Central administration of Ang II decreases prolactin plasma levels in OVX-estradiol primed rats, an effect blocked by peripheral administration of domperidone. This suggests that the Ang II effect on anterior pituitary hormone release is probably mediated by the tuberoinfundibular dopamine system (79). Also, DA inhibits the Ang II-stimulated prolactin release, and inositol phosphate (IP) production in anterior pituitary cells (80). This negative regulation of IP production, mediated by D_2 receptors, is one of the mechanisms by which DA controls hormonally stimulated prolactin release (81).

Behaviour

Behaviourally, icv administration of Ang II potentiates apomorphine-induced stereotypy (82), an effect blocked by both DA antagonists (82) and Ang II receptor antagonists (82,83). Also, Ang II decreases the number of punished responses in a conflict paradigm, while at the same time decreasing DA levels in the hippocampus and hypothalamus (84). Dopamine antagonists also block the cognitive effects of Ang II (85), and potentiate chemically-stimulated and electroconvulsive seizures by icv Ang II acutely (86).

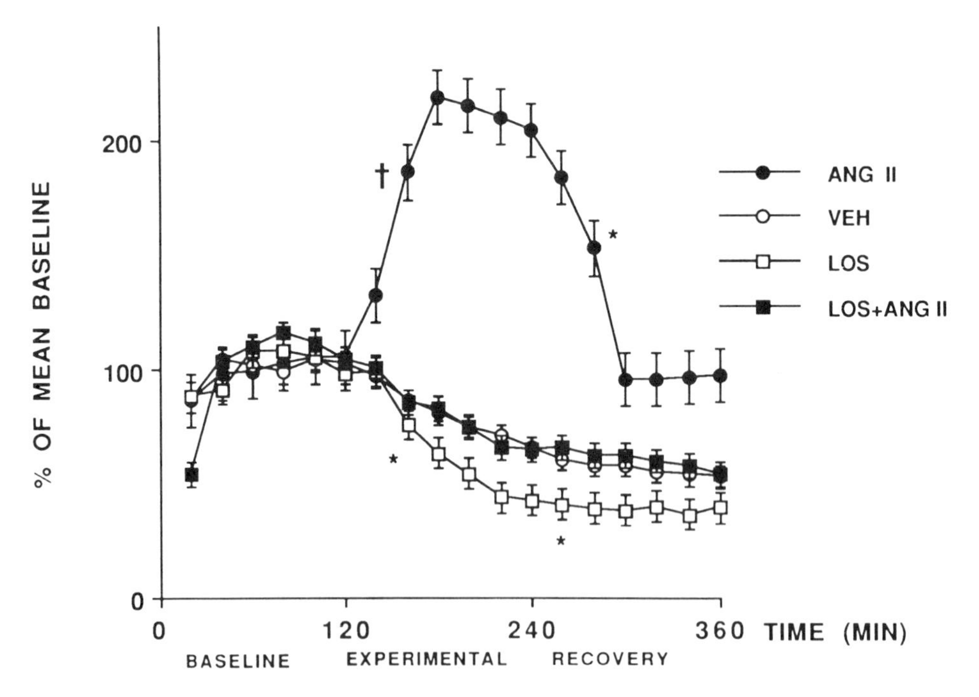

Figure 2. DOPAC outputs from right striatum, standardised to the mean of the first six baseline values. All rats recieved artificial CSF during baseline and recovery periods. During the experimental time period, rats recieved either Ang II (1μM) (●), vehicle (aCSF) ( ), Losartan (1μM) ( ), or Ang II + Losartan (1μM) (■) through the microdialysis probe. Values are mean ± standard error. Analysed by two way ANOVA: †$p<0.001$ compared to baseline, *$p<0.05$ compared to baseline.

In Vitro

In vitro, Ang II stimulates the release of newly synthesized DA from rat striatal slices (87), while losartan inhibits release (88). In accord with this, Ang II has been shown to mimic the electrophysiological effects of DA on cells in the rat striatum (89).

In vitro studies using hypothalamic slices have shown that Ang II enhances both DA and dihydroxy phenyl acetic acid (DOPAC) efflux (57). Neuronal hypothalamic cultures display increased neuronal DA levels in response to Ang II at high concentrations in the medium (90). Dopamine also elicits significant decreases in [^{125}I]Ang II specific binding to neuronal cultures in normal rats via adrenergic receptors (66).

Renin, via formation of angiotensin peptides, increases DA turnover in the tubero-infundibular DA neurons in the medial and lateral palisade zones of the hypothalamus (73).

In Vivo

In vivo, icv Ang II increases striatal DA levels (91), while peripherally administered losartan decreases them (88). Using *in vivo* microdialysis we have shown that Ang II induces a marked increase in extracellular DOPAC release, the major DA metabolite, in the striatum. When Ang II infusion was stopped, the concentration returned to control levels. The Ang II induced DOPAC increase is completely blocked by coadministration of the AT_1 specific antagonist, Losartan, indicating that the response is mediated via AT_1 receptors. Losartan alone led to a significant depression of DOPAC output relative to vehicle, suggesting that dopamine release is under a tonic facilitatory influence of Ang II via the Ang II AT_1 receptor (92) (Figure 2).

In rats with unilateral 6-OHDA lesions in the nigrostriatal pathway, injection of angiotensin II into the unlesioned striatum elicits tight dose-dependant rotations ipsilateral

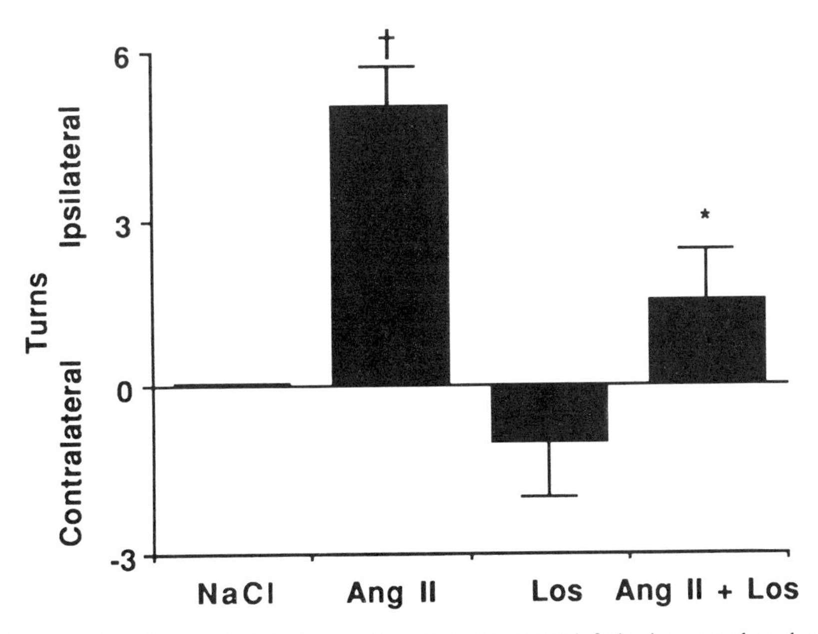

Figure 3. Turns elicited by intrastriatal injection after unilateral 6-OHDA infusion into contralateral substantia nigra. Results are mean number of ipsilateral turns ± standard errors in a contralateral or ipsilateral direction, relative to the lesion. Kruskal-Wallis test: †Ang II vs. NaCl $p<0.01$, *Ang II vs. Ang II + Los $p<0.05$.

to the lesion side. This rotation is blocked by coadministration of the angiotensin AT_1 receptor antagonist, Losartan, which has no significant effect when injected alone (Figure 3). Also, pretreatment with the D_2 receptor blocker, haloperidol, blocks this effect of Ang II, confirming the hypothesis that Ang II is intrinsically involved in modulating dopamine release in the striatum (Figure 4).

Microdialysis studies in the hypothalamus of conscious rats also show that Ang II potentiates K^+ stimulated DA release (93), while in the septum, icv Ang II increases septal DA levels (94).

Conclusion

Our autoradiographic studies show a close anatomical association of Ang II receptor binding with DA neurones in the nigrostriatal pathway in the human brain, while our *in vivo* studies using microdialysis and 6-OHDA lesions demonstrate that Ang II has an acute effect on DA turnover in the striatum, suggesting there is indeed an important interaction between brain Ang II and DA systems.

SUMMARY

There is a large body of evidence to support the concept of a relationship between brain Ang II and catecholamine systems. This interaction may participate in some central actions of Ang II such as cardiovascular control, dipsogenesis, and complex behaviours. It also extends to the nigrostriatal dopaminergic system which bear AT_1 receptors, both on their cell bodies in the substantia nigra presynaptically, and on their terminals in the striatum,

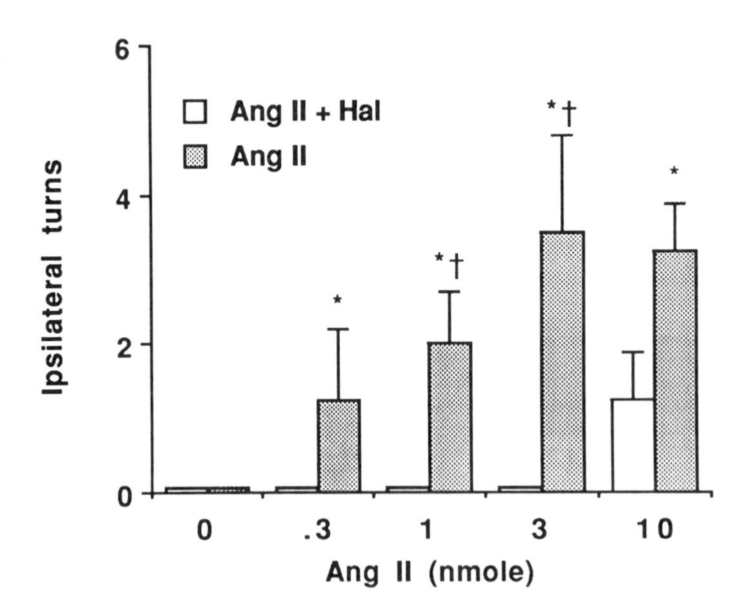

Figure 4. Pretreatment with haloperidol (2mg/kg i.p. one hour before test) (open bars) on intrastriatal Ang II stimulated rotation (stippled bars) after unilateral 6-OHDA infusion into contralateral substantia nigra. Results are mean number of ipsilateral turns ± standard errors. *Dose effect of Ang II on turning - ANOVA $p<0.05$.

where Ang II can markedly potentiate DA release. This observation suggests that drugs which modulate central Ang II may be useful in regulating central dopaminergic activity.

ACKNOWLEDGMENTS

This work was supported by funds from the National Health & Medical Research Council, National Heart Foundation of Australia, Austin Hospital Medical Research Foundation and The Ian Potter Foundation. T.A.J. is funded by a NH&MRC Dora Lush Postgraduate Scholarship. A.M.A. is the holder of a NH&MRC R.D. Wright Postdoctoral Fellowship. S.Y.C. is supported by a NH&MRC Australian Postdoctoral Fellowship. D.P.M. is funded by a NH&MRC Medical Postgraduate Scholarship.

REFERENCES

1. Saavedra, J.M. (1992). Pharmacol. Rev. *13*(2):329-380
2. Phillips, M.I. (1987). Ann. Rev. Physiol. *49*:413-435
3. Mendelsohn, F.A.O., Quirion, R., Saavedra, J.M., Aguilera, G. and Catt, K.J. (1984). Proc. Natl. Acad. Sci. USA *81*:1575-1579
4. Mendelsohn, F.A.O., Allen, A.M., Clevers, J., Denton, D.A., Tarjan, E. and McKinley. M.J. (1988). J. Comp. Neurol. *270*:372-384
5. Speth, R.C., Walmsley, J.K., Gehlert, D.R., Chernicky, C.L., Barnes, K.L. and Ferrario, C.M. (1984). Brain Res. *326*:137-143
6. McKinley, M.J., Allen, A.M., Clevers, J., Denton, D.A. and Mendelsohn, F.A.O. (1986). Brain Res. *375*:373-376
7. Millan, M.A., Jacobowitz, D.M., Catt, K.J. and Aguilera, G. (1990). Peptides *11*:243-253
8. Allen, A.M., Chai, S.Y., Clevers, J., McKinley, M.J., Paxinos, G. and Mendelsohn, F.A.O. (1988). J. Comp. Neurol. *269*:249-264
9. Allen, A.M., Paxinos, G., McKinley, M.J., Chai, S.Y. and Mendelsohn, F.A.O. (1991). J. Comp. Neurol. *312*:291-298
10. McKinley, M.J., Allen, A.M., Clevers, J., Paxinos, G. and Mendelsohn, F.A.O. (1987). Brain Res. *420*:375-379
11. Allen, A.M., Paxinos, G., Song, K. and Mendelsohn, F.A.O. (1992) in. Handbook of Chemical Neuroanatomy, Vol. 11: Neuropeptide receptors in the CNS (eds.Bjorklund, B., Hokfelt, T., Kuhar, M.J.) Elsevier Science Publishers, pp1-37.
12. Allen, A.M., MacGregor, D.P., Chai, S.Y., Donnan, G.A., Kaczmarczyk, S.J., Richardson, K., Kalnins, R., Ireton, J. and Mendelsohn, F.A.O. (1992). Ann. Neurol. *32*:339-344
13. MacGregor, D.P., Murone, C., Song, K., Allen, A.M., Paxinos, G, and Mendelsohn, F.A.O. (1995). Brain Res. (in press)
14. Lind, R.W., Swanson, L.W. and Ganten D. (1985). Neuroendocrinology *40*:2-24
15. Covenas, R., Fuxe, K., Cintra, A., Aguirre, J.A., Goldstein, M. and Ganten, D. (1990). Neurosci. Lett. *114*:160-166
16. Fuxe, K., Ganten, D., Hokfelt, T. and Bolme P. (1976). Neurosci. Lett. *2*:229-234
17. Ferrario, C.M. (1983). Hypertension *5*(suppl. V):V73-V79
18. Sasaki,S. and Dampney, R.A.L. (1990). Hypertension *15*:274-283
19. Unger, T., Becker, H., Petty, M., Demmert, G., Schneider, B., Ganten, D. and Lang, R.E. (1985). Circ. Res. *56*:563-575
20. Chevillard, C., Duchene, N., Pasquier, R. and Alexandre, J-M. (1979). Eur. J. Pharmacol. *58*:203-206.
21. Scholkens, B., Jung, W., Rascher, W., Dietz, R. and Ganten, D. (1982). Experientia *38*:469-470
22. Hoffman, W.E., Phillips, M.I. and Schmid P. (1977). Neuropharmacology *16*:563-569
23. Gordon, F.J., Brody, M.J., Fink, G.D., Buggy, J. and Johnson, A.K. (1979). Brain Res. *178*:161-173
24. Bellin, S.I., Landas, S.K. and Johnson, A.K. (1987). Brain Res. *416*:75-83
25. Pirola, C.J., Scheucher, A., Balda, M.S., Dabsys, S.M., Finkielman, S. and Nahmod VE. (1987). Neuropharmacology *26*(6):561-566
26. Walters, D.E. and Speth, R,C. (1989). Brain Res. Bull. *22*:283-288

27. Scholkens, B.A., Jung, W., Rascher,W., Schomig, A. and Ganten, D. (1980). Clin. Sci. *59*:53s-56s
28. Lewis, S.J., Allen, A.M., Verberne, A.J.M., Figdor, R., Jarrot, B. and Mendelsohn, F.A.O. (1986). Eur. J. Pharmacol. 125:305-307
29. Rettig, R., Healy, D.P. and Printz, M.P. (1986). Brain Res. *364*(2):233-240
30. Unger, T., Rascher, W., Schuster, C., Pavlovitch, R., Schomig, A., Dietz, R. and Ganten, D. (1981). Eur. J. Pharmacol. *71*:33-42
31. Allen, A.M., Dampney, R.A.L. and Mendelsohn, F.A.O. (1988). Am. J. Physiol. *255*:H1011-H1017
32. Li, Y.-W. and Guyenet, P.G. (1995) Am. J. Physiol. *268*:R272-R277
33. Fior, D.R., Yang, S.N., Hedlund, P.B., Narvaez, J.A., Agnati, L.F. and Fuxe, K. (1994). Eur. J. Pharmacol. *262*:271-282
34. Allen, A.M., Mendelsohn, F.A.O., Gieroba, Z.J. and Blessing, W.W. (1990). J. Neuroendocrinol. *2*:867-874.
35. Fitzsimons, J.T. and Setler, P.E. (1975). J. Physiol. *250*:613-631
36. Bellin, S.I., Bhatnagar, R.K. and Johnson. A.K. (1987). Brain Res. 403:105-112
37. Cunningham, J.T. and Johnson, A.K. (1989). Brain Res. *480*:65-71
38. Cunningham, J.T. and Johnson, A.K. (1991). Brain Res. *558*:112-116
39. McRae-Degeuree, A., Bellin, S.I., Landas, S.K. and Johnson, A.K. (1986). Brain Res. *374*:162-166
40. Ferrari, A.C., de Arruda Cambargo, L.A., Saad, W.A., Renzi, A., De Luca, L.A.J. and Menani. J.V. (1991). Brain Res.; 560:291-296.
41. Severs, W.B., Summy-Long, J., Daniels-Severs, A. and Connor, J.D. (1971). Pharmacology *5*:205-214
42. Fregley, M.J., Rowland, N.E.a nd Greenleaf, J.E. (1984). Brain Res. *298*:321-327
43. Bastos, R., Saad, W.A., Menani, J.V., Renzi, A., Silveira, J.E.N. and Camargo, L.A. (1994). Brain Res. *636*:81-86
44. Fregley, M.J., Rowland, N.E. and Greenleaf, J.E. (1984). Brain Res. Bull. *12*:393-398
45. Fregley, M.J. and Rowland, N.E. (1992). Proc. Soc. Exp. Biol. Med. *199*:158-164
46. Renaud, L.P. (1988) in Neurotransmitters and cortical function (eds. Avoli, M. and Reader, T.A.), Plenum Publishing Corporation, New York, pp 495-515
47. Veltmar, A., Qadri, F., Culman, J., Rascher, W. and Unger, T. (1991). J. Hypertens. *9*:S56-S57
48. Baranowska, D., Braszko, J.J. and Wisniewski, K. (1983). Psychopharmacology *81*:247-251
49. Braszko, J.J. and Wisniewski, K. (1976). Pol. J. Pharmacol. Pharm. *28*:667-672
50. Braszko, J.J., Wisniewski, K., Kupryszewski, G. and Witezuk, B. (1987). Behav. Brain Res. *25*:195-203
51. Braszko, J.J. and Wisniewski, K. (1988). Peptides *9*:475-479
52. Braszko, J.J. and Wisniewski, K. (1990). Psychoneuroendocrinology *15*(4):239-252
53. Wisniewski, K. and Braszko, J.J. (1984). Clin. and Exp. Hyper. -Theory and Practice *A6*:2127-2131
54. Cierco, M. and Israel, A. (1994). Eur. J. Pharmacol. *251*:103-106
55. Sumners, C., Phillips, M.I. and Raizada, M.K. (1983). Neurosci. Lett. *36*:305-309
56. Myers, L.M. and Sumners, C. (1989). Am. J. Physiol. *257*:C706-C713
57. Diz, D.I. and Pirro, N.T. (1992). Hypertension *19*(suppl.II):II-41-II-48
58. Meldrum, M.J., Xue, C., Badino, L. and Westfall, T.C. (1984). J. Cardiovas. Pharmacol. *6*:989-995.
59. Garcia-Sevilla, J.A., Dubocovich, M.L. and Langer, S.Z. (1979). Eur. J. Pharmacol. *56*:173-176.
60. Haulica, I., Coculescu, M., Ghinea, E., Stratone, A., Petrescu, G., Rosca, V. and Oprescu, M. (1980). Life Sci. *27*:809-813
61. Hirose, S., Yokosawe, H., Inagami, T. and Workman, J. (1980). Brain Res. *191*:489-499
62. Finocchiaro, L.M.E., Goldstein, D.J., Finkielman, S. and Nahmod, V.E. (1990). Endocrinology *126*:59-66.
63. Sumners, C. and Raizada, M.K. (1986). Am. J. Physiol. *250*:C236-244
64. Raizada, M.K., Muther, T.F. and Sumners, C. (1984). Am. J. Physiol. *247*:C364-372
65. Bottiglieri, D., Sumners, C. and Raizada, M.K. (1987). Brain Res. *403*:167-171
66. Sumners, C., Watkins, L.L. and Raizada, M.K. (1986). J. Neurochem. *47*:1117-1126
67. Sumners, C. (1992). Brain Res. Bull. *28*:411-415
68. Sumners, C. and Phillips, M.I. (1983). Am. J. Physiol. *244*:R257-R263
69. Tomaszewwicz, M., Micossi, L.G., Bielarczyk H., Luszawska, D., Santarelli, I. and Szutowicz, A. (1991). J. Neurochem. *56*:729-732
70. Stadler, T., Veltmar, A., Qadri, F. and Unger, T. (1992). Brain Res. *569*:117-122
71. Qadri, F., Culman, J., Veltmar, A., Maas, K., Rascher, W. and Unger, T. (1993). J. Pharmacol. Exp. Ther. *267*:567-574
72. Qadri, F., Badoer, E., Stadler, T. and Unger, T. (1991). Brain Res. *563*(1-2):137-141
73. Andersson, K., Fuxe, K., Agnati, L.F., Ganten, D., Zini, I., Eneroth, P., Mascagni, F. and Infantillina, F. (1982). Acta. Physiol. Scand. *116*:317-320

74. Simonnet,G. and Vincent, J.D. (1982). Neurochem. Int. *4*(2/3):149-155.
75. Kawabe, H., Brosnihan. B,, Diz, D.I. and Ferrario, C.M. (1986). Inter-Am. Soc. Proc. *8*(4):I84-I89
76. Sumners, C., Woodruff, G.N., Poat, J.A. and Munday, K.A. (1979) Psychopharmacology *60*:291-294.
77. Sumners, C., Woodruff. G.N. and Poat, J.A. (1981). Psychopharmacology *73*:180-183
78. Burt, D.R., Creese, L. and Snyder, S. (1977). Science *196*:326-328
79. Steele, M.K., McCann, S.M. and Negro-Vilar, A. (1982). Endocrinology *111*:722-729
80. Enjalbert, A., Sladeczek, F., Guillon, G., Bertrand, P., Shu, C., Epalbaum, J, Garcia-Sainz, A., Jard, S., Lombard, C., Kordon, C. and Boekaert, J. (1986). J. Biol. Chem. *261*:4071-4075
81. Journot, L., Homburger V., Pantaloni, C., Priam, M., Bockaert, J. and Enjalbert, A. (1987). J. Biol. Chem. *262*:15106-15110
82. Georgiev, V., Gyorgy, L., Getova, D. and Markovska, V. (1985). Bulg. Acad. Sci. *11*(4):19-26
83. Banks, R.J.A. and Dourish, C.T. (1991). Br. J. Pharmacol. *104*:63P
84. Georgiev, V., Stancheva, S., Kambourova, T. and Getova, D. (1990). Bulg. Acad. Sci. *16*(1):32-37
85. Hyttel, T. and Christensen, A.V. (1983). J. Neural. Trans. *18*(suppl.):157-164
86. Georgiev, V. and Kambourova, T. (1984). CR Acad. Bulg. Sci. *37*:391-393
87. Simonnet, G. and Giorguieff-Chesselet, M.F. (1979). Neurosci. Lett. *15*:153-158
88. Dwoskin, L.P., Jewell, A.L. and Cassis, L.A. (1992). Naunyn Schmiedeberg's Arch. Pharmacol. *345*:153-159
89. Simonnet, G., Giorguieff-Chesselet, M.F., Carayon, A., Bioulac B., Cesselin, F., Glowinski, W.and Vincent, J.D. (1981). J. Physiol. Paris *77*:71-79
90. Sumners, C., Myers, L.M., Kalberg, C.J. and Raizada, M.K. (1990) Prog. Neurobiol. *34*:355-385
91. Braszko, J.J., Kupryszewski, G., Witczuk, B. and Wisniewski, K. (1992). Pharmacol. Res. *25(*suppl2):9-10
92. Mendelsohn, F.A.O., Jenkins, T.A. and Berkovic, S.F. (1993) Brain Res. *613*:221-229
93. Badoer, E., Wurth, H., Turck, D., Qadri, F., Itoi, K., Dominiak, P. and Unger, T. (1990). Naunyn Schmiedeberg's Arch. Pharmacol. *340*:31-35
94. Nakao, K., Katsuura, G., Morii, N., Itoh, H., Shiono, S., Yamada, T., Sugawara, A., Sakamoto, M., Saito, Y., Eigyo, M., Matsushita, A. and Imura, H. (1986). Eur. J. Pharm. *131*:171-177

11

REGULATION OF THE HYPOTHALMIC-PITUITARY-ADRENAL AXIS AND VASOPRESSIN SECRETION

Role of Angiotensin II

Greti Aguilera and Alexander Kiss

Section on Endocrine Physiology, Developmental Endocrinology Branch
National Institute of Child Health and Human Development
NIH, Bethesda, Maryland 20892

INTRODUCTION

Angiotensin II (ANG II) has a number of actions in the central nervous system, including regulation of water intake, blood pressure and the secretion of pituitary hormones such as ACTH, prolactin and gonadotropins (1,2). With respect to ACTH, several in vitro studies in rodents and primates have shown that ANG II directly stimulates ACTH secretion and potentiates the stimulatory effect of corticotropin releasing hormone (CRH) through ANG II receptors in the pituitary corticotroph (3). However, the physiological importance of the direct pituitary action seen in vitro is not clear, and in most experimental conditions the stimulatory effect of ANG II on ACTH secretion in vivo appears to be mediated by central mechanisms (4). The stimulatory effect of peripherally administered ANG II on ACTH secretion is mediated, at least in part, by CRH, as demonstrated by the ability of CRH antiserum to block the effect (5), and by the transient increases in irCRH in the median eminence following ANG II administration (4). Circulating ANG II may centrally activate CRH release by acting directly on the median eminence or through the circumvetricular organs, areas containing abundant ANG II receptors (2,6).

Although the above evidence indicates that ANG II is involved in regulating the HPA axis, the physiological role of the peptide in the responses to acute or chronic stress is not understood. It is possible that ANG II is involved in stress-specific responses of the HPA axis under different stressors. Activation of the magnocellular system following chronic osmotic stimulation inhibits the HPA axis, whereas repeated physical-psychological stress results in hyperresponsiveness to a novel stress, with or without desensitization of the response to the primary stimulus, depending on the stress paradigm (7). Thus, the question remains as to whether differential activation of ANG II receptors in the hypothalamic paraventricular nucleus (PVN) by various stressors influences the response of the CRH neuron.

Recent Advances in Cellular and Molecular Aspects of Angiotensin Receptors
Edited by Mohan K. Raizada et al., Plenum Press, New York, 1996

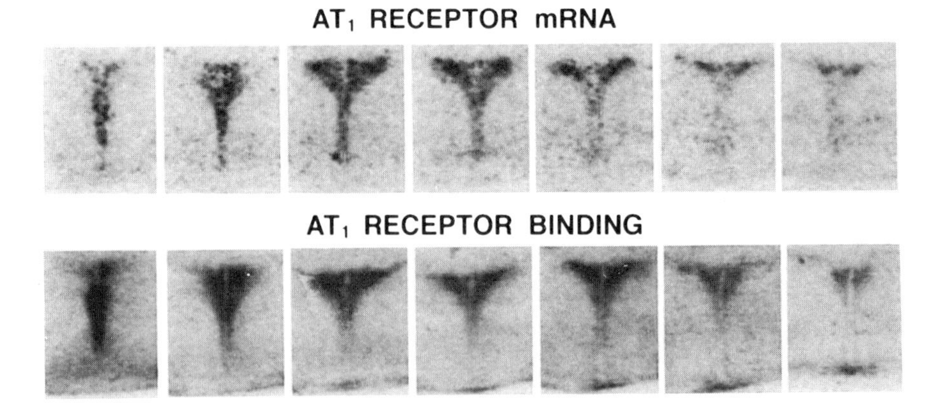

Figure 1. Distribution of AT_1 receptor mRNA and AT_1 receptor binding in the rat PVN. AT1 receptor mRNA was measured by in situ hybridization and AT1 receptors by autoradiographic analysis of the binding of $^{125}I[Sar^1,Ile^8]AII$ in serial coronal sections of the PVN. Film exposure time was 21 days for the in situ hybridization and 4 days for the binding study. Figure is representative of the results in 5 rats.

ANG II is also involved in the central regulation of vasopressin (VP) secretion. I.c.v. injection of ANG II causes release of VP from the posterior pituitary lobe (1,2). At least part of the effects of ANG II on VP secretion are indirect, mediated through catecholamine release by activation of ANG II receptors located in afferent terminals innervating magnocellular PVN neurons (8,9). On the other hand, the presence of ANG II binding sites and AT_1 receptor mRNA in the magnocellular PVN (10), suggests that ANG II directly modulates the function of VP neurons.

Ang II Receptor Binding and AT_1 Receptor mRNA in the PVN

Autoradiographic analysis of the binding of $^{125}I[Sar^1,Ile^8]ANG$ II in serial coronal sections of the PVN reveals high density staining throughout the anteroposterior axis of the PVN, with localization in the periventricular and parvicellular pars (Figure 1). Periventricular binding was distinct in the most anterior sections, where light microscopic analysis of the stained sections showed few lateral parvicellular or magnocellular cells. Binding was also present in the magnocellular areas, but it was less well defined and of lower density. In addition, high binding is clearly present in nerve fibers associated with the dorsolateral area of the PVN, and in the median eminence (2,6,10). Binding of the radiolabeled ligand is completely inhibited by the AT_1 receptor antagonist, Losartan, indicating that ANG II receptors in the PVN are type 1 (AT_1).

The topographic distribution of AT_1 receptor mRNA is similar to that of the AT_1 receptor binding, with the highest levels in the periventricular and parvicellular portions of the PVN. Very low hybridization was found in the dorsolateral magnocellular region. No AT_1 receptor mRNA is observed in the median eminence (Figure 1) indicating that the receptors are likely associated with nerve terminals rather than glial cells. The demonstration of AT_1 receptor mRNA in CRH perikarya (see below) suggests that the nerve terminals containing ANG II receptors originate in CRH cells, and that these AT_1 receptors mediate the stimulatory effect of peripherally administered ANG II on the release of CRH into the pituitary portal circulation (5,11).

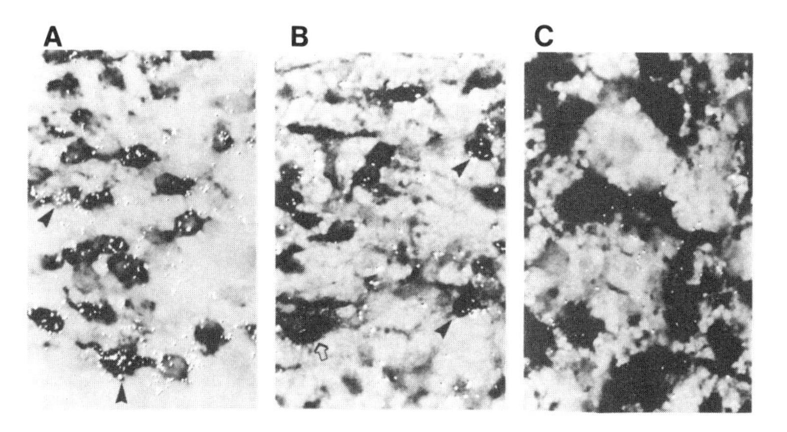

Figure 2. Double staining in situ hybridization for AT_1 receptor mRNA, and CRH, VP and oxytocin mRNAs in the PVN. Transcripts for the AT_1 receptor were present in CRH (darkly stained cells in (A, dark arrows) and parvicellular VP stained cells (B, dark arrows). No AT_1 receptor mRNA transcripts are observed in magnocellular VP stained cells (B, clear arrows) or oxytocyn stained cells (C).

Cellular Localization of AT_1 Receptor mRNA in the PVN

The question of the cell type expressing AT_1 receptors in the PVN has been addressed by simultaneous double staining in situ hybridization studies in hypothalamic sections using ^{35}S-labelled AT_1 receptor cRNA probe and digoxigenin labeled cRNA probes for CRH, VP, oxytocin or TRH (10). As shown in Figure 2, abundant AT_1 receptor mRNA hybridization overlying CRH stained cells and parvicellular VP stained cells. In contrast, the number of grains associated with magnocellular VP and oxytocin labeled cells and TRH stained cells are no different from the background.

The presence of high levels of AT_1 binding as well as AT_1 receptor mRNA in the periventricular area, indicates the presence of receptors in parvicellular perikarya other than CRH. This region of the PVN contains mainly TRH, somatostatin, and dopamine cells (12). While the lack of colocalization of AT_1 receptor mRNA and TRH mRNA in the periventricular area renders it unlikely that AT_1 receptors are in TRH cells, it is possible that they are associated with dopaminergic neurons, as suggested by the ability of central administration of ANG II to increase dopaminergic turnover in the hypothalamus (13) and decrease prolactin secretion (14). Central administration of ANG II has also been shown to reduce plasma levels of growth hormone (14), and it is likely that AT_1 receptors located in somatostatin cells mediate this effect.

The absence of detectable ANG II receptors in VP and oxytocin cells supports the view that ANG II regulates magnocellular function indirectly through stimulation of catecholamine release. With respect to VP, it has been shown that α-adrenergic antagonists prevent the stimulatory effect of central ANG II administration on VP secretion (8). Microdialysis studies have shown that i.c.v. administration of ANG II causes norepinephrine release from the PVN (9). The autoradiographic pattern of AT_1 binding adjacent to the dorsolateral portions of the PVN and the absence of AT_1 receptor mRNA in these areas, strongly suggest that the receptors are located in nerve fibers converging on the PVN. The PVN receives inputs from a number of well-defined pathways originating in noradrenergic areas of the brain stem, the bed nucleus of the stria terminalis and circumventricular organs, including the subfornical organ and organum vasculosum of the lamina terminalis (15). The presence of AT_1 receptor mRNA in cell bodies of nuclei originating in such pathways (1,6) suggests that these receptors are synthesized and transported to afferent terminals in the PVN where they influence the release of catecholamines or other neurotransmitters involved PVN function.

While most evidence indicates that the primary action of ANGII on VP secretion is indirect, the lack of AT_1 receptor mRNA in VP and oxytocin neurons in the colocalization studies is intriguing. First, in situ hybridization studies clearly show low levels of AT_1 receptor mRNA in the dorsal magnocellular PVN (10), and second, electrical recording of single magnocellular neurons showns increases in electrical activity when ANG II is applied to the perifusion bath (16). Therefore, it is not possible to rule out that magnocellular neurons express low levels of AT_1 receptor mRNA, which are undetectable with the techniques employed (10), and further studies in animals under osmotic stimulation will be needed to elucidate this problem. It is also conceivable that AT_1 receptors in the magnocellular system are located in glial cells. This is supported by the demonstration of high levels of AT_1 receptors in glial cell cultures (17). During stimulation of the hypothalamic neurohypophyseal system, astrocytes become juxtaposed, decreasing the surface coverage of magnocellular perikarya and dendrites (18), and ANG II receptors may be involved in this process.

Effect of i.c.v. Injection of ANG II on CRH and VP mRNA Levels

Central ANG II administration increases CRH mRNA levels in the PVN in studies using in situ hybridization (10) and Northern blots (19). In contrast, i.c.v. ANG II injection had no effect on VP mRNA levels in the PVN or SON, whereas the expected increases were observed after i.p. hypertonic saline injection (10).

In experiments using i.c.v. injection ANG II may influence the PVN through activation of distal receptors. The regulation of hypothalamic CRH expression and secretion is multifactorial and probably involves a number of neurotransmitters and neuropeptides (7,20). It is noteworthy that catecholamines, a major regulator of hypothalamic CRH secretion (20), appear not to be involved in the early increases in CRH mRNA following stress, suggesting the participation of other factors (21). In view of the ability of a single dose of ANG II to increase CRH mRNA in the PVN, activation of AT_1 receptors in CRH neurons should be considered as a mechanism for the increases in CRH mRNA levels observed 2 to 4 hr after acute stress (7).

Effect of Stress on AT_1 Receptor mRNA and ANG II Binding in the PVN

The specific localization of AT_1 receptors in CRH neurons of the parvicellular PVN strongly suggests a role for ANG II in the regulation of the HPA axis during stress (10). In fact, it has been shown that binding of ANG II to the PVN is glucocorticoid dependent and increases following repeated immobilization stress in the rat (2). An important question is whether ANG II is involved in the differential responsiveness of the HPA axis to various stress paradigms, and whether a correlation exists between AT_1 receptor expression in the PVN and the sensitivity of the HPA axis during stress (7).

In situ hybridazation studies show that acute physical stress (immobilization for one hr or i.p. hypertonic saline injection) causes rapid increases in AT_1 receptor mRNA in the dorsomedial and ventromedial areas of the PVN. AT_1 receptor mRNA levels in the PVN remain significantly elevated following repeated stress (3 times in 24 hr at 8 h intervals, or once a day for 14 days), though after repeated stress values are lower than those observed after a single immobilization. Both stress paradigms caused a similar pattern of increase of AT_1 receptor mRNA in the ventromedial and dorsomedial portions of the PVN, despite the strong osmotic component of i.p. hypertonic saline injections. The most unexpected finding is that water deprivation for 60 hr or 2% saline intake for 12 days, both predominant osmotic

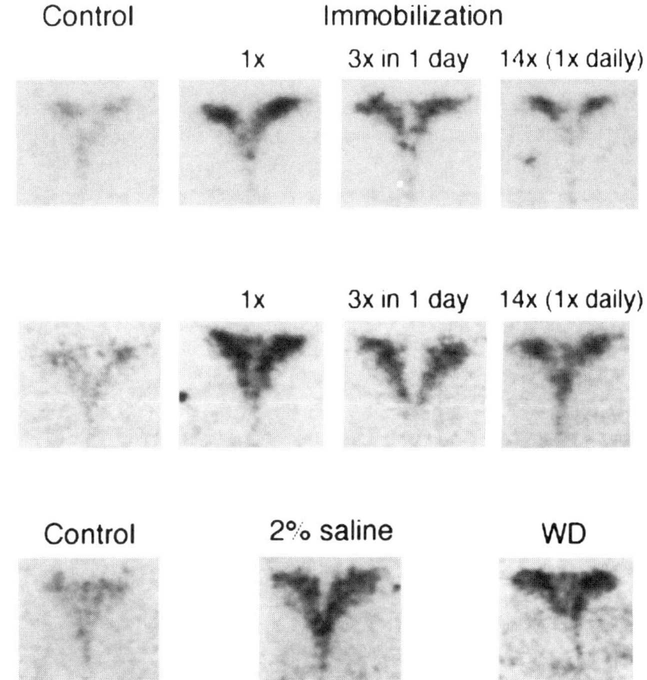

Figure 3. Effect of stress on AT_1 receptor mRNA in the PVN. Rats were subjected to a) immobilization (physical psychological stress) for one hr or b) i.p. injection of 5 ml of 1.5 M NaCl (painful stress with an osmotic component) and killed 3 hr later (1x); or repeated stress, 3 times in 24 hr at 8 hr intervals (3 x in day), or daily for 14 days (14 x (1 x daily); c) 2% NaCl intake for 12 days (2% saline), or water deprivation for 60 hr (WD)(osmotic stress models). Figure represent in situ hybridization of AT_1 receptor mRNA in representative PVN sections from each stress group.

stimuli, also result in marked increases in AT_1 receptor expression in the ventromedial and dorsomedial PVN (parvicellular), but not the magnocellular PVN or in the SON (Fig 3). The lack of significant changes in AT_1 receptor mRNA or ANG II binding in the magnocellular system of rats under osmotic stimulation is consistent with the failure to detect any AT_1 receptor mRNA colocalized in magnocellular vasopressin or oxytocin neurones (10). This supports the view that the recognized stimulatory effects of central ANG II on VP release are mediated by ANG II receptors located in afferent inputs to the PVN rather than by direct effects on magnocellular neurons (8,9).

These studies show that stress does indeed cause rapid and sustained increases in AT_1 receptor mRNA and ANG II receptors in the PVN. However, this response is not specific for the type of stimulation, with all stress paradigms resulting in comparable changes in AT_1 receptors, irrespective of their effect on the responsiveness of the HPA axis. Thus, a similar pattern of increase in AT_1 receptor expression in the parvicellular PVN was observed in both, osmotic stress (water deprivation or 2% saline intake) which is associated with inhibition od ACTH responses, and physical-psychological stress (immobilization and i.p. hypertonic saline injection) which is associated with hyperrresponsiveness of the HPA axis (7). Moreover, AT_1 receptor mRNA changes were similar in the two physical-psychological stress models, despite their different patterns of adaptation of the ACTH response to repeated stimulation.

There are conflicting reports regarding the effect of central ANG II blockade on ACTH responses to stress. In sheep, i.c.v. administration of a converting enzyme inhibitor attenuates ACTH responses to hemorrhage (22), whereas ACTH responses to ether stress in the rat (23) are not modified by central ANG II antagonists or converting enzyme inhibitors. While there is little evidence indicating that ANG II is responsible for the differential responses of the HPA axis to the various stress paradigms, it is possible that AT_1 receptors in the PVN are involved in the sympathetic responses to stress. Although neither i.c.v. nor peripheral administration of ANG II elevates basal plasma catecholamine levels (24,25), ANG II may have a modulatory or permissive role during stress-induced sympathetic activation. Physical-psychological and metabolic stress paradigms are clearly associated with elevated plasma catecholamine levels (24). Similarly, sympathetic activation is observed during osmotic stimulation, conditions in which basal plasma epinephrine and norepinephrine levels are normal, but there is a marked potentiation in the response to acute stress (26). CRH neurons of the PVN are involved in regulating sympathetic outflow (12), and appear to be critical for the plasma epinephrine responses to stress (27). Therefore, AT_1 receptors in CRH neurons could be implicated in the central control of sympathetic activity. In support of this possibility, i.c.v. injection of the AT_1 receptor antagonist, Losartan, attenuates the increase in splanchnic nerve activity during hyperthermia in the rat (28), and recent studies of our laboratory have shown that central AT_1 receptor blockade reduces plasma catecholamine responses to acute immobilization and i.p. hypertonic saline (D. Jezova, A Kiss and G. Aguilera, unpublished).

Effect of Adrenalectomy and Glucocorticoid Administration on AT_1 Receptor mRNA Levels and ANG II Binding in the PVN

AT_1 receptor mRNA levels and ANG II binding in the PVN decrease by about 30% compared to sham-operated rats by 18 hr, and remain at similar levels after 60 hr, 4 and 6 days adrenalectomy. This effect is prevented with glucocorticoid replacement with corticosterone in the drinking water or injection of dexamethasone. These effects indicate that

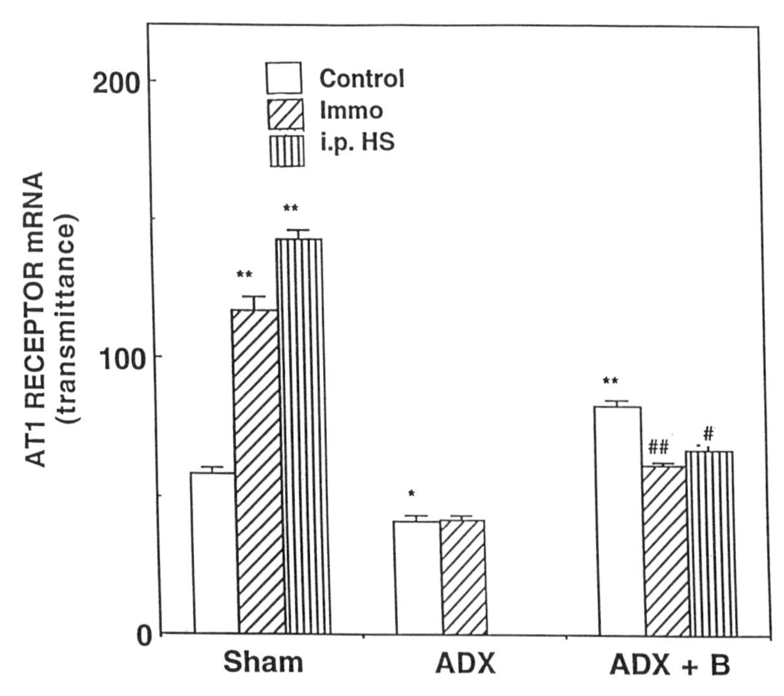

Figure 4. Effect of a single immobilization (Immo) or i.p. hypertonic saline injection (i.p. HS) on AT_1 receptor mRNA levels in the PVN of sham operated rats or adrenalectomized rats receiving 0.9% saline (ADX) or 30 μg/ml corticosterone (ADX + B) in the drinking water. Bars are the mean and SEM of the values obtained by in situ hybridization in 5 rats per group. **, $p < 0.01$ vs sham control; *, $p < 0.05$ vs sham; ##, $p < 0.01$ vs control ADX + B; #., $p < 0.05$ vs ADX + B.

adrenal steroids are necessary to maintain basal AT_1 receptor mRNA levels in the PVN. Glucocorticoids may have a direct stimulatory effect on AT_1 receptor mRNA transcription or stability, acting through glucocorticoid receptors located in the PVN (29). In addition, glucocorticoids could influence AT_1 receptor mRNA levels indirectly through regulating other factors, including ANG II itself, locally in PVN neurons and/or afferent innervation to the PVN (12). It has been shown that glucocorticoids regulate the expression of components of the renin-angiotensin system in the brain and periphery (2). Dexamethasone administration increases angiotensinogen mRNA in several areas of the brain, including the PVN, whereas removal of glucocorticoids by adrenalectomy increases irANG II in the PVN.

The effect of glucocorticoids on stress-induced AT_1 receptor mRNA has been studied in rats subjected to acute stress after pretreatment with dexamethasone, 100 μg 14 and 1 hr prior to stress. While dexamethasone treatment of control rats results in a small but significant increase in AT_1 receptor mRNA, it markedly potentiates the stimulatory effect of immobilization and i.p. hypertonic saline injection, compared with vehicle-treated stressed rats (Fig. 4).

The role of the stress-induced glucocorticoid surge on the AT_1 receptor mRNA response to stress has been addressed by experiments in adrenalectomized rats with and without glucocorticoid replacement (Fig. 4). In contrast to the marked stimulatory effect of stress on AT_1 receptor mRNA levels in the PVN in intact rats, stress is without effect in adrenalectomized rats with no glucocorticoid replacement. Corticosterone replacement in the drinking water prevents the inhibitory effect of adrenalectomy, and surprisingly, in these rats stress not only fails to increase, but significantly decreases AT_1 receptor mRNA expression in the PVN compared with the levels in unstressed adrenalectomized glucocorticoid-treated rats (Fig. 4).

It is clear from the evidence above that glucocorticoids are involved in the mechanism by which stress induces AT_1 receptors in the PVN. Thus, the inability of stress to increase AT_1 receptor mRNA in the PVN of adrenalectomized rats indicates that glucocorticoids are necessary for this effect. However, the failure of glucocorticoid replacement to restore the stimulatory effect of stress on AT_1 receptors in the PVN of adrenalectomized rats, indicates that basal glucocorticoid levels are not sufficient to increase AT_1 receptor mRNA in the PVN, and that higher stress levels are required. Notwithstanding the requirement of the surge, glucocorticoids cannot fully account for the increases in AT_1 receptor mRNA in the PVN following stress. Firstly, dexamethasone injection in intact rats fails to reproduce the magnitude of the increase in AT_1 receptor mRNA observed after acute stress, and secondly, dexamethasone potentiated the stimulatory effect of stress, indicating that increased circulating glucocorticoids must interact with other stress-dependent factors. In this regard it should be noted that in adrenalectomized rats with glucocorticoid replacement, stress not only had no effect, but inhibited AT_1 receptor mRNA levels in the PVN. This strongly suggests that stress has both inhibitory and stimulatory influences on AT_1 receptor expression in the PVN, and that a major effect of the glucocorticoid surge during acute stress is to suppress an inhibitory pathway activated by stress.

CONCLUSIONS

The colocalization of AT_1 receptor mRNA in CRH cells, in conjunction with the increases in CRH mRNA following central ANG II injection suggest that ANG II modulates CRH neurons directly through receptors in PVN perikarya. In contrast, the lack of AT_1 receptor mRNA in VP and oxytocin cells, and the autoradiographic pattern of ANG II binding, suggest that regulation of magnocellular function is largely indirect via receptors associated with afferent innervation to the PVN.

AT_1 receptor expression in the parvicellular but not in the magnocellular division of the PVN, is positively regulated by glucocorticoids. Regulation of AT_1 receptor expression in the PVN during stress appears to be under dual regulatory influences: stress-activated inhibitory pathways and the stimulatory effect of glucocorticoids. The lack of specificity of the changes in AT_1 receptor expression in the PVN following stress paradigms with opposite effects on ACTH secretion (osmotic and physical-psychological stress) does not support a role for ANG II as a major determinant of the response of HPA axis during stress. In addition, the strong correlation between AT_1 receptor mRNA and ANG II binding suggest that control of mRNA levels plays an important role in the regulation of AT_1 receptor levels in the PVN.

REFERENCES

1. Phillips, M.I. (1980) In: Enzymatic release of vasoactive peptides. Gross,F and Vogel, G., eds. Raven Press, New York, pp 335-370.
2. Saavedra, J.M. (1992) Endocrine Rev 13: 329-380.
3. Aguilera, G., Harwood, J.P, Wilson, J.X, Morell, J., Brown, J.H. and Catt KJ. (1983) J. Biol. Chem. 258: 8039-8045
4. Ganong, W.F. and Murakami, K. (1987) Ann NY Acad Sci 512: 176-186.
5. Rivier, C. and Vale, W. (1983) Reg Peptides 7:253-258.
6. Mendelsohn, F.A.O., Quirion, R, Saavedra, J.M., Aguilera, G. and Catt, K.J.(1984) Proc. Natl. Acad. Scie. (USA) 81: 1575-1579.
7. Aguilera, G. (1994) Front Neuroendocrinol 15: 321-350.
8. Veltmar, A., Culman, J., Quadri, F., Rasher, W. and Unger, T. (1992) J Pharmacol Exp Ther 263: 1253-1260.
9. Stadler, T., Veltmar, A., Quadri, F. and Unger, T. (1992) Brain Res 569:117-122.
10. Aguilera, G., Young, W.S., Kiss, A. and Bathia, A. 1995 Neuroendocrinology 61: 437-444.
11. Plotsky, P.M., Sutton, S.W., Bruhn, T.O. and Ferguson, A.V. (1988) Endocrinology 122:538-545.
12. Swanson, L.W., Sawshenko, P.E., Lind, R.W. and Rho, J-H. (1987) Ann NY Acad Sci 512: 12-23.
13. Fuxe, K., Anderson, K., Ganten, D., Hokfelt, T. and Encroth, P. (1980) In: Enzymatic release of vasoactive peptides, F Gross, G Vogel, eds, Raven press, NY, pp 161-170
14. Steele, M.K., McCann, S.D. and Negro-Vilar, A. (1982) Endocrinology 111: 722-729.
15. Sawshenko, P.E. and Swanson, L.W. J. Comp. Neurol. (1983) 218: 121-144.
16. Yang, C.R., Phillips, M.I., Renaud, L.P. (1992) Am J Physiol 32: R1333-R1338
17. Raizada, M.K., Phillips, M.I., Crews, F.T. and Sumners, C. (1987) Proc Natl Acad Sci USA 84: 4655-4658
18. Theodosis, D.Y. and Poulain DA. (1989) Brain Res 484: 361-366).
19. Sumitomo, T., Suda, T., Nakano, Y., Tozawa, F., Yamada, M. and Demura, H. (1991) Endocrinology 128: 2248-2252.
20. Al-Damluji, S. (1993) Clin Endocrinol Metab 7: 355-392
21. Kiss, A. and Aguilera, G. The Endocrine Soc. 75th Annual Meeting, Las Vegas (1993) abs 1085
22. Cameron, V.A., Espiner, E.A., Nicholls, M.G., MacFarlane, M.R. and Sadler, W.A. (1986) Life Sci 38: 553-559.
23. Buckner, F.S., Chen, F.N., Wade, CE. and Ganong, W.F. (1986) Neuroendocrinology 42: 97-101.
24. Kvetnansky, R., Jezova, D., Oprsalova, Z., Foldes, O., Michajlovskij, N., Dobrakovova, M., Lichardus, B. and Makara GB (1990) In: Circulating regulatory factors and neuroendocrine function. Porter JC, Jezova D, eds., Plenum Press, New York, pp 113-134.
25. Brown, M.R., Fisher, L.A. (1985) Fed Proc 44: 243-248.
26. Kiss, A., Jezova, D. and Aguilera G (1994) Brain Res 663: 84-92
27. Kvetnansky, R., Dobrakovova, M., Jezova, D., Oprrsalova, Z., Lichardus, B. and Makara G (1989) GR VanLoon, R Kvetnansky, R McCarty, J Axelrod, eds. Gordon and Breach Sci Publish., New York. pp 549-570.
28. Kregel, K.C., Stauss, H. and Unger, T. (1994) Am J Physiol 35: R1985-R1991.
29. Aronsson, M., Fuxe, K., Dong, Y., Okret, S. and Gustafsson, J-A (1988) Proc Natl Acad Sci USA 85: 9331-9335.

12

RELATIONSHIP BETWEEN THE DRINKING RESPONSE TO ANGIOTENSIN II AND INDUCTION OF FOS IN THE BRAIN

Neil E. Rowland, Melvin J. Fregly, Anny K. Rozelle, and Annie Morien

Departments of Psychology and Physiology
University of Florida
Gainesville, Florida 32611-2250

INTRODUCTION

One of the prominent actions of angiotensin (Ang) II in the central nervous system is the stimulation of water, and in some cases NaCl, intake (1,2). Either peripheral or central administration of Ang II induces water intake in water-replete animals. The dipsogenic effect of peripheral Ang II is abolished by either lesions of, or Ang II receptor antagonism in, the subfornical organ (SFO). The SFO is a circumventricular organ, and the fenestrations between capillary endothelia allow leakage of circulating Ang II into the parenchyma in the core and posterior parts of the SFO. The SFO has a high density of Ang II receptors most or all of which are subtype 1 (AT-1R) in rats (3,4). The SFO has efferent connections to other forebrain areas known to be involved in either fluid intake or neuroendocrine action, including organum vasculosum of the lamina terminalis (OVLT), median preoptic nucleus (MnPO), supraoptic nucleus (SO) and hypothalamic paraventricular nucleus (PVH). Some of these pathways are reciprocal, and some may use Ang II as a transmitter (5,6).

What has not been determined is the normal flow of information among these structures, including the impact of lesions and/or pharmacological antagonists on this flow. Recently, mapping of induced immediate early genes has proven useful for localizing functional activation of discrete brain regions in response to defined stimuli, including those relating to fluid balance (7,8). Fos is the protein product of c-fos, a gene that is expressd in many cell types in response to a variety of second messengers. Fos in turn is a nuclear transcription factor for other genes of unknown significance in brain. Thus, the detection of Fos-like immunoreactivity (Fos-IR) in the brain is a putative marker of neuronal activity (7,8). We have used this metric to determine some of the sites of action of Ang II, and whether either the number of cells or the apparent intensity of Fos-IR in one or more structures is correlated with water intake.

Recent Advances in Cellular and Molecular Aspects of Angiotensin Receptors
Edited by Mohan K. Raizada et al., Plenum Press, New York, 1996

FOS-IR AFTER EITHER PERIPHERAL OR CENTRAL ADMINISTRATION OF ANG II

Infusion of Ang II into rats through an indwelling jugular vein catheter (100 ng/rat/min for 1 hr) produced condensed Fos-IR in the perinuclear region of cells in the core and posterior parts of the SFO, in a midline strip of cells in the MnPO stretching to the medial part of the OVLT, and in the SO and PVH (9,10). The Fos-IR cells are most likely to be neurons, but we cannot rule out the possibility that some are glia. Control rats infused with isotonic NaCl vehicle had essentially no Fos-IR cells in these regions. A few regions, such as pyriform cortex and dorsomedial thalamus, showed moderate Fos-IR (*ie,* constitutive) in both control and Ang II-infused rats. Regions such as the central nucleus of the amygdala (CNA) showed moderate Fos-IR following infusion of either Ang II or an equi-pressor dose of the alpha-adrenergic agonist, phenylephrine, so this Fos-IR is presumably a nonspecific action of Ang II. Fos-IR was also induced by Ang II in cells scattered along the medial forebrain bundle, in a diffuse pattern that resembles the distribution of Ang II-IR nerve terminals (11).

Subcutaneous (SC) injection of Ang II (200-400 ug/kg) also produced Fos-IR in these same regions, but with lower intensity and less consistency than with intravenous infusion (unpublished data). The dipsogenic heptapeptide, Ang III (Ang 2-8), also was associated with Fos-IR in these regions, although 2-4-fold higher molar doses were neeeded to achieve comparable intensities to Ang II (10).

Cerebroventricular pulse (pCVT) injection of Ang II (100 ng) also produced strong Fos-IR in SFO, MnPO, OVLT, SO and PVH. Within the SFO the Fos-IR after pCVT injection was now restricted to the anterior and ventricular layers of the SFO (12). There was almost no overlap with the core and posterior regions stimulated in the intravenous studies (9). Interestingly, pCVT injection of the dipsogenic cholinomimetic, carbachol, induced prominent Fos-IR with almost the same distribution as pCVT Ang II. The pharmacological independence of Ang II and cholinergic drinking responses (13) suggests either that the same cells (in SFO) have both Ang II and cholinergic receptors, or that separate but overlapping neuronal populations with similar efferents are involved.

ROLE OF SFO

It is well-known that complete lesions of the SFO permanently abolish water intake induced by peripheral injection of Ang II. This has led to the idea that the SFO is the primary transduction site in brain for peripherally-administered Ang II. However, the studies of drinking to pCVT Ang II in rats after SFO lesions do not agree. There are several differences, but one relates to the time between lesion and behavioral testing: rats tested one week after either electrolytic lesion or transection of the efferent fibers showed greatly impaired drinking (14,15), while rats tested 3 mo after lesion responded almost normally (16) to pCVT Ang II. We used the Fos-IR mapping method to address these questions further.

The first study was designed to examine the primacy of the SFO in Fos-IR using intravenous infusion of Ang II (9). Male rats were given electrolytic lesions of the SFO or its ventral stalk. Those with a demonstrated absence of drinking to SC Ang II (about 2 weeks postlesion) showed no Fos-IR in OVLT, MnPO, SO or PVH after Ang II infusion (9). This is consistent with the view that the SFO is the unique transduction site and that either peripheral Ang II does not access the OVLT (at least at doses that are effective in SFO), or that it does not induce Fos-IR in this region. In either case, the Fos-IR in OVLT with peripheral Ang II presumably must result from transduction in the SFO.

The second (unpublished) study was designed to examine the role of the SFO in behavior and Fos-IR using pCVT injection of Ang II, and possible recovery or reorganization of function with time after the lesion. Adult male rats received SFO lesions as before, or sham surgery, and were screened for drinking to SC Ang II 1 week after surgery. On the basis of these results, they were divided into two behaviorally-matched groups, one destined for immediate Fos studies and the other for study 9-10 weeks later. All rats additionally received a lateral cerebroventricular cannula 1 week before a behavioral and (2 days later) Fos-IR study. [In the rats studied one week postlesion, the cannula was implanted at the time of lesion surgery].

The results obtained **1 week postsurgery** were very clear:

(i) Lesioned rats failed to drink to peripheral Ang II (200 ug/kg); M±SE intakes were 4.7±1.4 ml (controls) and .6±.4 ml (SFO lesion).
(ii) Lesioned rats also failed to drink to pCVT Ang II (10 ng); M±SE intakes were 8.2±1.7 ml (controls) and .7±.4 ml (SFO lesion).
(iii) Lesioned rats (with verified ICV cannulas) had no pCVT Ang II (100 ng)-induced Fos IR in MnPO (35±5 cells/section in controls, 0 in lesion rats), SON (84±7 vs 0), or PVH (60±15 vs 0). All of these differences were statistically reliable ($P<0.05$).

They suggest that, at 1 week postlesion, CVT-injected Ang II does not engage (by Fos-IR or drinking) any remaining site in the brain.

The results 9-10 weeks postsurgery were quite different. Five rats with SFO lesions had failed to drink (.7±.4 ml) to SC Ang II given 1 week after the lesion and so, based on the 1 week group data given above, would not have been expected to drink to CVT Ang II (since they did not have cannulas at that time, they were not so tested). However, when they were tested with pCVT Ang II after 9-10 weeks, they drank 3.8±1.4 ml, close to the 4.8±.2 ml consumed by two control rats so tested. Thus, there was substantial recovery of pCVT Ang II-induced drinking with time after SFO lesion. Fos-IR also showed good recovery to 63±1, 25±6 and 22±6 cells/section in MnPO, SON and PVH, respectively (N=3). These data are preliminary, since relatively small Ns are involved, but they confirm and apparently resolve the disparate published reports (14-16) by suggesting that recovery of function and/or reorganization may occur along the LT following SFO lesion. Importantly, the drinking behavior parallels the appearance of Fos-IR in the ventral LT and magnocellular nuclei.

In closing this section, we should note that relatively few cells in the area postrema/nucleus of the solitary tract (AP/NTS) show Fos in response to either central or peripheral Ang II. We also examined this in spontaneously hypertensive rats (SHR), with the hypothesis that they may show greater Fos reactivity in some regions of the brain than normotensive rats. However, we found no difference in Ang II-induced Fos between SHR and Wistar rats. What was unexpected was a much reduced induction of Fos in the SO and PVH of the WKY "controls", despite equal Fos in the SFO (17). This raises an interesting question concerning differences between WKY and outbred Wistar rats, that is beyond the scope of this paper.

EFFECT OF FLUID INTAKE ON FOS-IR

All of the above Fos-IR studies were conduced in euhydrated rats tested in the middle of the day without access to water after administration of Ang II (*ie* the behavioral and Fos studies were separate). However, we did find the occasional rat that showed little or no Fos-IR in SO and PVH; in most of these cases, the rat had been known to consume a substantial amount of water a few hours prior to the Ang II stimulus. This suggests that some aspect of water inhibits Fos-IR in SO and PVH magnocellular regions. In fact, allowing rats to drink water following pCVT Ang II prevents the appearance of Fos-IR in SO and PVH without affecting induction of Fos-IR in SFO or LT (18). We confirmed this result, and showed that intragastric infusion of water also was highly effective (19); Xu and Herbert (20) have shown that hypoosmolality is associated with this block of Fos-IR. However, using sham-drinking rats with open gastric fistulas, we also found an indication of orally-mediated inhibition of Ang II-induced Fos-IR, especially in the posterior SO (19). Thus, either direct hypoosmolality in SO and PVH, or input from extracranially-transduced information (eg by tongue, gut, liver) is sufficient to modulate the Ang II-related afferent information from SFO.

Hyperosmolarity, produced by intravenous infusion of moderate (but highly dipsogenic) doses of hypertonic NaCl (2 ml 2M), induces very strong FLI in the SO and PVH, but almost none along the LT (21). At still higher doses, that we consider extreme, we and others have found Fos-IR in SFO and in the dorsal cap region of the OVLT (21-23). However, because lower doses are able to give almost maximal Fos-IR in the SO and PVH without detectable stimulation of structures of the LT, we question the proposition that the LT is a primary osmoreceptive structure. The high sensitivity of the PVH and SO is further demonstrated by detection of Fos-IR in these nuclei of rats following normal-sized meals that evidently have effects on body fluid balance that are within the physiological range (24). Very recently, portal vein hypernatremia without concurrent change in systemic osmotic pressure has been shown to be an adequate stimulus of Fos-IR in SO and PVH (25). We thus believe that normal-sized meals activate portal osmoreceptors that project (presumably indirectly) to the SO and PVH. Activation of this system may then underlie prandial drinking.

PHARMACOLOGICAL SPECIFICITY OF ANG II ACTION

We have previously reported that either peripheral or central administration of the AT-1R antagonist, losartan potassium, blocks the dipsogenic effects of Ang II (26). Because the permeability of peripheral losartan into brain was not clear at the time of our initial Fos-IR studies, we used pCVT administration of losartan (50 ug) followed by pCVT Ang II (100 ng). We found a complete block of Ang II-induced Fos-IR in all brain regions (12). In behavioral studies, 1 ug losartan is sufficient to inhibit completely the water intake evoked by 10 ng pCVT Ang II (26). The specificity of pCVT losartan has been shown in other studies in which neither water intake (27) nor Fos-IR (unpublished data) induced by hyperosmolarity was affected. The primary site of action of pCVT losartan to block pCVT Ang II-induced Fos-IR is in the SFO {unpublished data cited in (28)}. These data are consistent with other evidence that the SFO contains primarily AT-1R subtype (3,4).

It was thus surprising when we found that pCVT administration of the AT-2R antagonist, PD 123319, also dose-dependently inhibited Ang II-induced water intake (26,29). The maximally effective dose was about 10-fold higher than for losartan. One possibility, then, was that PD 123319 is not totally specific for the type 2 receptor, and that the behavioral effects observed were due to nonselective blockade of the AT-1R. By use of the Fos-IR method, we have been able to show that this is an extremely unlikely explanation. Thus, pCVT PD 123319 (100 ug) was completely ineffective in attenuating Fos-IR induced in the SFO (or MnPO and OVLT) by pCVT Ang II (12). As noted above, the most likely site of access and action of both Ang II and the antagonists is in the SFO; the failure of PD 123319 to displace Ang II in SFO is not surprising given the apparent absence of AT-2R in that organ (3,4), but does not explain its antidipsogenic effect. However, the Fos-IR study did show an extensive inhibition of Ang II-induced activity by PD 123319 in both the SO and PVH. Thus, the profile observed after Ang II and PD resembles that seen after Ang II and water intake, with selective inhibition in the magnocellular neurosecretory nuclei. Ligand binding studies have indicated the presence of AT-2R in PVH (4), and the use of the AT-2R antibody (30) indicates a much more widespread presence in brain than is evident from the ligand displacement work. Whether PD123319 is either diffusing directly to the SO and PVH or acting at synapse(s) on afferent(s) to the PVH (eg via a pathway that normally carries hydrational information) cannot be ascertained from our data.

RELATIONSHIP BETWEEN FOS-IR AND DRINKING AFTER LOSARTAN

In the previous section, we noted that pCVT injection of a high dose of losartan completely abolished pICV Ang II-induced Fos-IR. It has also been shown that ICV Ang II-induced water intake can be inhibited by peripheral losartan, suggesting that this agent penetrates the SFO or other brain regions. In the present series of studies, we examined the inhibitory action of peripherally-administered losartan on Ang II-induced drinking and Fos-IR.

We previously reported (26) that the drinking response to pCVT Ang II (10 ng) was about 75% attenuated by 10 mg losartan/kg, SC. Because we have been using 100 ng Ang II as the stimulus in the Fos studies, we examined the inhibitory effects of losartan on drinking to this higher central dose of Ang II. We found about 25% and 50% attenuations with 10 and 20 mg losartan/kg SC, respectively. In the Fos measures using the same doses, Ang II-induced Fos-IR was completely inhibited in all regions (SFO, OVLT, SO, PVH) by 20 mg losartan/kg, and was attenuated by 50% (in PVN) to 80% (in SFO) of control levels with 10 mg losartan/kg SC. Thus, the reductions in Fos parallel the reductions in drinking, although the quantitative aspects of this relationship have not been established fully.

HYPOTENSION AND OTHER FORMS OF ANGIOTENSIN-RELATED THIRST

Peripheral administration of agents such as the B-adrenergic agonist, isoproterenol, serotonin (5HT), or the 5HT agonist, 5-carboxyamidotryptamine (5CT), is accompanied by release of renal renin, increase in circulating Ang II, and water intake. Several lines of evidence, and notably those with kininase II inhibitors, have indicated that the drinking response is either completely or partially dependent on circulating Ang II. In order to verify this further, we have mapped the expression of Fos-IR in the brain following administration of isoproterenol and 5CT (31,32). The most prominently-activated regions included the SFO (posterior part), SO and PVH, consistent with a role for circulating Ang II. However, neither the drinking response nor the Fos-IR induced by either 5CT or isoproterenol was reversed by losartan pretreatment (27,31). Further, in the case of 5CT, SFO lesions that abolished the drinking to SC Ang had no effect on 5CT-induced water intake {unpublished data, cited in (32)}. Since both 5CT and isoproterenol are hypotensive agents, it is likely that Ang II and baroreceptor inputs both play a role in the behavioral and Fos responses.

Hypovolemia, induced by either hemorrhage or polyethylene glycol (PEG), is accompanied by prominent Fos activation of the SO and PVH (33,34). However, relatively low levels of Fos were seen along the LT after PEG (33). It is already known that PEG-induced thirst is largely independent of Ang II, and so it is likely that baroreceptor input (most likely, venous/atrial) is sufficient to activate many cells in the SO and PVH without concurrent Ang II stimulation. PEG also induces sodium appetite, the latency of which is shortened by prior exposure to sodium-free diet (35). Three days' sodium-free diet prior to PEG did not, however, produce Fos patterns different from PEG alone (33). In contrast, sodium depletion by furosemide plus 22 hr sodium-free diet produced very pronounced Fos activation of the OVLT, MnPO and SFO. Since plasma Ang II is elevated in these situations, the contribution of Ang II to the Fos patterns needs to be examined. Of relevance to this issue is the finding that losartan (10 mg/kg SC) does not attenuate either depletion-induced salt appetite (26) or PEG thirst (27).

CONCLUSIONS

We have summarized recent data, mostly from our own labs, in which we have examined the patterns of Fos expression in rat brain after various known dipsogenic or natriorexigenic stimuli. With a few exceptions, the patterns of Fos expression are useful for indicating that circulating Ang II is transduced in the core and posterior parts of the SFO, via an AT-1 receptor, and these cells in turn activate cells in MnPO, OVLT, SO and PVH. An AT-2 receptor also seems to be implicated in the SO and PVH, possibly via an inhibitory pathway that is normally activated by either water intake or hypotonicity. Other dipsogenic stimuli, including hypertonicity and hypovolemia, produce prominent Fos activation in SO and PVH, But these can occur without concurrent induction of Fos in the SFO and OVLT. The Fos method appears to be a promising means of unravelling the specific pathways and the integration of information in the circuitry underlying regulation of fluid balance. In particular, the combination of Fos with the use of tissue lesion, pharmacological antagonist, retrograde transport, and other double-labelling methods, should allow a more detailed appreciation of the functional integration of thirst and sodium appetite.

REFERENCES

1. Rowland NE, Fregly MJ. *Appetite* 11 (1988) 143-178.
2. Johnson AK, Edwards GL. *Curr. Topics Neuroendocrinology* 10 (1990) 149-190.
3. Rowe BP, Saylor DL, Speth RC. *Neuroendocrinol* 55 (1992) 563-573.
4. Tsutsumi K, Saavedra JM. *Am. J. Physiol* 261 (1991) R209-216.
5. Bains JS, Potyok A, Ferguson AV. *Brain Res* 487 (1992) 223-229.
6. Oldfield BJ, Hards DK, McKinley MJ. *Brain Res* 558 (1991) 13-19.
7. Hoffman GE, Smith MS, Verbalis JG. *Front. in Neuroendocrinol.* 14 (1993) 173-213.
8. Sharp FR, Sagar SM, Swanson RA. *Crit. Rev. Neurobiol.* 7 (1993) 205-228.
9. McKinley MJ, Badoer E, Oldfield BJ. *Brain Res.* 594 (1992) 295-300.
10. Rowland NE, Li, BH, Rozelle AK, Fregly MJ, Garcia M, Smith GC. *Brain Res. Bull.* 33 (1994) 427-436.
11. Lind RW, Swanson LW, Ganten D. *Neuroendocrinol* 40 (1985) 2-24.
12. Rowland NE, Li BH, Rozelle AK, Smith GC. *Am. J. Physiol.* 267 (1994) R792-798.
13. Saad WA, Menani JV, Camargo LAA, Abrao-Saad W. *Braz. J. Med. Biol. Res.* 18 (1985) 37-46.
14. Buggy J, Fisher AE, Hoffman WE, Johnson AK, Phillips MI. *Science* 190 (1975) 72-74.
15. Lind RW, Johnson AK. *J. Neurosci.* 2 (1982) 1043-1051.
16. Lind RW, Thunhorst RL, Johnson AK. *Physiol. Behav.* 32 (1984) 69-74.
17. Rowland NE, Li BH, Fregly MJ, Smith GC. *Brain Res.* (in press).
18. Herbert J, Fosling ML, Howes SR, Stacey PM, Shiers HM. *Neuroscience* 51 (1992) 867-882.
19. Morian KR, Rowland NE. *Regul. Peptides* (in press)
20. Xu Z, Herbert J. *Brain Res.* 659 (1994) 157-168.
21. Han L, Rowland NE. *Brain Res.* (submitted).
22. Hamamura M, Nunez DJR, Leng G, Emson PC, Kiyama H. *Brain Res.* 572 (1992) 42-51.
23. Oldfield BJ, Bicknell RJ, McAllen RM, Weisinger RS, McKinley MJ. *Brain Res.* 561 (1991) 151-156.
24. Rowland NE. *Neurosci. Lett.* 189 (1995) 1-3.
25. Morita H, Yamashita Y, Tanaka K, Nishida Y, Hosomi H. *FASEB J.* 9 (1995) A630 (abstract).

26. Rowland NE, Rozelle AK, Riley PJ, Fregly MJ. *Brain Res. Bull.* 29 (1992) 389-393.
27. Fregly MJ, Rowland NE. *Proc. Soc. Exp. Biol. Med.* 199 (1992) 158-164.
28. Culman J, Hohle S, Qadri F, Edling O, Blume A, Lebrun C, Unger Th. *Clin Exper. Hypertens.* 17 91995) 281-293.
29. Rowland NE, Fregly MJ. *Brain Res. Bull.* 32 (1993) 391-394.
30. Reagan LP, Flanagan-Cato LM, Yee DK, Ma LY, Sakai RR, Fluharty SJ. (*this symposium*)
31. Rowland NE, Fregly MJ, Li BH, Smith GC. *Brain Res.* 654 (1994) 34-40
32. Rowland NE, Li BH, Fregly MJ, Smith GC. *Brain Res.* 664 (1994) 148-154.
33. Han L, Rowland NE. *Neurosci. Lett.* (in press)
34. Shen E, Dun SL, Ren C, Bennett-Clarke C, Dun NJ. *Brain Res.* 593 (1992) 136-139.
35. Stricker EM. *J. Comp. Physiol. Psychol.* *97 (1983) 725-737.*

13

IDENTIFICATION OF AT_1 RECEPTORS ON CULTURED ASTROCYTES

E. Ann Tallant, Debra I. Diz, and Carlos M. Ferrario

Hypertension Center
Bowman Gray School of Medicine
Medical Center Boulevard
Winston-Salem, North Carolina 27157-1032

INTRODUCTION

It is well established that the brain contains an endogenous renin-angiotensin system (RAS) that participates in the regulation of blood pressure through the control of water intake and sodium appetite, the secretion of ACTH and vasopressin and the regulation of sympathetic tone (1). While high affinity binding sites for Ang II were initially characterized in brain homogenates (2,3), the technique of *in vitro* receptor autoradiography allowed visualization of Ang II binding sites to specific brain regions including the circumventricular organs, hypothalamus, dorsal and ventral medulla, spinal cord and pituitary gland (1,4,5). These brain regions correspond to areas at which the peptide exerts its physiological effects (5) or where the responses are inhibited by Ang II receptor antagonists (6).

^{125}I-Ang II binding to brain homogenates or brain slices may represent receptors on neurons, astrocytes on vascular elements. Functional Ang II receptors have been identified on cultured neuronal (7,8), glial (9-11) and cerebral microvessel endothelial cells (12,13). Although much of the early work on the brain RAS focused on neuronal elements, recent evidence suggests that astrocytes may be a major source of angiotensinogen production in the brain. Thomas et al. (14) reported that angiotensinogen is present in pure neuronal cultures while we showed that cultured human astrocytes contain the mRNA for angiotensinogen (15). Deschepper et al. (16) found that cultured rat astroglial cells also contain angiotensinogen mRNA and secrete angiotensinogen constituitively. Furthermore, Stornetta and coworkers (17) co-localized angiotensinogen mRNA with the astrocyte-specific protein glial fibrillary acidic protein (GFAP) in rat brain using *in situ* hybridization histochemistry. In recent studies by Gyurko et al. (18), intracerebroventricular injection of antisense oligonucleotides to angiotensinogen resulted in a reduction in central levels of Ang II and a concomitant reduction in blood pressure. Thus the amount of angiotensinogen produced and secreted by astrocytes may play an important role in regulating the amount of Ang II that is available to interact with its receptors.

Recent Advances in Cellular and Molecular Aspects of Angiotensin Receptors
Edited by Mohan K. Raizada et al., Plenum Press, New York, 1996

Specific high affinity binding sites for Ang II were identified on rat C6 glioma cells (19) and in explant cultures of astrocytes from various regions of rat brain and spinal cord (20,21). In explant cultures, labeling was more intense in the spinal cord and brainstem than in cerebellum and cortex (20), suggesting that subpopulations of astrocytes may preferentially express Ang II receptors. Since astrocytes in select brain regions may have unique functions (i.e., production of precursor proteins such as angiotensinogen, participation in the blood brain barrier, or control of local neuronal environment), Ang II receptors on different brain region astrocytes may produce divergent types of responses. In this chapter, we review our studies on Ang II receptors on cultured human astrocytes and present new evidence showing that Ang II receptors on astrocytes are heterogeneously distributed in rat brain.

ANG II RECEPTORS ON CULTURED HUMAN ASTROCYTES

To identify astroglial Ang II receptors and their signaling pathways, we used cultured human astrocytoma cells of the CRTG3 cell line. This cell line was established by explant culture of neoplastic brain tissue isolated from a patient undergoing excision of a tumor mass histopathologically diagnosed as a grade IV astrocytoma [glioblastoma multiforme] (22). The majority of cells in these cultures stained positively with the glial specific marker glial fibrillary acid protein (GFAP) and negatively for the neuronal marker neuron-specific enolase (15,23).

To identify and characterize Ang II receptors, we measured binding of 125I-Ang II to intact CRTG3 cells, as described in the legend to Figure 1. Under these binding conditions, CRTG3 cells contained a single high affinity Ang II binding site with an affinity of 4.0 ± 0.9 nM and a maximal number of binding sites of 117.4 ± 1.2 fmol/mg protein. A representative

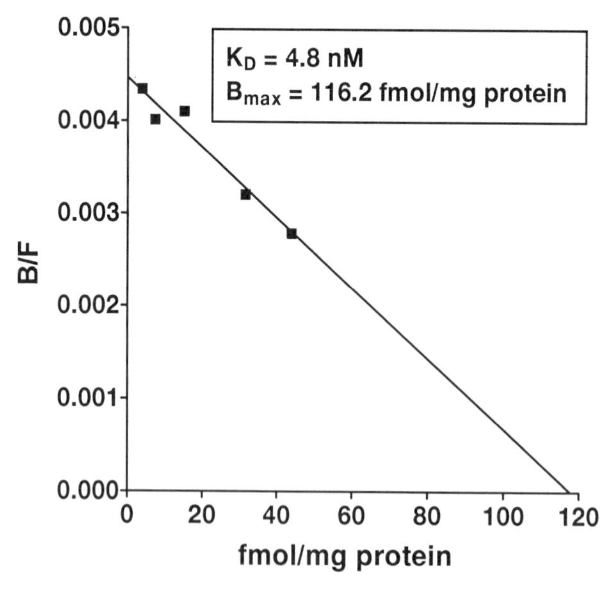

Figure 1. Scatchard plot of ^{125}I-Ang II binding to cultured human astrocytes of the CRTG3 cell line. Binding of increasing concentrations of ^{125}I-Ang II to intact CRTG3 cells growing as monolayers in 24-well dishes was measured in phosphate-buffered saline containing 0.1% bovine serum albumin. Binding was measured at 4 °C for 90 min to prevent internalization of the radioligand. Non-specific binding was measured in the presence of 10 μM unlabeled Ang II. The data were analyzed using the computer program RADLIG (Elsevier-Biosoft).

Scatchard plot is also shown in Figure 1. Competition curves using increasing concentrations of unlabeled Ang II also suggested that CRTG3 cells contain a high affinity Ang II binding site with an IC_{50} in the nanomolar range. In competition with various angiotensin peptides, the order of binding was Ang II > Ang I > Ang-(1-7) > Ang-(2-8). The receptor subtype was characterized by the binding of 125I-Ang II to intact cells in the presence of increasing concentrations of AT_1 and AT_2 selective antagonists or a non-selective sarcosine analog of Ang II. The maximal competition by 10 μM concentration of the AT_1 antagonists losartan, EXP 3174 or L-158,809 was 12.8, 29.1, and 26.3%, respectively, of the specific binding. A similar concentration of the AT_2 selective antagonists PD123177 or CGP 42112A also competed for 18.8 and 28.6%, respectively, of the specific binding, suggesting that both AT_1 and AT_2 receptors are present on CRTG3 cells. In contrast, 10 μM Sar1Thr8-Ang II competed for 72% of the specific binding sites. Since the sarcosine analog competed for more of the Ang II binding sites than can be accounted for by AT_1 and AT_2 selective antagonists, an additional subtype of Ang II receptor may also be present on CRTG3 cells. Finally, 28% of the specific 125I-Ang II bound was not displaced by Sar1Thr8-Ang II, suggesting that other binding sites may also be present Thus human astrocytes of the CRTG3 cell line contain high affinity Ang II binding sites of both the AT_1 and AT_2 subtype as well as possible additional binding sites.

AT_1 receptors in different tissues are coupled to both activation of a phosphoinositide-specific phospholipase C and inhibition of adenylate cyclase (24-26). To identify the cellular pathways activated by Ang II binding to CRTG3 cells, we measured phospholipase C activity as the production of inositol 1,4,5-trisphosphate (IP_3) using a specific radioreceptor assay. Fifteen seconds after the addition of 100 nM Ang II, the IP_3 levels were increased by 145.5 ± 4.6% of basal levels (27). To determine whether the rise in IP_3 caused subsequent mobilization of intracellular Ca^{2+}, CRTG3 cells were loaded with Fura 2 by incubation for 30 min in 2 μM Fura 2/AM and Ca^{2+} mobilization was measured using a spectrofluorometer. Incubation with 100 nM Ang II caused a consistent increase in the intracellular concentration of Ca^{2+}, averaging 190 ± 18 nM above basal (27). The increase in intracellular Ca^{2+} was transient, peaking at approximately 15 sec and returning to baseline by 60 sec. Ca^{2+} mobilization in response to Ang II was unchanged in the absence of extracellular Ca^{2+} or after incubation with the Ca^{2+} channel blockers verapamil or nifedipine, indicating that the increase was due to mobilization of intracellular Ca^{2+} by IP_3. In addition, Ca^{2+} mobilization by Ang II was completely blocked by the AT_1 selective antagonists losartan while the AT_2 antagonist CGP 42112A was ineffective (11). Thus Ang II activates AT_1 receptors on human astrocytes that are coupled to the activation of a phospholipase C and the mobilization of intracellular Ca^{2+}.

Ang II also stimulated a dose-dependent release of prostaglandins (PGs) from CRTG3 cells, measured as the production of PGE_2 or 6-keto $PGF_{1\alpha}$, a stable metabolite of prostacyclin [PGI_2] (27). Ang II stimulated PGE_2 and PGI_2 release in a dose-dependent manner. Losartan, the AT_1-selective antagonist, blocked the Ang II-induced release of PGE_2 (11); however, losartan stimulated PGI_2 synthesis, in the presence or absence of Ang II (28). In contrast, CGP 42112A, the AT_2-selective antagonist, totally blocked the Ang II-induced PGI_2 synthesis but only partially attenuated PGE_2 release. These data suggest that, in human astrocytes, Ang II activates AT_1 receptors participating in phospholipase C activation and PG release as well as AT_2 receptors coupled to the release of PGs (11).

Ang II also stimulated the release of both PGE_2 and PGI_2, with essentially the same order of potency, in two other human astrocytoma cell lines—STTG1 and WITG2—and rat C6 glioma cells (10,27). In contrast to the finding in CRTG3 cells, Ang II did not activate phospholipase C or cause an increase in intracellular Ca^{2+} in STTG1 and WITG2 cells (27) or C6 glioma cells (unpublished observation). However, bradykinin significantly increased cytosolic Ca^{2+} in all three astrocyte cell lines, indicating that they all had the potential to respond. These results suggest that astrocytes are heterogeneous in their responses to Ang II.

ANG II RECEPTORS ON NEONATAL RAT BRAIN ASTROCYTES

To study the heterogeneity of Ang II, we isolated astrocytes from brain regions known to participate in cardiovascular regulation (medullary and hypothalamic astrocytes) as well as region-specific astrocytes involved in sensory and motor function (cortical and cerebellar astrocytes). Primary cultures of astrocytes were prepared from the brains of one day old neonatal Sprague-Dawley rats. Brain tissue was dissected into cortical, hypothalamic, cerebellar and medullary regions and physically dissociated. Cells were grown to confluence, physically shaken to detact oligodendrocytes and passaged prior to use (29). Isolated cells stained positively with an antibody against glial fibrillary acidic protein and negatively with markers for neurons, fibroblasts or oligodendrites.

To identify Ang II binding sites on region-specific astrocytes, we measured the binding of ^{125}I-Ang II to isolated membranes from astrocytes from each region. As shown in the representative Scatchard plots in Figure 2, astrocytes from both the medulla and cerebellum contained high affinity Ang II binding sites. Although cortical astrocytes also contained a significant amount of ^{125}I-Ang II binding, a large majority of this binding was non-specific. Interestingly, astrocytes harvested from hypothalamic tissue contained no significant amount of ^{125}I-Ang II binding sites in binding studies in isolated membranes. Using subtype selective receptor antagonists, we determined the relative amount of AT_1 and AT_2 binding sites on astrocytes isolated from the medulla and cerebellum. In the medulla,

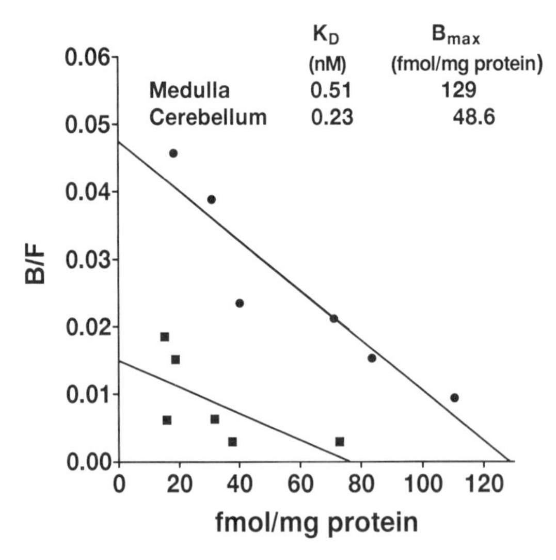

Figure 2. Scatchard plot of ^{125}I-Ang II binding to cultured astrocytes isolated from neonatal Sprague-Dawley rat medulla and cerebellum. Binding of increasing concentrations of ^{125}I-Ang II to membranes isolated from cell homogenates by centrifugation (37,000 x g for 20 min at 4 °C) was measured in Tris-buffered saline containing 0.1% bovine serum albumin and a cocktail of protease inhibitors (1 mM phenylmethylsulfonyl fluoride, 0.1 mg/ml soybean trypsin inhibitor, 10 μM amastatin, 10 μM bestatin, 2 μM leupeptin, 10 μM Z prolyl prolinal and 10 μM SCH 39370). Binding was measured at 22 °C for 60 min and membranes were harvested by filtration. Non-specific binding was measured in the presence of 10 μM unlabeled Ang II. The data were analyzed using the computer program RADLIG (Elsevier-Biosoft).

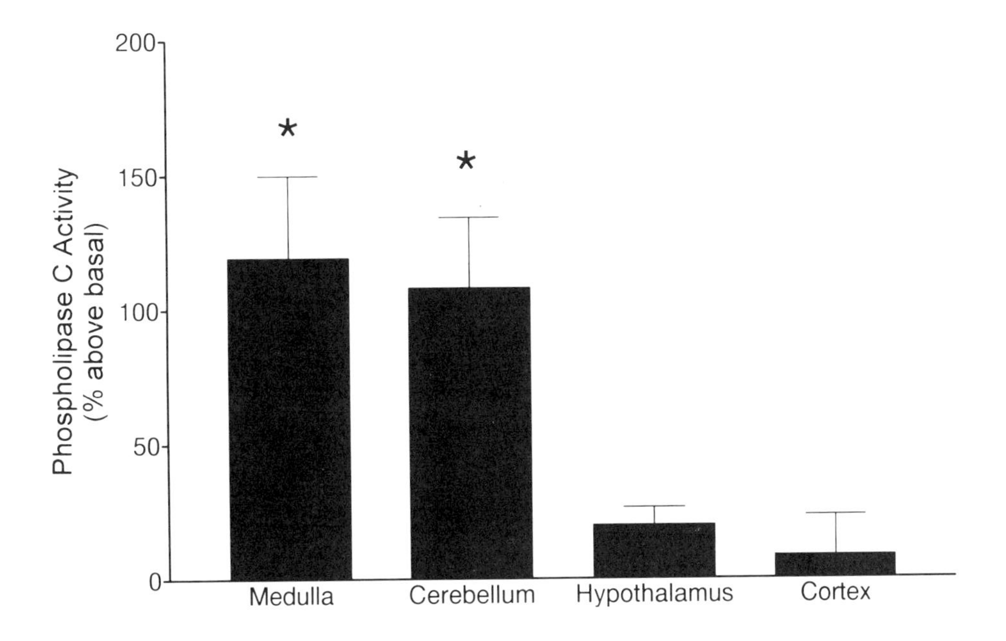

Figure 3. Ang II activation of phospholipase C in astrocytes isolated from rat brain. Astrocytes isolated from each region of one day old Sprague-Dawley rat brain were cultured as described in the text, loaded with ^{3}H-myoinositol by incubation for 72 hrs in inositol-free media, and stimulated with 1 μM Ang II for 30 min at 37 °C. Radiolabeled inositol phosphates were extracted with cold perchloric acid and separated by ion exchange chromatography, as described by Berridge (30). Phospholipase C activity was measured as the release of total inositol phosphates and is presented as the percentage above basal, in the presence of buffer alone, from astrocytes isolated from at least 4 different litters of rat pups.

the majority of the ^{125}I-Ang II binding sites were of the AT_1 subtype and could be effectively competed for by 1 μM losartan or EXP 3174 (90% and 81%, respectively). In contrast, the AT_2 selective antagonists PD123119 and CGP 42112A at a concentration of 1 μM competed for 22% and 39% of the total specific ^{125}I-Ang II binding sites in medullary astrocytes. In cerebellar astrocytes, losartan and EXP 3174 competed for essentially all of the specific ^{125}I-Ang II binding sites (99% and 98%, respectively). However, PD123119 and CGP 42112A at 1 μM concentration were also competitors for these binding sites in cerebellar astrocytes, competing for 65% and 73%, respectively, of the specific binding sites. These results suggest that astrocytes, especially those isolated from the cerebellum, contain AT_1 binding sites that also appear to be sensitive to inhibition by the AT_2 antagonist PD 123119.

To determine whether the AT_1 receptors on astrocytes isolated from neonatal rat brain are also coupled to inositol phosphate metabolism and the mobilization of intracellular Ca^{2+}, we measured phospholipase C activity in response to 1 μM Ang II in astrocytes from each brain region. Ang II caused a significant increase in the production of inositol phosphates in astrocytes isolated from the medulla and cerebellum of Sprague-Dawley rats, as shown in Figure 3. Phospholipase C activated in response to Ang II in both medullary and cerebellar astrocytes was inhibited by pre-incubation with the AT_1 selective antagonist losartan, while the AT_2 selective antagonist PD 123119 was ineffective (data not shown). Ang II also caused the release of prostacyclin from astrocytes isolated from the medulla or cerebellum, as shown

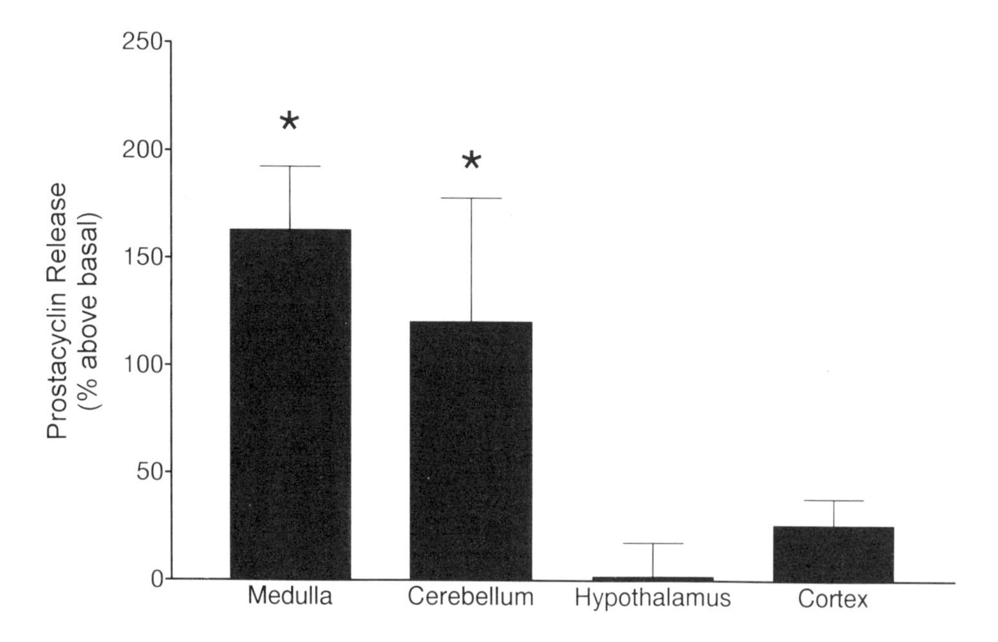

Figure 4. Ang II activation of prostacyclin release from astrocytes isolated from rat brain. Astrocytes isolated from each region of one day old Sprague-Dawley rat brain were cultured as described in the text. Cells were stimulated with 1 μM Ang II for 15 min and prostacyclin released into the media was measured as the production of its stable metabolite, 6-keto-prostaglandin $F_{2\alpha}$, using a commercial radioimmunoassay kit (Advanced Magnetics). Prostacyclin release is presented as the percentage above basal, in the presence of buffer alone, from astrocytes isolated from at least 4 different litters of rat pups.

in Figure 4. In contrast, a similar concentration of Ang II did not activate phospholipase C or release prostacyclin in astrocytes isolated from the hypothalamus or cortex, as shown in Figures 3 and 4, respectively.

CONCLUSIONS

Astrocytes isolated from human brain as well as specific regions of rat brain contain AT_1 receptors that are coupled to the activation of a phosphoinositide-specific phospholipase C and the production of prostaglandins. Ang II activation of phospholipase C was previously reported by Sumners et al. in astrocytes isolated from whole rat brain (31) or from the combined hypothalamus and medulla of rat brain (32). The production of inositol phosphates, specifically IP_3, causes a subsequent mobilization of intracellular Ca^{2+}, as we showed in human astrocytes. Activation of a Ca^{2+}-dependent phospholipase A_2 in response to the rise in intracellular Ca^{2+} may account for the PG release in response to Ang II that is coupled to an AT_1 receptor. However, in human astrocytes of the CRTG3 cell line, PG release was also partially due to activation of an AT_2 receptor and in the STTG1 and WITG2 cells lines, Pgs were released by Ang II in spite of no Ang II activation of PLC or mobilization of intracellular Ca^{2+}. These later results suggest that Ang II causes PG release in human astrocytes by at least two pathways—an AT_1 pathway that involves activation of a Ca^{2+}-dependent phospholipase A_2 and a Ca^{2+}-independent pathway. The subtype of receptor coupled to PG release in astrocytes isolated from neonatal rat brain is currently under investigation.

We identified Ang II receptors on astrocytes isolated from the medulla and cerebellum of neonatal rat brain. Although a small number of receptors was also present on cortical astrocytes, we did not observe any significant Ang II binding to astrocytes isolated from the hypothalamus either measuring binding to isolated membranes or by *in vitro* receptor autoradiography (data not shown). These results are reminiscent of earlier studies by Sumners et al. (32) who identified functional Ang II receptors on isolated astrocytes from the combined medulla and hypothalamus and by Hosli et al. (20) who showed major ^{125}I-Ang II binding to astrocytes isolated from the brainstem and less intense binding to astrocytes isolated from the cortex and cerebellum. However, our results contrast with those of Bottari et al. (21) who identified a large number of Ang II binding sites on astrocytes isolated from the diencephalon and pons. These later discrepancies are most likely due to differences in the isolated brain regions from which the astrocytes were cultured, the techniques for dissociating the brain tissue or the binding assays themselves.

Ang II receptors on astrocytes isolated from the neonatal rat medulla were predominantly of the AT_1 subtype. In contrast, cerebellar astrocytes contained both AT_1 receptors coupled to phosphoinositide metabolism as well as AT_2 receptors. Receptors of both the AT_1 and AT_2 subtype were identified in both the young and adult rat brain by *in vitro* receptor autoradiography (33). In the adult rat brain, the majority of the Ang II receptors were of the AT_1 subtype although AT_2 receptors were also identified in specific areas such as the inferior olive. This contrasts with the presence of a large number of AT_2 receptors in the young (2 week-old) rat brain in various regions including the cerebellar cortex (33). The presence of large numbers of AT_2 receptors in developing rat brain and their absence in the adult brain suggests that these receptors may be involved in the development of the central nervous system. Although AT_2 receptors may be present on neurons in the cerebellum, our results suggest that at least part of these receptors are present on astrocytes which may participate in neural development.

What role do Ang II receptors on astrocytes play in the central regulation of blood pressure? Astrocytes are a major neural source of angiotensinogen based on the co-localization of angiotensinogen mRNA and positive immunoreactivity for GFAP in various regions of rat brain (17). Evidence that the central level of angiotensinogen regulates the level of Ang II in the brain comes from a recent study where antisense oligonucleotides directed to the translation start site of angiotensinogen were injected into the ventricles of spontaneously hypertensive rats [SHRs] (18). A reduction in mean arterial pressure was accompanied by a small but significant decrease in the central levels of Ang II, suggesting that the levels of angiotensinogen may be a rate-limiting step in Ang II synthesis in the brain. In the periphery, plasma levels of angiotensinogen are positively regulated by Ang II (34). Ang II also regulates the expression of the mRNA for angiotensinogen in the brain. We showed that elevation of Ang II following its infusion into the ventricles (35) or after aortic coarctation (36) results in a significant increase in angiotensinogen mRNA expression in the medulla but not in the hypothalamus. Thus Ang II receptors on astrocytes may regulate the production of angiotensinogen in response to changes in endogenous concentrations of Ang II.

Ang II receptors on astrocytes may also participate in cardiovascular regulation by releasing PGs. We showed that Ang II receptors on astrocytes from both human and rat brain are coupled to the release of PGs and many effects of PGs on central regulation of cardiovascular function parallel those of Ang II. For example, intracerebroventricular injections of PGE_2 or PGI_2 produce pressor and tachycardic effects similar to Ang II given by the same route (37). In fact, the mechanisms mediating the elevation in pressure for both Ang II and PGs include increases in activity of the peripheral sympathetic nervous system (1,37). Furthermore, recent studies indicate that Ang II-induced vasopressin release may be mediated by PGs (38,39). PGs in the central nervous system may regulate cerebral blood flow and blood brain barrier permeability (40-42), actions which are also attributed to Ang II (41). Thus Ang II receptors on astrocytes may regulate cardiovascular function by controlling central Ang II levels or releasing PGs.

In summary, Ang II receptors of the AT_1 subtype are present on astrocytes isolated from both human brain and neonatal rat brain and are coupled to the production of inositol phosphates and the mobilization of intracellular Ca^{2+}. Astrocytes from the neonatal cerebellum also contain AT_2 receptors that may contribute to the development of the central nervous system. In neonatal rat brain, Ang II receptors are heterogeneously distributed and are present in areas that participate in cardiovascular regulation as well as in sensory and motor function. Since astrocytes are a major neural source of angiotensinogen and PGs, Ang II receptors on astrocytes may control the availability of angiotensinogen for Ang II production and the release of PGs which regulate cardiovascular function.

ACKOWLEDGMENTS

This work was supported in part by NS 31664 (EAT, DID), HL-38535 (DID) and HL-51952 (EAT, DID and CMF) from the National Institutes of Health and a Grant-in-Aid from the American Heart Association (EAT). Part of this work was carried out during an Established Investigatorship of the American Heart Association to DID. We acknowledge the generous gifts of Losartan (Losartan potassium; DuP 753), EXP 3174 and PD 123177 from Drs. P. Timmermans and R. Smith of du Pont (Wilmington, DE), CGP 42112A from Dr. M. DeGasparo of CIGA-GEIGY (Basle, Switzerland), PD 123319 from Parke-Davis (Ann Arbor, MI) and L-158,809 from Merck Sharp and Dohme Research Laboratories (West Point, PA). We thank June Tessiatore and Susan Bosch for excellent technical assistance.

REFERENCES

1. Ferrario, C.M., Ueno, Y., Diz, D.I. and Barnes, K.L. (1986) in Handbook of Hypertension, Vol.8: Pathophysiology of Hypertension— Regulatory Mechanisms. (eds. A. Zanchetti and R.C. Tarazi), Elsevier Science Publishers B.V. pp 431-54
2. Simonnet, G. and Vincent, J.D. (1982) Neurochem. Int. *4*:149-55
3. Bennett, J.P. and Snyder, S.H. (1976) J. Biol. Chem. *251*:7423-30
4. Speth, R.C., Wamsley, J.K., Gehlert, D.R., Chernicky, C.L., Barnes, K.L. and Ferrario, C.M. (1985) Brain Res. *326*:137-43
5. Brosnihan, K.B., Diz, D.I., Schiavone, M.T., Averill, D.B. and Ferrario, C.M. (1987) in Brain Peptides and Catecholamines in Cardiovascular Regulation (eds. J.P. Buckley and C.M. Ferrario), Raven Press, New York: pp 313-28
6. Campagnole-Santos, M.J., Diz, D.I., Santos, R.A., Khosla, M.C., Brosnihan, K.B. and Ferrario, C.M. (1988) Hypertension *11 (Suppl)*:I-167-71
7. Tallant, E.A., Diz, D.I., Khosla, M.C. and Ferrario, C.M. (1991) Hypertension *17*:1135-43
8. Fluharty, S.J. and Reagan, L.P. (1989) J. Neurochem. *52*:1393-400
9. Raizada, R.K., Phillips, M.I., Crews, F.T. and Sumners, C. (1987) Proc. Natl. Acad. Sci. USA *84*:4655-9
10. Jaiswal, N., Diz, D.I., Tallant, E.A., Khosla, M.C. and Ferrario, C.M. (1991) Am. J. Physiol. Regul. Integr. Comp. Physiol. *260*:R1000-6
11. Jaiswal, N., Tallant, E.A., Diz, D.I., Khosla, M.C. and Ferrario, C.M. (1991) Hypertension *17*:1115-20
12. Speth, R.C. and Harik, S.I. (1985) Proc. Natl. Acad. Sci. USA *82*:6340-3
13. Guillot, F.L. and Audus, K.L. (1991) Peptides *12*:535-40
14. Thomas, W.G., Greenland, K.J., Shinkel, T.A. and Sernia, C. (1992) Brain Res. *588*:191-200
15. Milsted, A., Barna, B.P., Ransohoff, R.M., Brosnihan, K.B. and Ferrario, C.M. (1990) Proc. Natl. Acad. Sci. USA *87*:5720-3
16. Deschepper, C.F., Bouhnik, J. and Ganong, W.F. (1986) Brain Res. *374*:195-8

17. Stornetta, R.L., Hawelu-Johnson, C.L., Guyenet, P.G. and Lynch, K.R. (1988) Science *242*:1444-6
18. Gyurko, R., Wielbo, D. and Phillips, M.I. (1993) Regul. Pept. *49*:167-74
19. Printz, M.P., Jennings, C., Healy, D.P. and Kalter, V. (1986) J. Cardiovasc. Pharmacol. *8(Suppl. 10)*:S62-8
20. Hosli, E. and Hosli, L. (1989) Neuroscience *31*:463-70
21. Bottari, S.P., Obermüller, N., Bogdal, Y., Zahs, K.R. and, Deschepper, C.F. (1992) Brain Res. *585*:372-6
22. Barna, B.P., Chou, S.M., Jacobs, B., Ransohoff, R.M., Hahn, J.F. and Bay, J.W. (1985) J. Neuroimmunol. *10*:151-8
23. Estes, M.L., Ransohoff, R.M., McMahon, J.T., Jacobs, B.S. and Barna, B.P. (1991) J. Neurosci. Res. *27*:697-705
24. Smith, J.B. (1986) Am. J. Physiol. *250*:F759-69
25. Campanile, C.P., Crane, J.K., Peach, M.J. and Garrison, J.C. (1982) J. Biol. Chem. *257*:4951-8
26. Johnson, R.M. and Garrison, J.C. (1987) J. Biol. Chem. *262*:17285-93
27. Tallant, E.A., Jaiswal, N., Diz, D.I. and Ferrario, C.M. (1991) Hypertension *18*:32-9
28. Jaiswal, N., Diz, D.I., Tallant, E.A., Khosla, M.C. and Ferrario, C.M. (1991) Am. J. Hypertens. *4*:228-33
29. Cole, R. and de Vellis, J. (1989). in A Dissection and Tissue Culture Manual of the Nervous System (eds. A. Shahar, J. de Vellis, A. Vernadakis and B. Haber), Alan R. Liss,Inc. New York: pp 121-33
30. Berridge, M.J. (1983) Biochem. J. *212*:849-58
31. Sumners, C., Tang, W., Zelezna, B. and Raizada, M.K. (1991) Proc. Natl. Acad. Sci. USA *88*:7567-71
32. Sumners, C., Meyers, L.M., Kalberg, C.J. and Raizada, M.K. (1990) Prog. Neurobiol. *34*:355-85
33. Tsutsumi, K. and Saavedra, J.M. (1991) Am. J. Physiol. Regul. Integr. Comp. Physiol. *261*:R209-16
34. Khayyall, M., MacGregor, J., Brown, J.J., Lever, A.F. and Robertson, J.I.S. (1973) Clin. Sci. *44*:87-90
35. Kohara, K., Brosnihan, K.B., Ferrario, C.M. and Milsted, A. (1992) Am. J. Physiol. *262*:E651-7
36. Nishimura, M., Milsted, A., Block, C.H., Brosnihan, K.B. and, Ferrario, C.M. (1992) Hypertension *20*:158-67
37. Chiu, E.K. and Richardson, J.S. (1983) J. Gen. Pharmacol. *14*:553-63
38. Brooks, D.P., Share, L. and Crofton, J.T. (1986) Endocrinology *118*:1716-22
39. Inoui, M., Crofton, J.T., Toba, K. and Share, L. (1990). FASEB J. *4*:A683
40. White, R.P. and Hagen, A.A. (1982) Pharmacol. Ther. *18*:313-31
41. Wahl, A., Unterberg, A., Baethmann, A. and Schilling, L. (1988) J. Cereb. Blood Flow Metab. *8*:621-34
42. Palmer, G.C. (1986) Neurosci. Biobehav. Rev. *10*:79-101

14

STRUCTURE-ACTIVITY RELATIONSHIP OF THE AGONIST-ANTAGONIST TRANSITION ON THE TYPE 1 ANGIOTENSIN II RECEPTOR; THE SEARCH FOR INVERSE AGONISTS

Jacqueline Pérodin, Roger Bossé, Sylvain Gagnon, Li-Ming Zhou, Richard Bouley, Richard Leduc, and Emanuel Escher

Département de Pharmacologie
Faculté de Médecine
Université de Sherbrooke
Sherbrooke, Québec, JIH 5N4, Canada

Peptidic angiotensin II (Ang) antagonists have been mostly reported to behave in a more or less competitive fashion. Thus, reinforcing the view of competitive analogues being compounds which can reversibly bind without producing any biological response to the receptor as well as competing with an agonist (e.g. Ang or other peptidic agonists) for the same site. Recently, a new concept was introduced that changes the classical view of agonist-antagonist action. This concept presents the receptor as a dynamic structure capable of undergoing a conformational change between a biologically active and an inactive form. A bound ligand may shift the equilibrium to either side, according to its pharmacological character as an agonist or an antagonist: An antagonist favoring the inactive form of the receptor represents what is called an "inverse agonist". All peptidic Ang analogues bind to the same locus on the AT1 receptor but non-peptidic AT1 binding compounds (e.g. L-158,809 and DuP 753) seem to bind to different loci. Furthermore, it has also been shown that non-peptidic Ang antagonists do not possess the ability to recognize Ang receptors from amphibian or avian origins. In the present contribution, we attempt to fathom the molecular parameters that bring the transition from an agonistic to an antagonistic behaviour in order to select the compounds that display most profoundly these antagonistic features. We believe to possess the necessary tools (enlarged but planar aromatic side-chains in position 8) in order to explore the concept of inverse agonism on the mammalian AT1 receptor.

Recent Advances in Cellular and Molecular Aspects of Angiotensin Receptors
Edited by Mohan K. Raizada et al., Plenum Press, New York, 1996

INTRODUCTION

Classical pharmacology defines an antagonist as a compound that acts against a particular agonist. Within this definition several subclassifications as to the mechanism of action of antagonists have been made, for example of physiological antagonism. In this particular case, a given pharmacophore produces a physiological action that is opposed to that of another compound. A typical example is the vasoconstrictive action of adrenaline in the emergency treatment of the life-threatening vasodilation and extravasation induced by massive histamine liberation in anaphylactic reactions. Pharmacological antagonism, or more precisely, competitive antagonism was considered to be the action of a compound that competes with the agonist for access to the binding site. An antagonist of this definition lacks the capacity to induce receptor activation. Pure competitive antagonists were therefore viewed as compounds that had no activity by themselves but only prevented the respective agonists from binding and activating their targets.

The knowledge accumulated in the last twenty years did not modify considerably the basic antagonist concept: Signal transduction mechanisms, membrane receptor structure and regulation have received much attention and elucidation. The non-occupied membrane receptors, regardless of their belonging to a particular receptor family, were still regarded to be physiologically silent. Neither the ligand-gated ion channels, nor the tyrosine-kinase receptors of insulin and insulin-like growth factors, nor the superfamily of the G-protein coupled heptahelical receptors were considered to have an intrinsic activity in the resting non-occupied state.

Site-directed mutagenesis experiments and experimentally induced overexpression of catecholamine receptors in the last years have however profoundly changed this picture. Substitutions of alanine in position 293 in the α_{1B}-adrenergic receptor by site-directed mutagenesis and the expression of this mutated receptor in a recombinant cell line has produced a system with high stimulation of the second messengers even in the absence of agonist [1,2]. Similar experiments with the β_2-adrenergic receptors and rhodopsin have revealed many constitutively active forms of these receptors [3,4]. Simple overexpression of wild-type receptors in recombinant cell lines also revealed considerable basal second messenger activity due to intrinsic receptor activity (i.e. in absence of agonist) [5,6,7]. The surprising observation was that certain classical antagonists were able to reduce the intrinsic or basal activity of such a mutant or overexpressed wild-type receptor and that certain other antagonists did not or only weakly reduced basal activity. Pindolol, labetolol and timolol are all efficient antagonists in a classical competitive manner of adrenergic β_2 activation. Timolol however has a very high capacity of reducing spontaneous activity in overexpressed β_2-receptor preparations whereas other classical β_2-antagonists like pindolol and labetolol show only very weak inhibitory properties against this spontaneous activity [8]. On the other hand, labetolol and pindolol are also able in a competitive manner to block the inverse agonistic activity of timolol in this system.

The conventional classification of full agonists, partial agonists and so-called pure antagonists by their intrinsic activity (α_E=1 for agonists, 1> α_E >0 for partial agonists, and α_E=0 for pure antagonists) has therefore to be revised: Full agonists (α_E=1), partial agonists (1> α_E >0), neutral antagonists (α_E=0) (e.g. pindolol), partial inverse agonists (0> α_E >-1) and full inverse agonists (α_E=-1) (e.g. timolol).

According to these experimental findings, not only theoretical pharmacology will need profound reconsideration but so will therapeutics. Several cases of diseases due to the constitutive activity of mutant receptors are now described. Somatic mutation on the thyrotropin receptor may lead to thyroid adenoma [9]. A single-point mutation in the seventh transmembrane element of rhodopsin has been linked to progressive degeneration of the

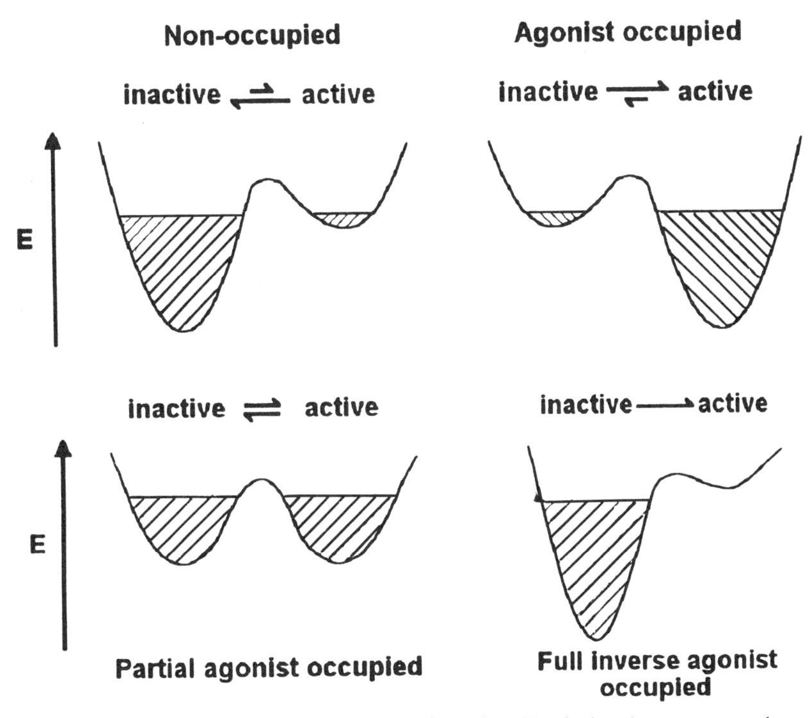

Figure 1. Theoretical energy landscapes of receptor conformation. Hatched regions represent the proportion of receptors under a particular state.

retina due to constitutive activity [4]. Also, a hereditary form of precocious puberty has been traced to a mutated form of the LH receptor with constitutive activity [10]. The list of diseases linked to constitutive activity is very likely to increase dramatically in the near future.

In this context, it should also to be brought to mind that some oncogenes have a receptor-like heptahelical structure [11], no known ligand but a strong intrinsic (or constitutive) activity. It is therefore very tempting to speculate that constitutive activity, either from somatic mutations or through transient overexpression in a particular physiological situation, may be a major physiological process involved in a variety of diseases, and especially in tumorigenesis. Research on inverse agonists is therefore a very promising avenue in the development of new therapeutic strategies.

From a mechanistic point of view, constitutive receptor activity is the result of an equilibrium of two receptor conformations. In the non-occupied state a predominant but not exclusive non-activiting conformation is in equilibrium with a small population of the activated form (upper left, figure 1). A trivial explanation on why constitutive activity has not been discovered earlier is probably that it was too small to be seen, for example, in smooth muscle pharmacology. It may also simply have been overlooked in other systems (e.g. secretory systems): Seemingly, nobody bothered to measure aldosterone production from adrenocortical cells in the presence of saralasin only in order to compare it to the basal activity since it was against the theory to observe anything to begin with!

The binding of an agonistic ligand shifts the equilibrium of the previously unoccupied receptor from the mostly inactive form to the active conformer allowing for successful G-protein coupling and hence, biological response. The higher the intrinsic activity of the agonist, the more complete is the equilibrium shift towards the active form. Agonists through their interaction with the receptor stabilize the active conformer and thus lowering the minimum energy of this conformation (upper right, figure 1). Partial agonists, for example of 50% intrinsic activity, would therefore produce an equilibrium in which the inactive and the active forms have the same minimum energy and the same receptor population (lower left, figure 1). Consequently, only a half-maximal response would be observed even under receptor saturating conditions assuming a linear relationship between receptor activation and biological response. On the other hand, a full inverse agonist would shift the equilibrium almost exclusively to the left, thus favoring solely the non-active form (lower right, fig.1).

On the angiotensin II (Ang) system, research on constitutive activity has yet to be produced. The physiological significance as well as the biological pathways of the mammalian AT_1 receptor are quite well characterized [12,13,14]. On the other hand, the second type of mammalian Ang receptors, AT_2, has only recently been cloned, and its biological significance is still rather controversial [14,15,16,17]. Other mammalian Ang receptor types have been suggested recently but their roles and functions are even far more sketchy [18,19,20]. So far, the non-mammalian Ang receptors have received very little attention [21,22,23].

Ang antagonists are among the longest-known peptide hormone antagonists. From a therapeutic point of view, they were however frustrating due to their very short half-lives and partial pressor activity in animal models of human hypertension. In the mean time, the clinically approved non-peptidic Ang antagonist, Losartan (Dup 753), has been developped. It is a highly selective AT_1 antagonist together with many other compounds of similar structure and pharmacology [14]. Although much effort was made to show congruency of parts of the Ang molecule with functional elements of these non-peptidic antagonists in order to rationalize receptor recognition [14], it has become clear in the light of site-directed mutagenesis studies that the binding loci of those non-peptidic antagonists are not the same as that of the Ang molecule [14,24,25].

All non-peptidic ligands of the AT_1 and of other peptidic receptors are of antagonist nature. Even a recently disclosed non-peptide analogue with apparently Ang related pressor activity [26] still has predominant antagonist properties [27]. The present study was therefore undertaken in order to identify the structural elements of peptidic Ang antagonists that lead to reduction of intrinsic activity and to select the most appropriate analogue for the search in inverse agonism. It was argued that inverse agonists should have close structural similarity to the agonists in order to have more or less the same binding locus and therefore the possibility to undergo the same or similar conformational changes once within the receptor. In the present work hypothesis, non-peptidic antagonists were assumed to be poor candidates due to their different binding locus and the high probability that they act only as access-deniers to the agonist Ang (neutral antagonists).

MATERIALS AND METHODS

Materials: Bovine adrenal glands were obtained from a nearby slaughterhouse. Ang II was from Sigma (St. Louis, MO). [^{125}I]Ang (2000 Ci/mmol) was prepared with iodogen (Pierce, Rockford, IL) as described by Fraker and Speck [28]. The product was purified to apparent homogeneity by HPLC (reverse-phase C-18), and the specific radioactivity was determined by self-displacement in the binding system.

Peptide synthesis: The peptides used in these experiments were synthesized in our laboratory (unless indicated otherwise) using solid phase synthesis as described by Merrifield [29]. Synthesis of the modified amino acids incorporated into the Ang analogues at position 8 was described in the following references: (3′, 5′-Br_2-4′-Cl) Phenylalanine (*m,m′*Br_2-*p*Cl-Phe), mono- (*pX*-Phe), penta-halogenated Phenylalanine (X_5-Phe), and D-Phenylalanine [30]; (4′-S-acetamidomethyl) mercapto-Phenylalanine (PheSAcm) [31]; Carboranylalanine (Car) [32]; Cymanthrenylalanine (Cym) and Ferrocenylalanine (Fer) [33].

Preparation of membranes: Bovine adrenal cortexes (dissected free of medullary tissue) were homogenized with eight strokes of a Dounce homogenizer (loose pestle) in a medium containing 110 mM KCl, 10 mM NaCl, 2 mM $MgCl_2$, 25 mM Tris-HCl pH 7.2, 5 mM KH_2PO_4, 1 mM dithiothreitol and 2 mM EGTA. After centrifugation at 500 x g for 15 min, the supernatant was centrifuged at 35,000 x g for 20 min. The pellet was resuspended in the same medium without EGTA supplemented with glycerol (14% v/v) and sorbitol (1.4% w/v), at a concentration of 10-15 mg of protein/ml. The membranes were stored at -70°C until used for binding studies.

Ang competition studies: Ang binding studies were performed in the presence of 25 mM Tris-HCl (pH 7.4), 100 mM NaCl, 5 mM $MgCl_2$, and 0.1% bovine serum albumin (BSA). PD123319 (5 μM) was also added to the medium in order to inhibit AT_2 binding. Membranes (200 μg/ml of protein) were incubated for 45 min at room temperature. Binding of [^{125}I]Ang (0.05 nM) was challenged by increasing concentrations of Ang or Ang analogues. Bound radioactivity was separeted from free ligand by filtration through A/E fiber filters presoaked in the binding buffer. The non-specific binding was mesured in the presence of 1 μM Ang. The receptor-bound radioactivity was analyzed by gamma-counting.

Table 1. Biological properties of Angiotensin analogues

Pos.8	D-xxx (pK_D)	pA_2	pD_2	α_E	L-xxx (pK_D)	pA_2	pD_2	α_E
Phe	8.65	8.18	---	0.00	9.27	---	8.67	1.00
Trp	8.77	8.9[35]	---	0.00[35]	8.52	8.36[35]	8.23	0.22
ß-Nal	9.69	8.08	---	0.00	8.49	8.27	7.71	0.34
Pyr	8.80	7.75	---	0.00	8.80	7.75	---	0.00
Ind[34](*)	8.82	8.92[(a)]	---	≈0	8.82	8.92[(a)]	---	≈0
Bip[34]	9.13	8.82[(a)]	---	≈0	8.76	8.66[(a)]	---	≈0
Cym	8.54	7.96	---	0.00	8.44	7.96	---	0.00
Fer	6.79	7.13	---	0.00	6.79	7.13	---	0.00
$TyrOC_{18}$	NA	NA	NA	NA	6.13	inactive	inactive	0.00
Car	NA	NA	NA	NA	7.13	6.12	6.31	0.15
PheSAcm	NA	NA	NA	NA	8.08	8.09	---	0.2
Ala	NA	NA	NA	NA	9.44	8.6[35]	---	0.00
Met	NA	NA	NA	NA	9.16	8.50	---	0.00
Ile	NA	NA	NA	NA	7.38	9.1[35]	---	0.00
Leu	8.09	8.11	---	0.00	9.39	8.25	---	0.00
Gly(*)	8.41	8.56	---	0.00	8.41	8.56	---	0.00
Lys	NA	NA	NA	NA	6.64	6.23	---	0.00
*p*Cl-Phe	8.55	8.49	---	0.00	NA	NA	NA	NA
*p*Br-Phe	8.69	7.12	8.35	0.14	8.85	---	8.34	0.88
*p*I-Phe	8.37	7.27	---	0.00	9.24	---	7.85	0.90
pN_3-Phe	8.51	9.0	---	0.00	9.00	---	8.72	0.95
pNO_2-Phe	7.63	6.77	---	0.00	8.69	9.43	9.07	0.46
m,m′Br_2*p*Cl-Phe	NA	NA	NA	NA	9.12	---	7.58	0.47
F_5-Phe	8.67	7.25	---	0.00	8.45	7.91	7.08	0.10
Cl_5-Phe	8.62	7.96	---	0.00	8.62	7.92	---	0.00
Br_5-Phe	8.72	7.78	---	0.00	8.72	7.83	---	0.00
Me_5-Phe	9.05	7.63	---	0.00	9.05	7.63	---	0.00

(*) indicates an achiral compound; (a) extrapolated from Hsieh *et al* [35]. pK_D values were determined on bovine surrenal cortex. pA_2, pD_2 and α_E were determined on rabbit aorta. Pyr, Fer and Me_5-Phe are racemic mixtures. *Abbreviations:* ß-Naphtalanine (ß-Nal); Pyrenylalanine (Pyr); Biphenylalaninine (Bip); 2-aminoindan-2-carboxylic acid (Ind); methyl group (Me); not available (NA).

Biological assay on rabbit aorta strips: New Zealand rabbits of either sex (1.5-2.5 kg) were killed by stunning and carotid exsanguination. The thoracic aorta was rapidly removed after a full-length thoracic-laparotomy and immersed in oxygenated Krebs solution. The excised aorta was mounted onto a glass rod, adipose tissue removed, then cut into 5 mm rings. The tissues were suspended in 5 ml organ baths containing continously oxygenated Krebs solution, changed every 15 min and maintained at 37°C. An initial tension of 2 g was applied and repeatedly readjusted until stabilised (averaging a 90 min incubation period). Tissue contractions were measured isometrically using force transducers (Grass FT03) and recorded in a Grass polygraph (Grass, Quincy, Mass).

RESULTS

The present study was carried out to compare peptide analogues in their affinity and in their ability to elicit biological response. The first parameter was determined by binding experiments on bovine adrenocortical membranes in presence of the AT_2-selective ligand PD123319 in order to measure the affinities to AT_1 only. The intrinsic activity (efficacy) was measured on rabbit aorta, a typical and classical functional assay of the AT_1 receptor. This assay was preferred to other more recent assays, e.g. IP_3 formation or cAMP inhibition since some analogues have been prepared and measured many years ago and are no more available, or they were results from other studies. The numerical results are presented in the table.

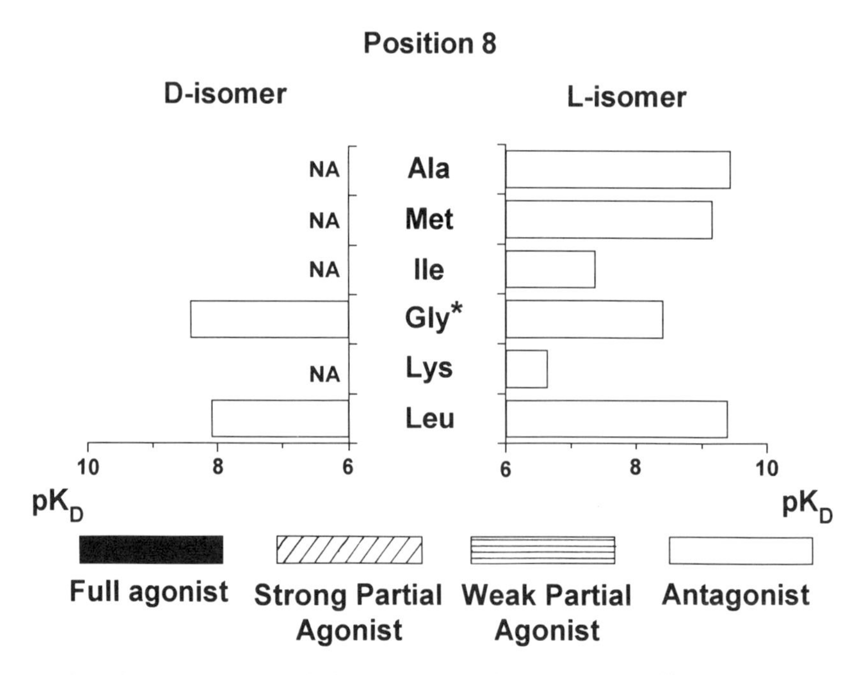

Figure 2. Effect of L- and D- isomer of aliphatic amino acids residus on the affinity and relative intrinsic activity of angiotensin analogues substituted in position 8.

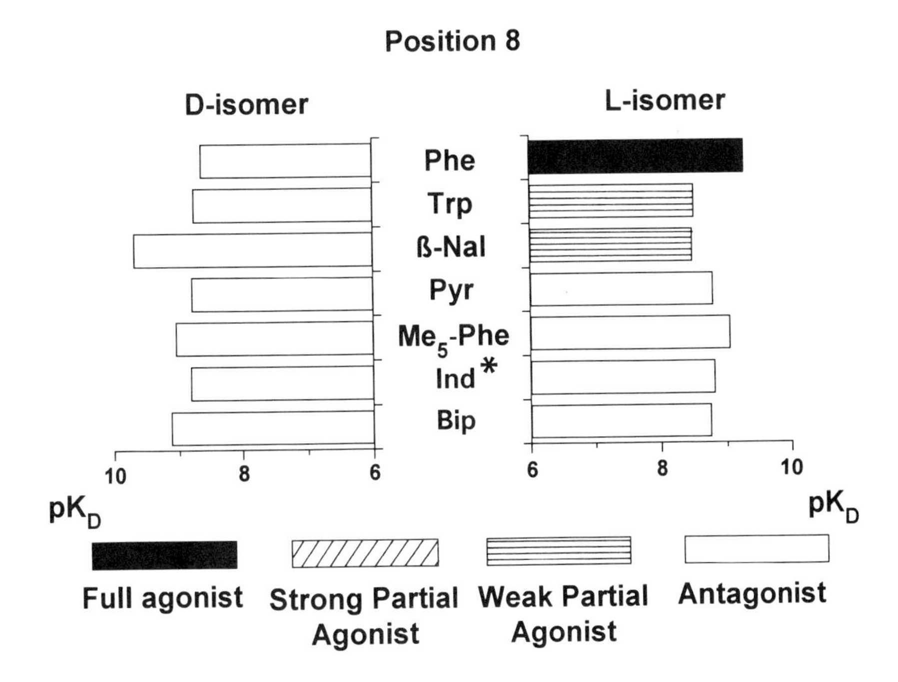

Figure 3. Effect of L- and D- isomer of aromatic amino acids residus on the affinity and relative intrinsic activity of angiotensin analogues substituted in position 8.

DISCUSSION

Rabbit aorta has been used among other functional bioassays in the past. There are several reasons why this bioassay was preferred over others, for example guinea pig ileum or rat uterus: Rabbit aorta develops quite stable contractions, Ang induced desensitization is slow and therefore the tissue allows experiments with cumulative concentration increases. This feature is extremely important since intrinsic activity could be explored with high precision and reproductibility. It also permitted to carry out cumulative concentration-response experiments for antagonist experiments, a feature particularily important in the present study since a great number of antagonists or predominantly antagonistic partial agonists had to be assayed. Assays on recombinant systems expressing the human receptor would have been preferable, but meaningful comparison to earlier results and those from the literature would have been impossible. It is however clear that future bioassays and especially those for measuring eventual inverse agonism have to be carried out on such systems.

Several groups of compounds have been synthesized and compared amongst themselves concerning affinity and efficacy. The standard was always Ang with sarcosine in position 1 ([Sar^1]Ang) instead of the natural Asp, and the permutation was in position 8. In the first group (fig. 2) are compared analogues with aliphatic side-chains in position 8, begininng with Gly, Ala, Leu, Ile, Met and Lys. Since only one peptide was produced with a D-amino acid in this position (gly beeing achiral), a meaningful L- vs D- diastereomer

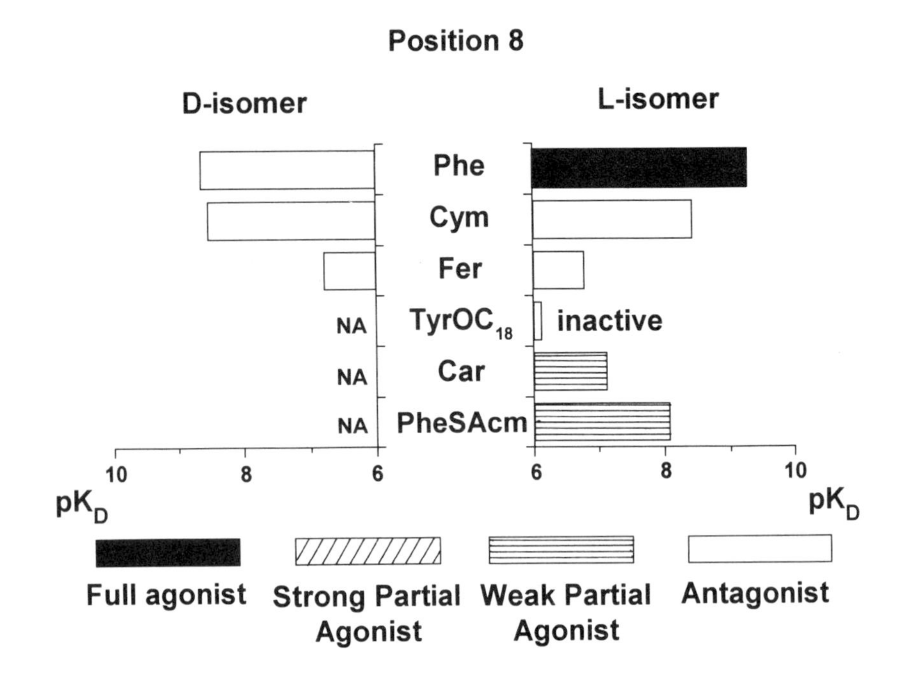

Figure 4. Effect of L- and D- isomer of complex aromatic amino acids residus on the affinity and relative intrinsic activity of angiotensin analogues substituted in position 8.

comparision has to be restricted to Leu. Besides the fact that all those analogues are pure antagonists in this bioassay, it is surprising that [Sar^1, Leu^8]Ang has practically a 20-fold higher affinity compared to its diastereomer [Sar^1, D-Leu^8]Ang. The affinity of [Sar^1, D-Leu^8]Ang is even lower than that of [Sar^1, Gly^8]Ang, as if the aliphatic side-chain in the D- configuration does not contribute to the total analogue binding energy if not in a slightly negative fashion. Two other compounds behave similarily, the parent molecule [Sar^1]Ang vs [Sar^1, D-Phe^8]Ang and its nitro derivative. In both cases a significantly but smaller affinity is observed for [Sar^1, D-Phe^8]Ang and [Sar^1, D-NO_2Phe^8]Ang if compared to their respective L- diastereomer. All non β-branched L-amino acids induce approximately the same affinity except Lys whose ε-amino-function probably perturbates the hydrophobic interior of the AT_1 receptor and thus reduces affinity by almost three orders of magnitude.

In the second set of compounds (fig 3), again position 8 diastereomers are compared but all have aromatic hydrocarbon side-chains, with the exception of Trp wich has an aromatic indole moiety. It is striking that even a minor extension of the aromatic moiety immediately reduces intrinsic activity (Trp and β-Nal, the latter is a synthetic non-natural amino acid with a β-naphtyl side-chain). On the other hand, affinity is not particularily influenced, even if quite large aromatic residues are introduced, as in the case of Pyr (pyrene-moiety, a -$C_{16}H_9$, a fused four-ring aromatic hydrocarbon) or pentamethylated Phe (Me_5Phe) where all ring protons are exchanged for methyl groups. Both residues have comparable ring diameters and identical affinities between the diastereomers. The observed high affinity of [Sar^1, D-βNal^8]Ang is however difficult to explain.

In the following set, extreme alterations of the aromatic side-chain were performed (fig 4). Introduction of a metallocenic moiety increased the "bulkiness" of a side-chain perpendicular to the aromatic nucleus. Cym (cymanthrenylalanine) is a manganese-tricarbonyl adduct to a cyclopentadienyl side-chain, whereas Fer stands for ferrocenylalanine, the classical ferrocene moiety (bis-cyclo-pentadienyl iron) in the side-chain. Cym maintains reasonable affinity but the even bulkier Fer indudes affinity loss close to two orders of magnitude.

A further extreme is the highly hydrophobic carboranyl-moiety in Car. This residue is $C_2B_{10}H_{11}$, a spherical pseudoaromatic [36]. The molecular diameter is that of a benzene ring, and its space requirements are those of a phenyl spinning around its 1-4 axis. The weak affinity of [Sar[1], Car[8]]Ang together with that of [Sar[1], Fer[8]]Ang shows that the Ang binding pocket has difficulties to accomodate such bulky structures that do not have the relative planarity of simple aromatic systems. It is also noteworthy that [Sar[1], Car[8]]Ang still has some partial agonistic properties wich underlines the relative resemblance of the carboranyl moiety with the simple phenyl system of phenylalanine. The next two analogues that have a substitution in the para-position of phenylalanine have a totally different hydrophobic behaviour. The first is an octadecanoyl ether of tyrosine ($TyrOC_{18}$), therefore lipid-like and extremely bulky. The second is acetamidomethyl substituted *p*-mercaptophenylalanine (PheSAcm), thus quite hydrophilic but less than the afore-treated lysine side-chain. The relative affinity of [Sar[1], PheSAcm[8]]Ang is tenfold below [Sar[1]]Ang but more than tenfold above

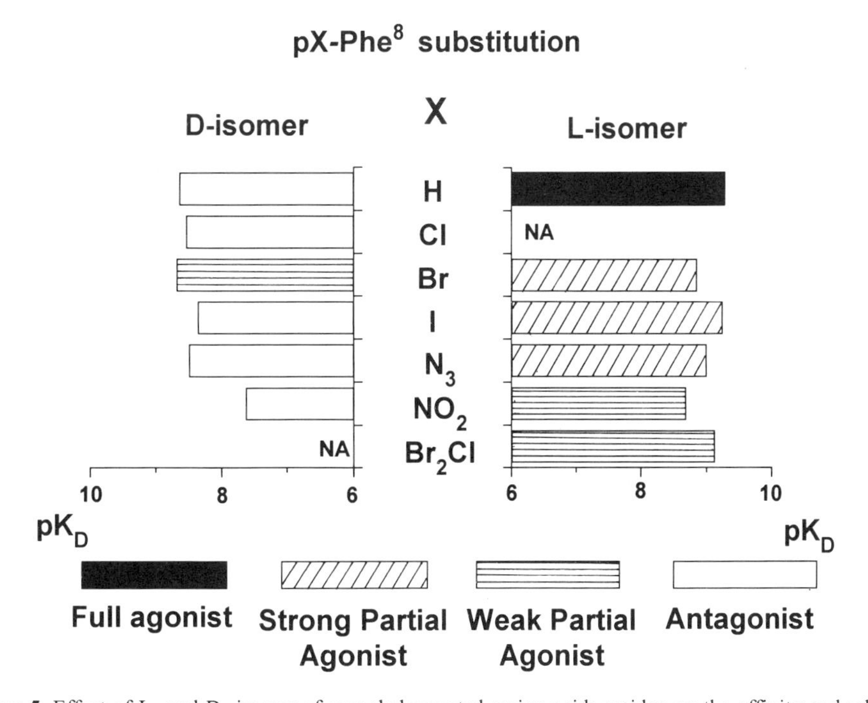

Figure 5. Effect of L- and D- isomer of monohalogenated amino acids residus on the affinity and relative intrinsic activity of angiotensin analogues substituted in position 8.

[Sar[1], Lys[8]]Ang and the partial agonism (α_E=0.2) of [Sar[1], PheSAcm[8]]Ang is therefore in good accordance with the other observed activity trends. [Sar[1], TyrOC$_{18}$]Ang is totally inactive, a not suprising feature considering the enormous space requirement necessary to accomodation such a large prosthetic group. The partitioning behaviour is probably placing this peptide into a membrane environment.

In figure 5, more subtle modifications affect the aromatic nucleus in position 8. None of the applied modifications increases the "thickness" of the aromatic residue and all substituents are in the para-position. The electronegativety of the substituents is in the following order: $Br_2Cl > NO_2 >> Br \approx Cl > I > N_3 > H$ [37]. Nevertheless, the space requirements are different: $Br_2Cl > I > N_3 > Br > NO_2 > Cl >> H$. The potency order of the L- series is however $H > I > Br_2Cl > Br > NO_2 > N_3$, while that of the D- series is $Br > H > Cl > N_3 > I > NO_2$. Especially in the case of the D- series, the differences are quite small and the affinities may be considered identical for H, Cl, Br, I and N_3 analogues due to the inherent variability of such affinity measurements wich are generally in the range of 0.2 log units. Some may need however more detailed analysis in the L- series: intrinsic activity is progressively affected, the two monohalogenated analogues (Br, I) and the pseudohalogen analogue N_3 have slightly reduced efficacy. The NO_2 and the tri-halogenated (Br_2Cl) analogue have already below half maximal intrinsic activity or efficacy. There is a clear trend to antagonism without affecting affinity if the phenyl ring in position 8 of the analogue is substituted also sidewards in the meta-position or by a branched substituent (NO_2 and SAcm). This is in good accordance with the polyaromatic analogue series (fig.3) where the

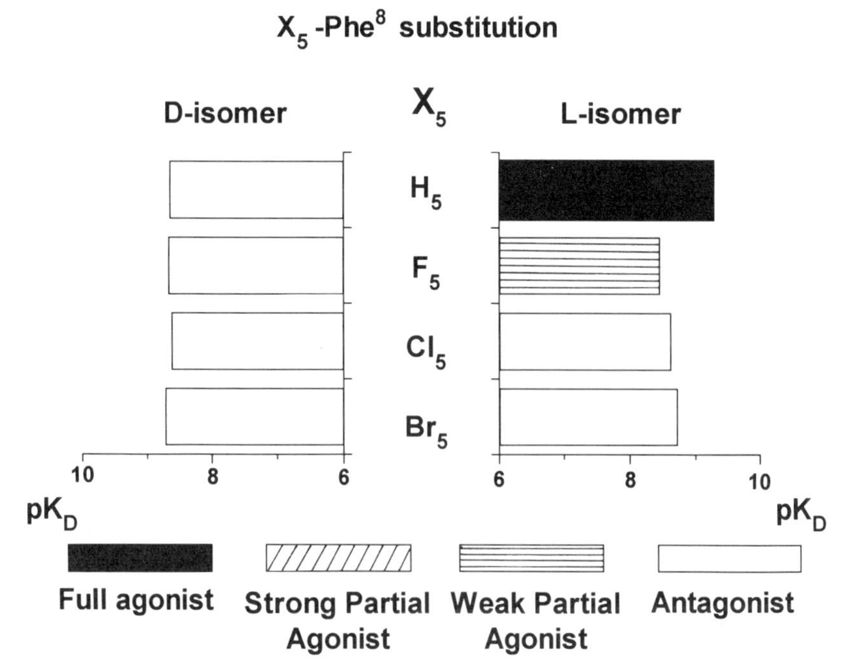

Figure 6. Effect of L- and D- isomer of polyhalogenated amino acids residus on the affinity and relative intrinsic activity of angiotensin analogues substituted in position 8.

Trp and ß-Nal analogues have only residual efficacy. The even larger Pyr, together with the penta-methylated Phe (Me_5-Phe), are pure antagonists analogues in the rabbit aorta assay. This feature is further explored in the pentahalogenated series found in figure 6. Another interesting feature is the difference of the L- or D- diastereoisomer, with exception of the D-BrPhe analogue that has some totally unexpected weak partial agonistic effect, all D-analogues have three to tenfold lower affinity than the corresponding L- diastereoisomer. This difference is disappearing if even larger aromatic residues are involved.

In the last group of analogues (fig.6), total substitution of the phenyl hydrogens in position 8 by halogens was explored. This penta-substitution, together with Me_5Phe analogue from figure 3 show a clear picture of a congruent affinity parallelism in both the L- and the D- series and conversion to pure antagonism except for the L-F_5Phe analogue that maintains some weak partial agonism. Since the diameters of fluorine and hydrogen are similar, fluorine being somewhat larger, the agonist-antagonist transition seems to rest in the subtle ring-enlargement induced by substitution with the halogens from the three first periods of the periodic system.

The most pronounced antagonists are therefore the analogues that display: a) affinity similarity between L- and D- diastereoisomer, b) contain substantial sidewards ring enlargement but also, c) maintain the normal aromatic thickness and planarity. The analogues that display those three properties in the most pronounced fashion are therefore the three penta-substituted analogues [Sar^1, Me_5Phe^8]Ang, [Sar^1, Cl_5Phe^8]Ang and [Sar^1, Br_5Phe^8]Ang together with the polyaromatic analogue [Sar^1, Pyr^8]Ang. Studies considering also duratrion of action place the two latter antagonists in a particular class because both analogues display unusually long duration of action, both *in vivo* and *in vitro* [38,39].

Studies are actually underway to screen several classical and new bioassays for constitutive AT_1 activity and eventual inverse agonism of the above mentioned peptide analogues.

ACKNOWLEDGMENTS

This study had been supported by grants to E.E. from the Canadian Medical Research Council, the Quebec Heart and Stroke Foundation and the Fonds pour la Recherche en Santé du Québec (FRSQ). R.L. benefits from a scholarship from the FRSQ. This work is part of the doctoral thesis of J.P., R.B., and part of the M.Sc. work of S.G.

REFERENCES

1. Kjelsberg, M.A., Cotecchia, S., Ostrowski, J., Caron, M.G. and Lefkowitz, R. J., Constitutive Activation of the α_{1B}-Adrenergic Receptor by All Amino Acid Substitution at a Single Site, J. Biol. Chem. 267 (1992) 1430-1433.
2. Allen, L.F., Lefkowitz, R.J., Caron, M. and Cotecchia, S., G-protein-coupled receptor genes as protooncogenes: Constitutively activating mutation of the α_{1B}-adrenergic receptor enhances mitogenesis and tumorigenecity, Proc. Natl. Acad. Sci. USA 88 (1991) 11354-11358.
3. Samama, P., Cotecchia, S., Costa, T. and Lefkowitz, R.J., A Mutation-induced Activated State of the $ß_2$-Adrenergic Receptor, J. Biol. Chem. 268 (1993) 4625-4636.
4. Robinson, P.R., Cohen, G.B. Zhukovsky, E.A. and Oprian, D.D., Constitutively active mutants of rhodopsin, Neuron 9 (1992) 719-725.
5. Schutz, W. and Freissmuth, M., Reverse intrinsic activity of antagonists on G protein-coupled receptors, Trends Pharmacol. Sci. 13 (1992) 376-380.
6. Costa, T., Ogino, Y., Munson, P.J., Onaran, H.O. and Rodbard, D., Drug efficacy at guanine nucleotide-binding regulatory protein-linked receptors: Thermodynamic interpretation of negative antagonism and receptor activity in the absence of ligand, Mol. Pharmacol. 41 (1992) 549-560.

7. Barker, E.L., Westphal, R.S. Schmidt, D. and Sanders-Bush, E., Constitutively active 5′-hydroxytryptamine 2C receptors reveal novel inverse agonist activity of receptor ligands, J. Biol. Chem. 269 (1994) 11687-11690.
8. Chidiac, P., Hebert, T., Valiquette, M., Dennis, M. and Bouvier, M., Inverse agonist activity of ß-adrenergic antagonists, Mol. Pharmacol. 45 (1994) 490-499.
9. Parma, J., Duprez, L., Van Sade, J., Cochaux, P., Gervy, P., Mockel, J., Dumont, J. and Vassart, G., Somatic mutations in the thyrotropin receptor gene cause hyperfunctioning thyroid adenomas, Nature 365 (1993) 649-651.
10. Shenker, A., Laue, L., Kosug, S., Merendino, J.J., Minegishi, T. and Cutler, J.B., A constitutively activating mutation of the luteinizing hormone receptor in familial male precocious puberty, Nature 365 (1993) 652-654.
11. Young, D., Waitches, G., Birchmeier, C., Fasano, O. and Wigler, M., Isolation and characterization of a new cellular oncogene encoding a protein with multiple potential transmembrane domains, Cell 45 (1986) 711-719.
12. Regoli, D., Park, W.K. and Rioux, F., Pharmacology of Angiotensin, Pharmacological Reviews 26 (1974) 69-123.
13. Griendling, K.K., Lassèque, B. and Alexander, R.W., The Vascular Angiotensin (AT_1) Receptor, Thrombosis and Haemostasis 70 (1993) 188-192.
14. Timmermans, P.B.M.W.M., Wong, P.C., Chiu, A.T., Herblin, W.F., Benfield, P., Carini, D.J., Lee, R.J., Wexler, R.R., Saye, J.A.M. and Smith, R.D., Angiotensin II Receptors and Angiotensin II Receptor Antagonists, Pharmacological Reviews 45 (1993) 205-247.
15. Dudley, D.T., Panek, R.L., Major, T.C., Lu, G.H., Burns, R.F., Klinkefus, B.A., Hodges, J.C. and Weishaar, R.E., Subclasses of angiotensin II binding sites and their functional significance, Mol. Pharmacol. 38 (1990) 370-377.
16. Bottari, S.P., King, I.N., Reichlin, S., Dahlstroem, I., Lydon, N. and DeGasparo, M., The angiotensin AT_2 receptor stimulates protein tyrosin phosphatase activity and mediates inhibition of particulate guanylate cyclase, Biochem. Biophys. Res. Commun. 183 (1992) 206-211.
17. Kambayashi,Y., Bardhan, S., Takahashi, K., Tsuzuki, S. Inui, H., Hamakubo and Inagami, T., Molecular cloning of a novel angiotensin II receptor isoform involved in phosphotyrosine phosphatase inhibition, J. Biol. Chem. 268 (1993) 24543-24546.
18. Bernier, S.G., Fournier, A. and Guillemette, G., A specific binding site recognizing a fragment of angiotensin II in bovine adrenal cortex membranes, Eur. J. Pharmacol. 271 (1994) 55-63.
19. Siemens, I.R., Reagan, L.P., Yee, D.K. and Fluharty, S.J., Biochemical characterizaton of two distinct angiotensin AT_2 receptor populations in murine neuroblastoma N1E-115 cells, J. Neurochem. 62 (1994) 2106-2115.
20. Servant, G., Escher, E. and Guillemette, G., Non-AT_1 and non-AT_2 binding sites observed in PC12 cells after confluency, (to be published).
21. Murphy, T.J., Nakamura, Y., Takeuchi, K. and Alexander, R.W., A cloned angiotensin receptor isoform from the turkey adrenal gland is pharmacologically distinct from mammalian angiotensin receptors, Mol. Pharmacol. 44 (1993) 1-7.
22. Ji, H., Sanberg, K., Zhang, Y., Catt, K.J., Molecualr cloning, sequencing and functional expression of an amphibian angiotensin II receptor, Biochem. Biophys. Res. Commun. 194 (1993) 756-762.
23. Nishimura, H., Walker, O.E., Patton, C.M., Madison, A.T., Chiu, A.T. and Keiser, J., Novel angiotensin receptor subtypes in fowl, Am. J. Physiol. 267 (1994) R1174-R1181.
24. Wong, P.C., Hart, S.D., Chiu, A.T., Herblin, W.F. Carini, D.J., Smith, R.D., Wexler, R.R. and Timmermans, P.B.M.W.M., Pharmacology of DuP 532, a Selective and Noncompetitive AT_1 Receptor Antagonist, J. Pharmacol. Experm. Therap. 259 (1991) 861-870.
25. Chang, R.S.L., Siegl, P.K.S., Clineschmidt, B.V., Mantlo, N.B., Chakravarty, P.K., Greenlee, W.J., Patchett, A.A. and Lotti, V.J., *In Vitro* Pharmacology of L-158,809, a New Highly Potent and Selective Angiotensin II Receptor Antagonist, J. Pharmacol. Experm. Therap. 262 (1992) 133-138.
26. Perlman, S. Schambye, H.T., Riviero, R.A., Greenlee, W.J., Hjorth, S.A. and Schwartz, T.W., Non-peptidic angiotensin agonist. Functional and molecular interaction with the AT_1 receptor, J. Biol. Chem. 270 (1995) 1493-1496.
27. Pérodin, J and Escher, E.(unpuplished results).
28. Fraker, P.J. and Speck, J.C., Protein and cell iodination with a sparingly soluble chloroamine, 1,3,4,6-tetrachloro-3a,6a-diphenylglycoloryl, Biochem. Biophys. Res. Commun. 80 (1978) 849-857.
29. Merrifield, R.B., Solid-phase peptide synthesis. I: The synthesis of a tetrapeptide, J. Am. Chem. Soc., 85 (1963) 2149-2154.

30. Leduc, R., Bernier, M. and Escher, E., Angiotensin-II Analogues. I: Synthesis and Incorporation of the Halogenated Amino Acids 3-(4′-Iodophenyl)alanine, 3-(3′,5′-Dibromo-4′-chlorophenyl)alanine, 3-(3′,4′,5′-Tribromophenyl)alanine, and 3-(2′,3′,4′,5′,6′-Pentabromophenyl)alanine, Helvetica Chimica acta 66 (1983) 960-970.
31. Escher, E., Bernier, M. and Parent, P., Angiotensin II Analogues. Part II. Synthesis and Incorporation of the Sulfur-Containing Aromatic Amino Acids: L-(4′-SH)Phe, L-(4′-SO_2NH_2)Phe, L-(4′-SO_3^-)Phe and L-(4′-S-CH_3)Phe, Helvetica Chimica acta 66 (1983) 1355-1365.
32. Quang, K.D., Thanei, P., Caviezel, M. and Schwyzer, R., The Synthesis of (S)-(+)-2-Amino-3-(1-adamantyl)-propionic Acid (L-(+)-Adamantylalanine, Ada) as a *"Fat"* or *"Super" Analogue of Leucine and Phenylalanine, Helvetica Chimica acta 62 (1979) 956-964.*
33. Tartar, A., Demarly, A., Sergheraert, C. and Escher, E., Metallocenic angiotensin II analogues, "Peptides '83" V. Hruby (Ed) Pierce Corp., Rockford, III, 1984, pp. 377-380.
34. Hsieh, K.H., LaHann, T.R. and Speth, R.C., Topographic Probes of Angiotensin and Receptor: Potent Angiotensin II Agonist Containing Diphenylalanine and Long-Acting Antagonists Containing Biphenylalanine and 2-Indan Amino Acid in Position 8, J. Med. Chem. 32 (1989) 898-903.
35. Samanen, J., Narindray, D., Adams Jr, W., Cash, T., Yellin, T. and Regoli, D., Effects of D-Amino Acid Substitution on Antagonist Activities of Angiotensin II Analogues, J. Med. Chem. 31 (1988) 510-516.
36. Leukart, O., Caviezel, M., Eberie, A., Escher, E., Tun-Kyi, A. and Schwyzer, R., L-o-Carboranyl, a Boron Analogue of Phenylalanine, Helvetica Chimica Acta 59 (1976) 2184-2187.
37. Hansch, C., Leo, A., Unger, S.H., Kim, K.H., Nikaitani, D. and Lien, E.J., "Aromatic" Substituent Constants for Structure-Activity Correlations, J. Med. Chem. 16 (1973) 1207-1216.
38. Holck, M., Bossé, R., Fischli, W., Gerold, H. and Escher, E., An Angiotensin II antagonist with strongly prolonged action, Biochem. Biophys. Res. Commun. 160 (1989) 1350-1356.
39. Bossé, R., Gerold, H., Fischli, W., Holck, M. and Escher, E., An angiotensin antagonist with prolonged action and antihypertensive properties, J. Cardiovasc. Pharmacol. 16 (1990) S50-S55.

15

MOLECULAR CLONING AND EXPRESSION OF ANGIOTENSIN II TYPE 2 RECEPTOR GENE

Toshihiro Ichiki, Yoshikazu Kambayashi, and Tadashi Inagami

Department of Biochemistry
Vanderbilt University School of Medicine
Nashville, Tennessee 37232

INTRODUCTION

Angiotensin II (Ang II) has been known as one of potent vasoactive peptides which also regulates fluid homeostasis, blood pressure, drinking behavior, aldosterone release, peripheral norepinephrine release, cell proliferation and hypertrophy (1-3). These diversity of specific physiological effects of Ang II has suggested possible presence of subtypes of Ang II receptor. While the recognition of reducing agent-sensitive and -insensitive Ang II receptors had suggested the possibility of multiple subtypes (4), the question remained unresolved until the development of new peptide and non-peptidic antagonists of the Ang II receptor (5-7). These compounds turned out to be isoform specific and powerful tools for the characterization and isolation of the Ang II receptors and receptor genes.

Another breakthrough in the attempt at resolving the problem of the multiplicity of the Ang II receptor came from the application of the expression cloning method which does not require the purification of a receptor protein. Combination of the expression cloning and isoform specific antagonists have allowed us to clone two isoforms of Ang II receptor designated as type 1 (AT_1) (8, 9) and type 2 (AT_2) (10, 11) receptors.

CLONING OF THE AT_2 RECEPTOR GENE

Kambayashi et al. (10) and Mukoyama et al. (11) used the expression cloning method to isolate the rat AT_2 receptor cDNA. mRNA from PC12W cells, which is a rat pheochromocytoma cell line, and rat fetus which also expresses the AT_2 receptor at a high level in mesenchymal tissue, were reverse transcribed and the resultant cDNA was cloned into the pcDNA I expression vector. COS-7 cells were transfected with this cDNA library. Transfected cells were selected by their binding to ^{125}I-labeled [Sar^1 , Ile^8]Ang II (Sarile) or CGP42112A which is a peptidic ligand specific to the AT_2 receptor. Selection of positive clone was repeated until a single clone which encodes the AT_2 receptor was obtained.

Recent Advances in Cellular and Molecular Aspects of Angiotensin Receptors
Edited by Mohan K. Raizada et al., Plenum Press, New York, 1996

The AT_2 receptor, stably expressed in COS-7 cells using another expression vector, pCR/CMV, showed high-affinity binding to [^{125}I]-Sarile and -CGP42112A with a dissociation constants of 0.23 and 0.11 nmol/L, respectively. The AT_2 receptor specific antagonist PD123319 competed effectively with [^{125}I]-Ang II with a IC_{50} of 1.7 nmol/L.

We (12) and Nakajima et al. (13) cloned mouse AT_2 cDNA from a mouse fetus cDNA library by means of a conventional plaque hybridization method and rat AT_2 cDNA as a probe. We then cloned genomic DNA of the mouse AT_2 gene (14). Nucleotide sequence of the 4.5 Kb Eco RI segment of the genomic DNA of the mouse AT_2 showed the presence of three exons. An entire coding region was included in the third exon without an interruption by an intron. Based on the finding that the coding region of the mouse AT_2 gene is not interrupted by an intron, Tsuzuki et al. (15) screened a human genomic DNA library with a rat AT_2 cDNA probe and cloned human AT_2 gene. Recently human AT_2 cDNA was cloned from a cDNA library made from human lung (16).

Although two closely related subtypes of the AT1 receptor designated as AT_{1a} and AT_{1b} encoded by a different gene were reported, southern blot analysis of genomic DNA of rat, mouse, and human suggest that a single AT_2 receptor gene is present in the genome.

Human AT_2 gene is localized to Xq22 in the X chromosome and rat AT_2 gene is localized to Xq34 (17).

RAT ANGIOTENSIN RECEPTOR AT_1 AND AT_2

```
AT1                       MALN  SSAEDGIKRI  QDDCPKAGRH  SYIFVMIPTL
AT2  MKDNFSFAAT  SRNITSSLPF  DNLNATGTNE  SAFNCSHKPA  DKHLEAIPVL   50

AT1  YSIIFVVGIF  GNSLVVIVIY  FYMKLKTVAS  VFLLNLALAD  LCFLLTLPLW
AT2  YYMIFVIGFA  VNIVVVSLFC  CQKGPKKVSS  IYIFNLAVAD  LLLLATLPLW  100
       TM-1                                               TM-2

AT1  AVYTAMEYRW  PFGNHLCKIA  SASVTFNLYA  SVFLLTCLSI  DRYLAIVHPM
AT2  ATYYSYRYDW  LFGPVMCKVF  GSFLTLNMFA  SIFFITCMSV  DRYQSVIYPF  150
                                 TM-3

AT1  KSRLRRTMLV  AKVTCIIIWL  MAGLASLPAV  IHRNVYFIEN  TNITVCAFHY
AT2  LSQRRNP-WQ  ASYVVPLVWC  MACLSSLPTF  YFRDVRTIEY  LGVNACIMAF  199
                    TM-4

AT1  ESRNSTLPIG  LGLT-KNILG  FLFPFLIILT  SYTLIWKALK  KAYEIQKNKP
AT2  PPEKYAQWSA  GIALMKNILG  FIIPLIFIAT  CYFGIRKHLL  KTNSYGKNRI  249
                        TM-5

AT1  RNDDIFRIIM  AIVLFFFFSW  VPHQIFTFLD  VLIQLGVIHD  CKISDIVDTA
AT2  TRDQVLKMAA  AVVLAFIICW  LPFHVLTFLD  ALTWMGIINS  CEVIAVIDLA  299
                       TM-6

AT1  MPITICIAYF  NNCLNPLFYG  FLGKKFKKYF  LQLLKYIPPK  AKSHSSLSTK
AT2  LPFAILLGFT  NSCVNPFLYC  FVGNRFQQKL  RSVFRVPITW  LQGKRETMSC  349
          TM-7

AT1  MSTLSYRPSD  NMSSSAKKPA  SCFEVE
AT2  RKSSSLREMD  TFVS                                          363
```

Figure 1. Amino acid sequences of rat angiotensin II AT_{1a} and AT_2 receptors deduced from the nucleotide sequence of the respective complementary DNAs. Sequence in shaded boxes are identical. Of the overall 32% sequence identity, the transmembrane regions show a somewhat higher degree of homology.

STRUCTURE OF THE AT_2 RECEPTOR

The deduced amino acid sequences encoded by the open reading frame of the AT_2 cDNA (rat (10, 11) and mouse (12, 13)) or the genomic DNA (human (15)) consist of 363 amino acid residues with an estimated molecular weight of 41 KDa. Rat AT_2 receptor has 32% amino acid sequence identity with rat AT_1 receptor (Fig. 1).

Hydropathy analysis of the amino acid sequences showed that the AT_2 receptor had seven putative transmembrane domains. The transmembrane regions show a somewhat higher degree of homology with the AT_1 than other domains. The Asp^{141}-Arg^{142}-Tyr^{143} sequence in the amino terminal region of the second cytosolic loop which are highly conserved among many seven-transmembrane type receptors is present. There are five potential N-linked glycosylation sites which are clustered exclusively in the amino terminal region of the receptor (Fig. 2).

There are two possible phosphorylation sites for protein kinase C (PKC) in the carboxyl-terminal tail. These residues are conserved in all species. Rat AT_2 receptor has 98.5% homology with mouse AT_2 receptor in amino acid sequence, human AT_2 receptor has 92% homology with rat AT_2 receptor. Differences of amino acid sequence between human and rodents AT_2 receptors are mainly found in the amino terminal region.

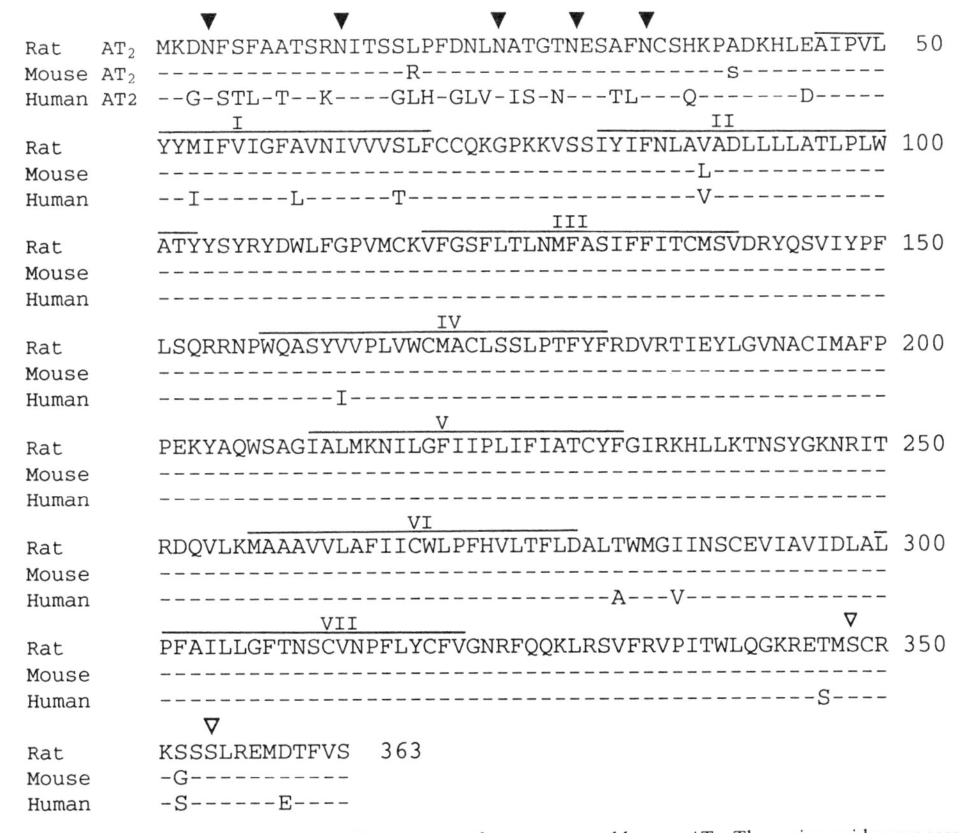

Figure 2. Comparison of the amino acid sequences of rat, mouse and human AT_2. The amino acid sequences of the AT_2 receptor deduced from the nucleotide sequences of complementary DNA (rat and mouse) and genomic DNA (Human) are shown. Putative transmembrane regions are indicated with lines (I-VII). The potential N-linked glycosylation sites are marked by filled triangles. Possible phosphorylation sites by protein kinase C are indicated by open triangles.

G-PROTEIN COUPLING AND SIGNALING OF THE AT_2 RECEPTOR

Although the AT_2 receptor seems to belong to the seven-transmembrane domain type receptor family, its ligand binding affinity is not affected by GTP-γS (18, 19) which is a nonhydrolyzable analog of GTP as other G-protein coupled receptors do. Furthermore GTP-γS did not affect the dissociation rate of Ang II from the AT_2 receptor. Tsutsumi et al. (20) described the presence of GTP-γS sensitive AT_2 receptor sites in ventral thalamic nucleus, medial geniculate nucleus and locus coeruleus of the brain. Binding of [^{125}I]-Ang II to these brain regions was pertussis toxin-sensitive suggesting the involvement of Gi- or Go-protein in the signaling pathway in these AT_2 receptor sites. Other reports examining the AT2 receptor sites in PC12W cells (18) or ovarian granulosa cells (19) failed to see the effect of GTP-γS. A few GTP-γS insensitive seven-transmembrane domain type receptors were reported which include somatostatin (SSTR1) (21) and dopamine (D3) (22) receptors. The AT_2 receptor and these receptors have common sequences in their third cytosolic loop which is generally believed to couple to G-protein (11). Therefore, the lack of the effect of GTP-γS on ligand-binding affinity does not necessarily indicate the absence of G-protein coupling. Rather, it may suggest that the AT_2 receptor belongs to a new family of seven-transmembrane domain type receptor.

Another interesting feature of the AT_2 receptor to note is the lack of agonist-induced internalization. Dudley et al. (23) and Pucell et al. (19) found that the AT_2 receptor sites were not decreased by Ang II in R3T3 cells and ovarian granulosa cells, respectively. Furthermore, the AT_2 receptor sites were increased by the agonist, Ang II or even by PD123319 or Sarile which are considered to be antagonists, a phenomenon which may be called homologous up-regulation (24).

It is generally agreed that signals that are mediated by the AT_2 receptor do not involve the change in cytosolic calcium, inositol 1, 4, 5, *tris*phosphate or cAMP (25, 26). Bottari et al. (27) observed that stimulation of the AT_2 receptor in PC12W cells activated phosphotyrosine phosphatase (PTPase) and reduced cGMP level. They observed the rapid dephosphorylation of tyrosine phosphrorylated protein in PC12W cells upon stimulation by Ang II. Whereas, we observed inhibition of PTPase activity in PC12W cells upon stimulation by Ang II (10). The PTPase activity was determined by using [^{32}P]-Raytide or p-nitrophenylphosphate as a substrate. The inhibition of PTPase by Ang II was abolished by preincubation of cells with pertussis toxin, which suggest the involvement of Gi-or Go-proteins in the signaling pathway of the AT_2 receptor. We failed to see any effect of Ang II on the cGMP level in PC12W cells. Another report using PC12W cells failed to see any effect of Ang II on tyrosine, serine or threonine phosphorylation (26).

In the neonatal rat hypothalamic neuronal cells, AT_2-mediated decrease in cGMP was reported by Sumners et al. (28) . Kang et al. reported that stimulation of the AT_2 receptor in neonatal hypothalamic cells increased the delayed rectifier potassium currents (29). They also found that the AT_2-mediated increase of potassium current was inhibited by Okadaic acid, an inhibitor of serine/threonine protein phosphatase but not by sodium orthovanadate, a PTPase inhibitor. The delayed rectifier potassium current activated by stimulation of the AT_2 receptor was inhibited by anti-type 2A protein phosphatase antibody or anti-Gi-protein antibody. Another report showed that the stimulation of the AT_2 receptor inhibited T-type calcium current in non-differentiated NG108-15 cells (30). Stimulation of the AT_2 receptor by Ang II or CGP42112A inhibited T-type calcium currents at membrane potential higher than -40 mV and shifted the current voltage curve at lower potentials. They also observed that hydrolysis of GTP took place when NG108-15 cells were stimulated with Ang II and the hydrolysis of GTP induced by Ang II was blocked by sodium orthovanadate which suggests the involvement of PTPase activation in the signaling pathway of the AT2 receptor. In these cells, the T-type calcium currents was not affected by microcystin-LR, another serine/threonine phosphates inhibitor. Lokuta et al. (31) examined an effect of Ang II on arachidonic acid release in cultured cardiac myocyte. Arachidonic acid release stimulated by Ang II was blocked by PD123319. Webb et al. (25) and Leung et al. (26), however, failed to see the release of arachidonic acid from PC12W cells upon stimulation by Ang II.

These results suggest that the AT_2 receptor may couple to different effector molecule in a tissue specific manner. Alternatively, multiple Ang II receptors which bind to PD123319 or CGP42112A are present. The result of the southern blot analysis, however, suggests the presence of a single AT_2 receptor gene in a genome. Studies using cross-linking experiments estimated the molecular weight of the AT_2 receptor at 100 KDa and 79 KDa in ovarian granulosa cells (19) and R3T3 cells (23), respectively, although the molecular weight deduced from the AT_2 cDNA is 43 KDa, And Siemens et al. (32) reported the presence of two populations of the AT_2 receptor in NIE-115 neuroblastoma cells. Both receptor populations had high affinity to CGP42112A and low affinity to losartan. Sensitivity to PD123319 and dithiothreitol was different. A difference in glycosylation may explain the apparent difference in the molecular weight of the AT_2 receptor, but it is difficult to believe that it affects the signaling mechanism. Another possibility is an alternative splicing of the AT_2 mRNA. In prostaglandin E receptor, alternative splicing of the 3′ area of the coding region results in translation products which couple to different G-protein (33). To date, however, no evidence of the alternative splicing of the AT_2 mRNA was reported.

BIOLOGICAL FUNCTION OF THE AT_2 RECEPTOR

No biological function for the AT_2 receptor has been established. Recently, two groups, however, showed that stimulation of the AT_2 receptor inhibit cell growth. Stoll et al. showed that the AT_2 receptor specific antagonist PD123177 increased the DNA synthesis in AT_1- and AT_2-expressing coronary endothelial cells stimulated by Ang II (34). This result may suggest that signals of the AT_2 receptor antagonize the mitogenic effects of Ang II via the AT1 receptor. Nakajima et al. introduced the AT_2 receptor gene driven by a heterologous promoter into injured artery (35). Stimulation of this transfected AT_2 receptor by Ang II suppressed neointimal formation and inhibition of neointimal formation was abolished by PD123319.

Several papers reported that the stimulation of the AT_2 receptor had some hemodynamic effects, although it is not yet confirmed. Naveri et al. reported that stimulation of the AT_2 receptor by CGP42112A increased cerebrovascular resistance (36). Whereas Scheuer et al. reported that the AT_2 receptor mediated a transient depressor effect of Ang II when rats were pretreated with losartan (37). This depressor effect of Ang II was blocked by PD123319. We have previously showed that PD123319 increased blood pressure when injected into cisterna magna (38). Taking the latter two reports together, the AT_2 receptor in central nervous system may play some role in reducing blood pressure.

EXPRESSION OF THE AT_2 RECEPTOR AND mRNA

The AT_2 receptor shows a unique tissue specific and ontogeny-dependent expression. In rat fetus, the AT_2 receptor is expressed highly in mesenchymal tissues (39) and certain brain nuclei such (40) as hypoglossal nucleus and oculomoter nucleus. Expression in these regions is rapidly shut off after birth. In adult rats, the AT_2 receptor is expressed in several brain nuclei (41), heart (42, 43), adrenal medulla (6) and myometrium (5). In human, mRNA

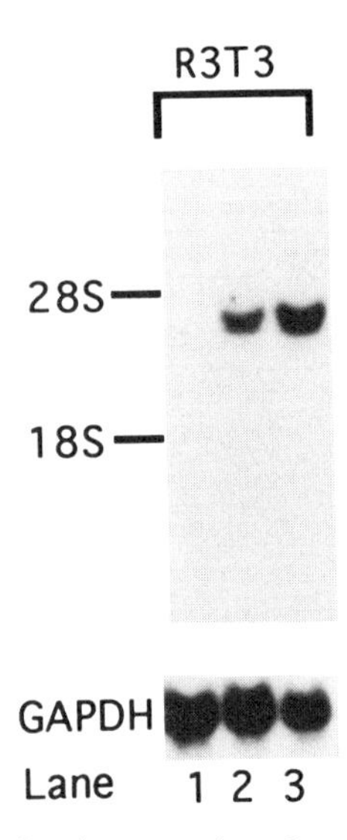

Figure 3. Northern blots showing the expression of mouse AT_2 mRNA in R3T3 cells.

of the AT_2 was detected in fetal lung and kidney by northern blot analysis (44). The AT_2 receptor sites in the heart was up-regulated in the hypertrophied left ventricle made by aortic banding, whereas the AT1 receptor sites were down-regulated (45). Nio et al. reported that AT_2 mRNA as well as AT1a mRNA was up-regulated in infarcted myocardium (46). De Gaspero et al. (47) and Cox et al. (48) showed that the AT_2 receptor sites were decreased during pregnancy in human and ovine uterus, respectively. These in vivo studies on the expression of the AT_2 receptor or mRNA may suggest that the Ang II plays some role in the development of fetus, reproduction and neuronal activity through the AT_2 receptor.

In in Vitro studies, R3T3 cells, a mouse fibroblast cell line, and PC12W cells, a rat pheochromocytoma cell line, were reported to express the AT_2 receptor sites but not AT1. The AT_2 receptor sites in PC12W cells were decrease by stimulation of nerve growth factor or dibutyryl cAMP (26). In R3T3 cells, the AT_2 receptor sites were increased after cells were confluent and cultured in the medium without serum (24). We have observed that AT_2 mRNA was increased after cells were confluent and serum were depleted (Fig. 3).

R3T3 cells were cultured in medium containing 10% fetal calf serum until cells were confluent (lane 1). Then serum was removed and cultured for additional 1 days (lane 2) and 2 days (lane 3). Total RNA was prepared by the acid guanidium- phenol-chloroform extraction method. Fifteen μg of total RNA was electrophoresed, blotted and hybridized with ^{32}P-labeled mouse AT_2 cDNA. Expression of the AT_2 mRNA was increased after cells were confluent and serum was removed.

The induction of the AT_2 mRNA seems to be parallel to the increase of the AT_2 receptor sites. Addition of serum or fibroblast growth factor to the medium reduced the AT_2 receptor sites in R3T3 cells. Dudley et al. also reported that AT_2 sites in R3T3 cells were increased when R3T3 cells were incubated with Ang II or antagonist such as PD123319 or Sarile (24).

To clarify the unique expression pattern and regulation of the AT_2 receptor, we have cloned the promoter region of the mouse AT_2 gene (49). About 1.5 Kb fragment including transcription initiation site of the mouse AT_2 gene was obtained from a mouse genomic DNA library and the nucleotide sequence was determined. Computer homology analysis of this

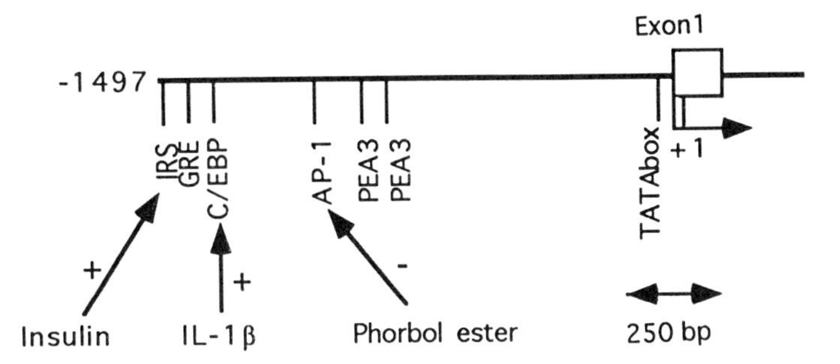

Figure 4. Molecular structure of the 5′ flanking region for mouse AT_2 gene.

segment revealed the presence of several potential cis DNA elements such as TATA box, CCAAT/enhancer binding protein (C/EBP) site and AP-1 (Fig. 4).

Relative locations for potential cis DNA elements are shown. Box indicates the first exon and arrows indicate transcription initiation site. IRS: Insulin response element, AP-1: protein kinase C response element, GRE: glucocorticoid response element, C/EBP: CCAAT/enhancer binding protein recognition site, PEA3: PEA3 recognition site. Possible stimulation of promoter activity by insulin, IL-1β and phorbol ester are shown.

The C/EBP region plays an important role in the differentiation of adipocyte (50) as well as transcriptional up-regulation of certain genes by interleukin (IL)-1β or IL-6 (51). The AP-1 site is bound by Fos/Jun heterodimeric transcription factors upon stimulation by PKC or phorbol ester, an activator of PKC (52). We have also found nucleotide sequence highly homologous to the insulin response sequence of the phosphoenolpyruvate carboxykinase gene promoter which is up-regulated by insulin (53). We found that stimulation of R3T3 cells by insulin or IL-1β enhanced the expression of the AT_2 mRNA whereas phorbol ester suppressed it. We have not determined whether the effect of these peptide hormones on the expression of the AT_2 mRNA occurs at the transcriptional level or post-transcriptional level. But these results suggest that the expression of the AT_2 receptor is regulated by multiple growth factors both in positive and negative directions. And the presence of the responsive elements to IL-1β (C/EBP) or insulin (IRS) or phorbol ester (AP-1) in the promoter region of the mouse AT_2 gene may indicate that these effects take place at the transcriptional level (Fig. 4). Promoter analysis to map the growth factor responsive elements may clarify the tissue-specific and ontogeny-dependent expression of the AT_2 gene.

CONCLUDING REMARK

Although gene encoding the AT_2 receptor has been cloned in rat, mouse and human and a seven-transmembrane domain structure was clarified, no consensus with regard to the signaling mechanism and physiological function of this receptor has emerged yet. Studies aimed at clarifying apparent heterogeneity of the AT_2 receptor and examining whether the diverse signals that are sensitive to PD123319 or stimulated by CGP42112A are (or are not) mediated by the same receptor, are urgently called for.

ACKNOWLEDGMENT

This work was supported by research grant HL35323 and HL14192 from National Institute of Health

REFERENCES

1. Peach, M.J. (1981). Biochem. Pharmacol. *30*:2745-2751
2. Timmermans, P.B.M.W.M., Wong, P.C., Chiu, A.T., et al (1993) Pharmacol. Rev. *45*:205-251
3. Bottari, S.P., de Gasparo, M.,Steckelings, U.M.,.et al (1993) Frontiers in Neuroendocrinol. *14*:123-171
4. Birabeau, M.A., Capponi, A.M., Vallotton, M.B. (1984) Mol. Cell. Endocrinol. *37*:181-189
5. Whitebread, S., Mele, M., Kamber, B., et al (1989) Biochem. Biophys. Res. Commun. *163*:284-291
6. Chiu, A.T., Herblin, W.F., McCall, D.E., et al (1989) Biochem Biophys Res Commun. *165*:196-203.
7. Chang, R.S.L. and Lotti, V.J. (1990) Mol. Pharmacol.*29*:347-351
8. Sasaki, K., Yamano, Y., Bardhan, S., et al. (1991) Nature *351*:230-233
9. Murphy, T.J., Alexander, R.W., Griendling, K.K., et al. (1991) Nature *351*:233-236
10. Kambayashi, Y., Bardhan, S., Takahashi, K., et al (1993) J. Biol. Chem. *268*:24543-24546
11. Mukoyama, M., Nakajima, M., Horiuchi, M., et al (1993) J. Biol. Chem. *268*:24539-24542
12. Ichiki, T., Herold, C.L., Kambayashi, Y., et al (1994) Biochem. Biophys. Acta. *1189*: 247-250
13. Nakajima, M., Mukoyama, M., Pratt, R.E., et al (1993) Biochem. Biophys. Res. Commun. *197*:1449-1454
14. Ichiki, T. and Inagami, T. (1995) Circ. Res. In Press
15. Tsuzuki, S., Ichiki, T., Nakakubo, H., et al (1994) Biochem. Biophys. Res. Commun. *200*:1449-1454
16. Martin, M.M., Su, B., and Elton, T.S. (1994) Biochem. Biophys. Res. Commun. *205*:645-651
17. Tissir, F., Riviere, M., Guo, D.F., et al (1995) Cytogenet. Cell Genet. In Press
18. Bottari, S.P., Taylor, V., King, I.N., et al (1991) Euro. J. Pharmacol. *207*:157-163.
19.. Pucell, A.G., Hodges, J.C., Sen, I., et al (1991) Endocrinol. *128*:1947-1958
20. Tsutsumi, K. and Saavedra, J.M. (1992) Mol. Pharmacol. *41*:290-297
21. Rens-Domiano, S., Law, S.F., Yamada, Y., et al (1992) Mol. Pharmacol. *42*:28-34
22. Sokoloff, P., Giros, B., Martres, M.P., et al (1990) Nature *347*:146-151
23. Dudley, D.T., Hubbell, S.E. and Summerfelt R.M. (1991) Mol. Pharmacol. *40*:360-367
24. Dudley, D.T. and Summerfelt, R.M. (1993) Regulatory Peptides *44*:199-206
25. Webb, M.L., Liu, E.C.-K., Cohen, R.B., et al (1992) Peptides *13*:499-508.
26. Leung, K.H., Roscoe, W.A., Smith, R.D., et al (1992) Eur. J. Pharmacol. *227*:63-70
27. Bottari, S.P., King, I.N., Reichlin, S., et al (1992) Biochem. Biophys. Res. Commun. *183*:206-211
28. Sumners, C., Tang, W., Zelezna, B., et al (1991) Proc. Natl. Acad. Sci. USA. *88*:7567-7571
29. Kang, J., Posner, P. and Sumners, C. (1994) Am. J. Physiol. *267*:C1389-C1397
30. Buisson, B., Laflamme, L., Bottari, S.P., et al (1995) J. Biol. Chem. *270*:1670-1674
31. Lokuta, A.J., Cooper, C., Gaa, S.T., et al (1994) J. Biol. Chem. *269*: 4832-4838
32. Siemens, I. V., Reagan, L.P., Yee, D.K., et al (1994) J. Neurochem. *62*:2106-2115
33. Namba, T., Sugimoto, Y., Negishi, M., et al (1993) Nature *365*:166-170
34. Stoll, M., Steckelings, M., Paul, M., et al (1995) J. Clin. Invest. *95*:651-657
35. Nakajima, M., Horiuchi, M., Morishita, R., et al (1994) Hypertension *24*:379 (abstract)
36. Naveri, L., Stromberg, C. and Saavedra, J.M. (1994) Regulatory Peptide *52*: 21-29
37. Scheuer, D.A. and Perrone, M.H. (1993) Am. J. Physiol. *264*:R917-R923
38. Yoshida, M., Hamakubo, T. and Inagami, T. (1994) Am. J. Physiol. *266*:R802-R808
39. Grady, E.F., Sechi, L.A., Griffin, C.A., et al (1991) J. Clin. Invest. *88*:921-933
40. Tsutsumi, K., Viswanathan, M., Stromberg, C., et al (1991) Eur. J. Pharmacol. *198*:89-92
41. Tsutsumi, K. and Saavedra, J.M. (1991) Am. J. Physiol. *261*:R209-R216
42. Sechi, L., Griffin, C.A., Grady, E.F., et al (1992) Circ. Res. *71*:1482-1489
43. Suzuki, J., Matsubara, H., Urakami, M., et al (1993) Circ. Res. *73*:439-447
44. Koike, G., Horiuchi, M., Yamada, T., et al (1994) Biochem. Biophys. Res. Commun. *203*:1842-1850
45. Lopez, J.J., Lorell, B.H., Ingelfinger, J.R., et al (1994) Am. J. Physiol. *267*:H844-H852
46. Nio, Y., Matsubara, H., Murasawa, S., et al (1994) J. Clin. Invest. *95*:46- 54
47. de Gasparo, M., Whitebread, S., Kalenga, M.K., et al (1994) Regulatory Peptide *53*:39-45
48. Cox, B.E., Ipson, M.A., Shaul, P.W., et al (1993) J. Clin. Invest. *92*:2240-2248
49. Ichiki, T. and Inagami, T. (1995) Hypertension In Press
50. Christy, R.J., Yang, V.W., Ntambi, J.M., et al (1989) Genes Dev. *3*:1323-1335
51. Akira, S., Isshiki, H., Nakajima, T., et al (1992) in Interleukins: Molecular Biology and Immunology, Chem. Immunol. (ed. T. Kishimoto), Karger, Basel, pp 299-322
52. Angel, P., Imagawa, M., Chiu, R., et al (1987) Cell *49*:729-739.
53. O'Brien, R.M. and Granner, D.K. (1991) Biochem. J. *278:609-619*

16

MOLECULAR CLONING OF THE HUMAN AT_2 RECEPTOR

Mickey M. Martin, Baogen Su, and Terry S. Elton

Brigham Young University
Department of Chemistry and Biochemistry
Benson Science Building
Provo, Utah 84602

INTRODUCTION

The peptide hormone angiotensin II (AII), the biologically active component of the renin-angiotensin system, exerts a wide variety of physiological effects on the cardiovascular, endocrine and central and peripheral nervous system (1). AII exerts its physiological effects in the target tissues through specific interactions with plasma membrane receptors (2). Given the diversity of AII-mediated events and the differential signal transduction mechanisms, it has been proposed that multiple AII receptor subtypes exist (3,4).

Recently, two major pharmacologically distinct AII receptor subtypes have been described by using specific antagonists (5-10). The AII type 1 receptor (AT_1R) has a high affinity for the nonpeptide antagonist DuP 753 (losartan), whereas the AII type 2 receptor (AT_2R) has a high affinity for the nonpeptide antagonist PD123319 and the peptide agonist CGP42112A and a low affinity for losartan (5-10).

Recent cloning of the AT_1R from adrenal glomerulosa and vascular smooth muscle cells has revealed that this receptor belongs to the seven-transmembrane, G-protein coupled receptor superfamily (11,12). To date, extensive pharmacological evidence indicates that all of the effects currently associated with AII seem to be mediated by the AT_1R, through its ability to couple to the phospholipase C/IP_3 pathway or to inhibit adenylate cyclase activity (5-12).

In contrast, much less is known about the structure and function of the AT_2R. Extensive studies investigating the tissue distribution of the AT_2R revealed its presence in a wide variety of tissues, including rat adrenal medulla (3,5), rat brain (8,13,14), rat ovarian granulosa cells (15), rabbit and human uterus (6,7) and mesenchymal tissues of the rat fetus (10). Since the receptor is transiently expressed at high levels in the late stages of embryonic development (10) and in the immature brain (14) it has been suggested that the AT_2R may play an important role in growth and development.

Recently, AT_2R cDNA clones have been isolated from rat PC12w cells (16) and from fetal tissues of the rat and mouse (17,18). These clones encode a receptor protein of 363

Recent Advances in Cellular and Molecular Aspects of Angiotensin Receptors
Edited by Mohan K. Raizada et al., Plenum Press, New York, 1996

```
-140                       GGATCCGCGCAGAATTCAAAGCATTCTGCAGCCTGAATTTTGAAGGAGTGTGTTTAGGCAC   -80
     TAAGCAAGCTGATTTATGATAACTGCTTTAAACTTCAACAACCAAAGGCATAAGAACTAGGAGCTGCTGACATTTCAAT     -1

                 ●                                   ●
     ATG AAG GGC AAC TCC ACC CTT GCC ACT ACT AGC AAA AAC ATT ACC AGC GGT CTT CAC TTC    60
     met lys gly asn ser thr leu ala thr thr ser lys asn ile thr ser gly leu his phe    20
                 ●                   ●               ●
     GGG CTT GTG AAC ATC TCT GGC AAC AAT GAG TCT ACC TTG AAC TGT TCA CAG AAA CCA TCA   120
     gly leu val asn ile ser gly asn asn glu ser thr leu asn cys ser gln lys pro ser    40

     GAT AAG CAT TTA GAT GCA ATT CCT ATT CTT TAC TAC ATT ATA TTT GTA ATT GGA TTT CTG   180
     asp lys his leu asp ala ile pro ile leu tyr tyr ile ile phe val ile gly phe leu    60

     GTC AAT ATT GTC GTG GTT ACA CTG TTT TGT TGT CAA AAG GGT CCT AAA AAG GTT TCT AGC   240
     val asn ile val val val thr leu phe cys cys gln lys gly pro lys lys val ser ser    80

     ATA TAC ATC TTC AAC CTC GCT GTG GCT GAT TTA CTC CTT TTG GCT ACT CTT CCT CTA TGG   300
     ile tyr ile phe asn leu ala val ala asp leu leu leu leu ala thr leu pro leu trp   100

     GCA ACC TAT TAT TCT TAT AGA TAT GAC TGG CTC TTT GGA CCT GTG ATG TGC AAA GTT TTT   360
     ala thr tyr tyr ser tyr arg tyr asp trp leu phe gly pro val met cys lys val phe   120

     GGT TCT TTT CTT ACC CTG AAC ATG TTT GCA AGC ATT TTT TTT ATC ACC TGC ATG AGT GTT   420
     gly ser phe leu thr leu asn met phe ala ser ile phe phe ile thr cys met ser val   140

     GAT AGG TAC CAA TCT GTC ATC TAC CCC TTT CTG TCT CAA AGA AGA AAT CCC TGG CAA GCA   480
     asp arg tyr gln ser val ile tyr pro phe leu ser gln arg arg asn pro trp gln ala   160

     TCT TAT ATA GTT CCC CTT GTT TGG TGT ATG GCC TGT TTG TCC TCA TTG CCA ACA TTT TAT   540
     ser tyr ile val pro leu val trp cys met ala cys leu ser ser leu pro thr phe tyr   180

     TTT CGA GAC GTC AGA ACC ATT GAA TAC TTA GGA GTG AAT GCT TGC ATT ATG GCT TTC CCA   600
     phe arg asp val arg thr ile glu tyr leu gly val asn ala cys ile met ala phe pro   200

     CCT GAG AAA TAT GCC CAA TGG TCA GCT GGG ATT GCC TTA ATG AAA AAT ATC CTT GGT TTT   660
     pro glu lys tyr ala gln trp ser ala gly ile ala leu met lys asn ile leu gly phe   220

     ATT ATC CCT TTA ATA TTC ATA GCA ACA TGC TAT TTT GGA ATT AGA AAA CAC TTA CTG AAG   720
     ile ile pro leu ile phe ile ala thr cys tyr phe gly ile arg lys his leu leu lys   240

     ACG AAT AGC TAT GGG AAG AAC AGG ATA ACC CGT GAC CAA GTC CTG AAG ATG GCA GCT GCT   780
     thr asn ser tyr gly lys asn arg ile thr arg asp gln val leu lys met ala ala ala   260

     GTT GTT CTG GCC TTC ATC ATT TGG TGC CTT CCC TTC CAT GTT CTG ACC TTC CTG GAT GCT   840
     val val leu ala phe ile ile cys trp leu pro phe his val leu thr phe leu asp ala   280

     CTG GCC TGG ATG GGT GTC ATT AAT AGC TGC GAA GTT ATA GC[illegible] GTC ATT GAC CTG GCA CTT   900
     leu ala trp met gly val ile asn ser cys glu val ile ala val ile asp leu ala leu   300

     CCT TTT GCC ATC CTC TTG GGA TTC ACC AAC AGC TGC GTT AAT CCG TTT CTG TAT TGT TTT   960
     pro phe ala ile leu leu gly phe thr asn ser cys val asn pro phe leu tyr cys phe   320

     GTT GGA AAC CGG TTC CAA CAG AAG CTC CGC AGT GTG TTT AGG GTT CCA ATT ACT TGG CTC  1020
     val gly asn arg phe gln gln lys leu arg ser val phe arg val pro ile thr trp leu   340
                                                         #
     CAA GGG AAA AGA GAG AGT ATG TCT TGC CGG AAA AGC AGT TCT CTT AGA GAA ATG GAG ACC  1080
     gln gly lys arg glu ser met ser cys arg lys ser ser ser leu arg glu met glu thr   360

     TTT GTG TCT TAAACGGAGAGCAAAATGCATGTAATCAACATGGCTACTTGCTTTGAGGCTCACCAGAATTATTTTT  1156
     phe val ser END                                                                   363

     AAGTGGTTTTAATAAAATAATAAAATTTCCCCTAATCTTTTCTGAATCTTCTGAAACCAAATGTAACTATGTTTATCGT  1235
     CCAGTGACTTTCAGGAATGCCCATTGTTTTCTGATATGTTTGTACAAGATTTCATTGGTGAGAC                 1299
```

Figure 1. Nucleotide and deduced amino acid sequence of the hAT_2R. The predicted 363 amino acid residue sequence is shown below the nucleotide sequence. Putative transmembrane domains as determined by hydrophobicity analysis are underlined. Putative N-glycosylation sites are indicated with closed circles at amino acid residues 4, 13, 24, 29 and 34. A consensus protein kinase C phosphorylation site is indicated by a (#). The Genbank accession number is U16957. (From Martin et al., Biochem. Biophys. Res. Commun., 205, 647, 1994. With permission.)

```
                                                                   TM-1
hAT2R    MKGNSTLATTSKNITSGLHFGLVNISGNNESTLNCSQKPSDKHLDAIPILYYII    54
rAT2R    ..D.FSF.A..R....S.P.DNL.AT.T...AF...H..A....E.......M.    54
mAT2R    ..D.FSF.A..R....SRP.DNL.AT.T...AF...H.......E.......M.    54

                                      TM-2
hAT2R    FVIGFLVNIVVVTLFCCQKGPKKVSSIYIFNLAVADLLLLATLPLWATYYSYRY   108
rAT2R    .....A......S.........................................   108
mAT2R    .....A......S....................L....................   108

                     TM-3                                   TM-4
hAT2R    DWLFGPVMCLVFGSFLTLNMFASIFFITCMSVDRYQSVIYPFLSQRRNPWQASY   162
rAT2R    ......................................................   162
mAT2R    ......................................................   162

                                                          TM-5
hAT2R    IVPLVWCMACLSSLPTFYFRDVRTIEYLGVNACIMAFPPEKYAQWSAGIALMKN   216
rAT2R    V.....................................................   216
mAT2R    V.....................................................   216

                                                   TM-6
hAT2R    ILGFIIPLIFIATCYFGIRKHLLKTNSYGKNRITRDQVLKMAAAVVLAFIICWL   270
rAT2R    ......................................................   270
mAT2R    ......................................................   270

                                         TM-7
hAT2R    PFHVLTFLDALAWMGVINSCEVIAVIDLALPFAILLGFTNSCVNPFLYCFVGNR   324
rAT2R    ...........T...I......................................   324
mAT2R    ...........T...I......................................   324

hAT2R    FQQKLRSVFRVPITWLQGKRESMSCRKSSSLREMETFVS                  363
rAT2R    .....................T............D....                  363
mAT2R    .....................T.....G......D....                  363
```

Figure 2. Comparison of the amino acid sequences between the human, rat and mouse AT_2R. The amino acid sequences of the hAT_2R, rAT_2R and mAT_2R are shown. Amino acids are represented by their single letter code. Putative transmembrane domains are indicated with a line and are identified as TM-1 through TM-7. Amino acids in the rAT_2R and mAT_2R sequences which are different from the corresponding hAT_2R sequences are shown. (From Martin et al., Biochem. Biophys. Res. Commun., 205, 648, 1994. With permission.)

amino acids which has 34% sequence identity with the AT_1R. Like the AT_1R, the AT_2R has a seven-transmembrane domain topology. In contrast to the AT_1R, the AT_2R is not sensitive to GTP analogs (15,17,19,20) and does not mediate IP_3 production or adenylate cyclase activity (15,17,19-21). The AT_2R has been shown to mediate prostaglandin synthesis (22), to modulate T-type Ca^{2+} channels (23), and to inhibit cGMP production (24-26). Finally, it has been demonstrated that the AT_2R mediates the inhibition of protein tyrosine phosphatase and that this action was dependent on a pertussis toxin-sensitive G-protein coupled mechanism (16,27).

Currently, data on the human AT_2R (hAT_2R) are very limited. Therefore, the cloning of the human AT_2R should provide further insight into the function of this receptor subtype in humans. With the cloning of the hAT_2R we will be able to begin to investigate the tissue distribution in human tissues and determine whether species specific differences exist in the expression of this receptor. In this paper, we will describe the cloning of the hAT_2R cDNA and gene, and examine the tissue distribution of its mRNA.

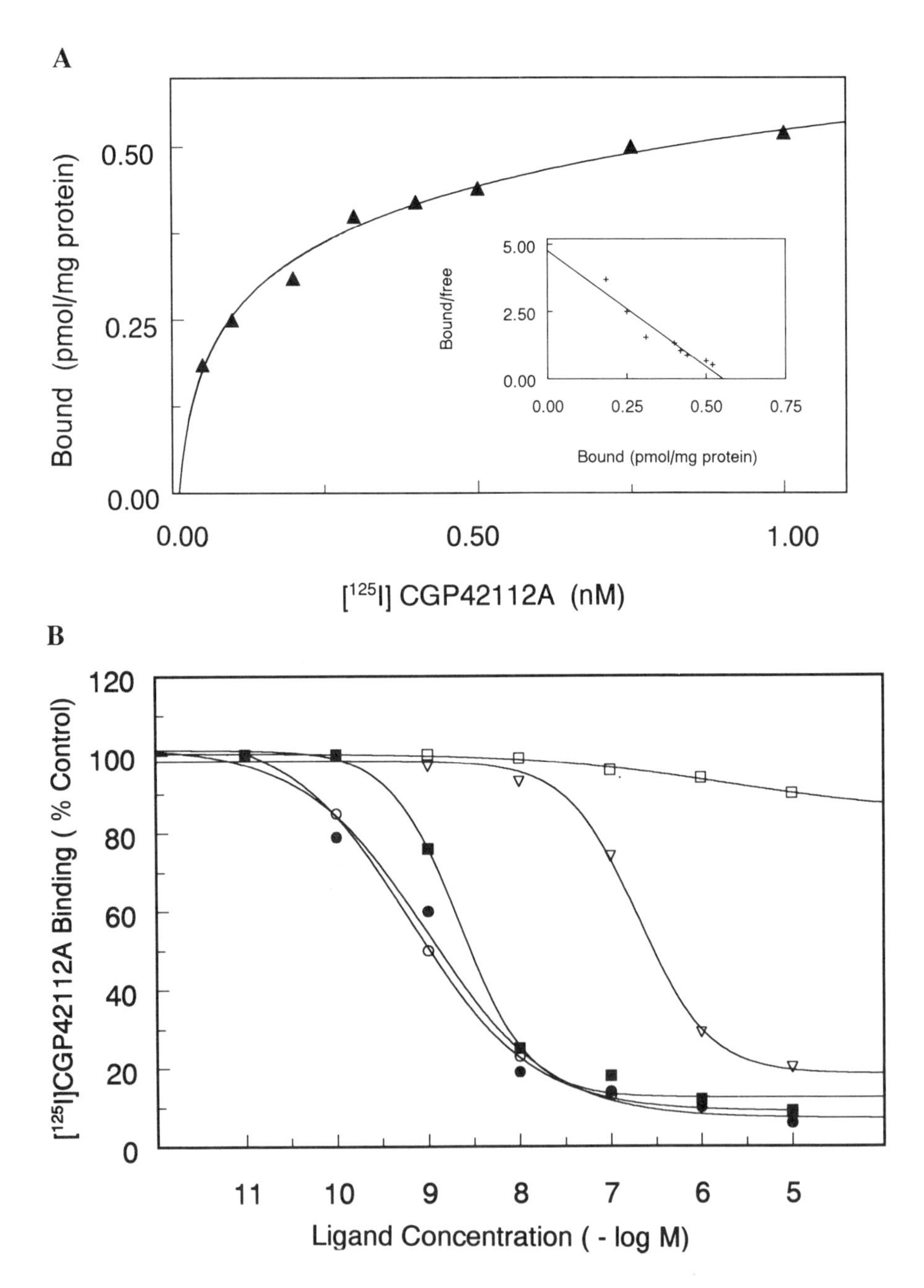

Figure 3. Binding characteristics of [^{125}I]CGP42112A to COS-7 cells transfected with the pcDNA3-hAT$_2$R construct. A. Saturation isotherm of the specific binding of [^{125}I]CGP42112A to membranes from COS-7 cells expressing the hAT$_2$R. Inset shows a Scatchard plot of the same data. B. Displacement of specific [^{125}I]CGP42112A binding in the hAT$_2$R cDNA transfected COS-7 cell membranes by unlabeled Sar^1,Ile^8-AII (○), CGP42112A (●), AII (■), AI ( ) and Losartan (□). The plotted data are from a typical experiment. (From Martin et al., Biochem. Biophys. Res. Commun., 205, 649, 1994. With permission.)

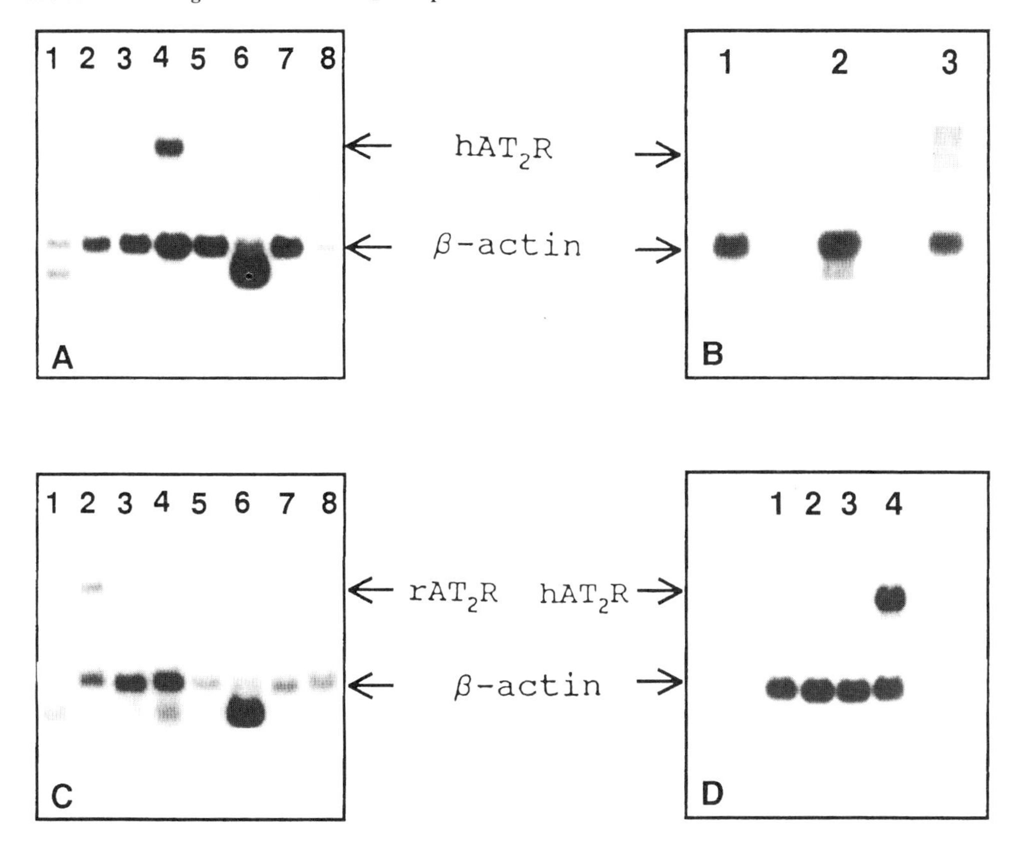

Figure 4. Northern blot analysis. A. The human adult multiple tissue Northern was sequentially probed with a radiolabeled hAT_2R cDNA probe and a β-actin control probe. Lanes 1-8 contained, in order, two µg of poly A^+ mRNA isolated from human adult heart, brain, placenta, lung, liver, skeletal muscle, kidney and pancreas. B. Northern blot was probed as described in A. Lanes 1-3 contained 5 µg of poly A^+ mRNA isolated from human adult adrenal, aorta and pituitary. C. The adult rat multiple tissue Northern blot was sequentially probed with a radiolabeled rAT_2R PCR probe and the control probe. Lanes 1-8 contained, in order, two µg of poly A^+ mRNA isolated from adult rat heart, brain, spleen, lung, liver, skeletal muscle, kidney and testis. D. Human fetal multiple tissue Northern blot was probed as described in A. Lanes 1-4 contain, in order, two µg of poly A^+ mRNA isolated from human fetal brain, lung, liver and kidney. (From Martin et al., Biochem. Biophys. Res. Commun., 205, 650, 1994. With permission.)

CLONING OF THE HUMAN AT_2R cDNA

It is evident that vast species differences exist in the proportion and distribution of the AT_2R (reviewed in 29). To determine the appropriate human tissue to utilize for cDNA cloning, the published rat cDNA sequences (16,17) were utilized to synthesize an amplimer set to generate an AT_2R specific PCR probe. This probe was subsequently radiolabeled and used for Northern analysis. Interestingly, Northern results demonstrated that the AT_2R mRNA was abundantly expressed in human lung tissue. Therefore, we concluded that a human lung cDNA library should be screened. A human adult lung 5′-stretch cDNA library constructed in the λDR2 vector (Clontech, Palo Alto, CA) was screened by standard techniques with the rat AT_2R specific PCR probe. Several positive clones were plaque purified and insert sizes

Figure 5. Sequence comparison between the hAT_2R cDNA and the coding region of a hAT_2R genomic clone. The hAT_2R cDNA sequence was obtained from Martin et al. (29). The hAT_2R genomic clone coding region and 5'-flanking region sequence was obtained from Koike et al. (39). Where the sequence is identical between the hAT_2R cDNA and hAT_2R gene has been arbitrarily designated exon B. The sequence shaded upstream from exon B is distinct between the hAT_2R cDNA and hAT_2R gene sequences. (From Martin, M.M. and Elton, T.S., Biochem. Biophys. Res. Commun., in press, 1995. With permission.)

were determined by PCR utilizing λDR2 insert screening amplimers. Lambda clone $hAT_2R/7$ had an insert size of approximately 2.7 kb and was selected for further characterization. Lambda clone $hAT_2R/7$ was converted into a plasmid, pDR2-$hAT_2R/7$, by cre-lox conversion. Both strands of the 2.7 kb hAT_2R cDNA clone were sequenced by the dideoxy sequencing method. The nucleotide and deduced amino acid sequence of the $hAT_2R/7$ cDNA are shown in Figure 1 (29). This 2.7 kb clone harbored a 1089 bp open reading frame with a deduced amino acid sequence of 363 amino acid residues. Hydropathy analysis of the deduced amino acid sequence revealed the presence of seven putative transmembrane domains. There are multiple serine and threonine amino acid residues present in the third intracellular loop and carboxyl tail domains of the hAT_2R that may be putative phosphorylation sites (31). There are also five consensus sites for N-glycosylation (31) in the amino-terminal domain of the hAT_2R. Recently, Servant et al. (32) demonstrated by photoaffinity labeling experiments, that the hAT_2R is N-glycosylated in at least three sites *in vivo*.

A comparison of deduced amino acid sequences of the human, rat and mouse AT_2R is shown in Figure 2. The amino acid sequence of the hAT_2R was 92% identical to the rat and mouse AT_2R cDNA sequence (16-18). Only the amino terminal extracellular domain of the hAT_2R shows divergence from the rat and mouse AT_2R (Fig. 2). All of the consensus sites for N-glycosylation and the putative phosphorylation sites are conserved in these three species, suggesting that these amino acid residues are functionally important.

To examine the binding properties of the hAT_2R cDNA, it was subcloned into the eukaryotic expression vector, pcDNA3 (Invitrogen, San Diego, CA). The new construct pcDNA3-$hAT_2R/7$ was transiently transfected into COS-7 cells by electroporation. The cells were harvested 72 hours post-transfection, membranes were prepared and utilized for radioreceptor binding assays. As shown in Figure 3A, hAT_2R cDNA expressed in COS-7 cells showed a single high affinity binding site. Plasma membrane preparation from these cells showed a K_d of 0.33 nM for the binding of [^{125}I]CGP42112A. Displacement of this binding is shown in Figure 3B. The binding of [^{125}I]CGP42112A was competed by unlabeled angiotensin analogs and AT_2R specific ligands in the following rank order: Sar^1,Ile^8-AII=CGP42112A>AII>AI. Losartan (10 μM), an AT_1 selective antagonist (5-10), did not inhibit [^{125}I]CGP42112A binding. These binding characteristics are in good agreement with those observed in the membranes from that reported for the mouse and rat AT_2R (16-18).

```
AATTATGTAGGTTGAAGGCTCCCAGTGGACAGACCAAACATATAAGAAGGAAACCAGAGATCTGGTGCTATTACGTCCCA   80
GCGTCTGAGAGAACGAGTAAGCACAGAATTCAAAGCATTCTGCAGCCTGAATTTTGAAGGTAAGTATGAAGCAATTTATA  160
TATAATTTACTTGGAAGGTAGAACATACATTAAATGAAAATATTTTTTATGGATGAACTTCTGTTTTTCCTGTGTTTTAA  240
CACTGTATTTTGCAAAACTCCTAAATTATTTAGCTGCTGTTTCTCTTACAGGAGTGTGTTTAGGCACTAAGCAAGCTGAT  320
TTATGATAACTGCTTTAAACTTCAACAACCAGTAAGTCTTCAAGTGGAATTTATTATTGATTCTTTTATGTTAATTTGTT  400
AGGTCAAAAGAAAAATCTTTAGAGCAAAATAAAAGTTTTGCTCTTTATTAGGAGGTTCTTTAGATATTACACTTTTAATT  480
GGGTAGCTTATTTGCATGTATTTTGAAACTATCTAAAGTAAATAGTGTTTCCTTTGTATGCTTATCTTTAGCTAATGTGT  560
TTTTTTTTTGGTTTAAATATGCTCTAGTGAAAAAATCACAAAACTCACACTGTAACGTTTGAGAGCAACGGCTATTCAGT  640
TCGGTTAAACCGAAAAGAGAGAAAAAGAAAAATGATAAGACTATATCATGATTAATCATTGATGTTCATTTCTCATTTCA  720
CTGAGCAATCTAAACTACTTCTACTTGAACCTAACATAGGTTTGGCTGTTTTTTTCTTTATCTGATTTAATTGAAAATGG  800
ATTGTTTAGATATTAAATAGACCAAAATTAGATATCTATATCTAAAACTAATATTCTCTATTAAAAGGAGAAGACTTAGA  880
AATATAAATATATTTCATAACTTCTTCTTAACTATATTATTATTGTGAAAACAATAGAAAAGTAAAACTAAAGGAATTAT  960
TTTTAATTATTAAAGACAAACCATGATTCCCCATCACAATATGTTCCTTTGGCATATTCTCTTTTAGTTTTTTATATACT 1040
TATTTTTTCCATTGCTTATGTTACTTATATTTTCCACAGCTGTAGTGATTTGTATGTAATTAGTTACTTAAATAGAGCAT 1120
TTCTCTATGCAATAGCAAGCAGCAGTGTGGTTTCTAGAAAGTATACAAACAGAGATATATGCTCAAGTGTTTTTTGAAAG 1200
ATTACTTTTCTTACCTCTGGGTTTCTTATTTGTTTCACACATTTCCTGCCCTGACTAATCTTTTGAAATATTAATTTTCT 1280
TTAAATCTTTCAAAAACATCATAGAAAAGATGGCGAAAAGGTAGTTTTATTCAGAACTTTGAATCTCTGAAATGTCTTTT 1360
TTCTACTATCAGGAAAATTATTTCTCCTTTTTCAAGAGGTAAAATGAATTCTTGTTTTACAAGCCATGGCTCTGAAACTT 1440
AATGTTTTCTATAATCACTCACTTTTTTTTTGCTTTTGACAAACATTCAAAATGCTAATGATTCAAGGATGTCCTCAGCT 1520
CTGTATGTGTTCTAAGAGTTCTATGTTTTTTCTCCACAGAAGGCATAAGAACTAGGAGCTGCTGACATTTCAATATGAAG 1600
GGCAACTCCACCCTTGCCACTACTAGCAAAAACATTACCAGCGGTCTTCACTTCGGGCTTGTGAACATCTCTGGCAACAA 1680
TGAGTCTACCTTGAACTGTTCACAGAAACCATCAGATAAGCATTTAGATGCAATTCCTATTCTTTACTACATTATATTTG 1760
TAATTGGATTTCTGGTCAATATTGTCGTGGTTACACTGTTTTGTTGTCAAAAGGGTCCTAAAAAGGTTTCTAGCATATAC 1840
ATCTTCAACCTCGCTGTGGCTGATTTACTCCTTTTGGCTACTCTTCCTCTATGGGCAACCTATTATTCTTATAGATATGA 1920
CTGGCTCTTTGGACCTGTGATGTGCAAAGTTTTTGGTTCTTTTCTTACCCTGAACATGTTTGCAAGCATTTTTTTTATCA 2000
CCTGCATGAGTGTTGATAGGTACCAATCTGTCATCTACCCCTTTCTGTCTCAAAGAAGAAATCCCTGGCAAGCATCTTAT 2080
ATAGTTCCCCTTGTTTGGTGTATGGCCTGTTTGTCCTCATTGCCAACATTTTATTTTCGAGACGTCAGAACCATTGAATA 2160
CTTAGGAGTGAATGCTTGCATTATGGCTTTCCCACCTGAGAAATATGCCCAATGGTCAGCTGGGATTGCCTTAATGAAAA 2240
ATATCCTTGGTTTTATTATCCCTTTAATATTCATAGCAACATGCTATTTTGGAATTAGAAAACACTTACTGAAGACGAAT 2320
AGCTATGGGAAGAACAGGATAACCCGTGACCAAGTCCTGAAGATGGCAGCTGCTGTTGTTCTGGCCTTCATCATTTGCTG 2400
GGTTCCCTTCCATGTTCTGACCTTCCTGGATGCTCTGGCCTGGATGGGTGTCATTAATAGCTGCGAAGTTATAGCAGTCA 2480
TTGACCTGGCACTTCCTTTTGCCATCCTCTTGGGATTCACCAACAGCTGCGTTAATCCGTTTCTGTATTGTTTTGTTGGA 2560
AACCGGTTCCAACAGAAGCTCCGCAGTGTGTTTAGGGTTCCAATTACTTGGCTCCAAGGGAAAAGAGAGAGTATGTCTTG 2640
CCGGAAAAGCAGTTCTCTTAGAGAAATGGAGACCTTTGTGTCTTAAACGTGAGAGCAAAATGCATGTAATCAACATGGCT 2720
ACTTGCTTTGAGGCTCACCAGAATTATTTTTAAGTGGTTTTAATAAAATAATAAAATTTCCCCTAATCTTTTCTGAATCT 2800
TCTGAAACCAAATGTAACTATGTTTTAACGTCCAGTGACTTTCAGGAATTGCCCATTGTTTTTCTGATATGTTTGTACAA 2880
GATTTTCATTGGTGAGACATATTTACAACCTAGAAGTAACTGGTGATATATCTCAAATTGTAATTAATAATAGATTGTGA 2960
ATAATGATTTGGGGATTCAGATTTCTCTTTGAAACATGCTTGTGTTTCTTAGTGGGGTTTTATATCCATTTTTATCAGGA 3040
TTTCCTCTTGAACCAGAACCAGTCTTTCAACTCATTGCATCATTTACAAGACAACATTGTAAGAGAGATGAGCACTTCTA 3120
AGTTGAGTATATTATAATAGATTAGTACTGGATTATTCAGGCTTTAGGCATATGCTTCTTTAAAAACGCTATAAATTATA 3200
TTCCTCTTGCATTTCACTTGAGTGGAGGTTTATAGTTAATCTATAACTACATATTGAATAGGGCTAGGAATATAGATTAA 3280
ATCATACTCCTATGCTTTAGCTTATTTTTACAGTTATAGAAAGCAAGATGTACTATAACATAGAATTGCAATCTATAATA 3360
TTTGTGTGTTCACTAAACTCTGAATAAGCACTTTTTAAAAAACTTTCTACTCATTTTAATGATTGTTTAAAGGTTTCTAT 3440
TTTCTCTGATACTTTTTTGAAATCAGTAAACACTGTGTATTGTTGTAAAATGTAAAGGTCACTTTTCACATCCTTGACTT 3520
TTTAGATGTGCTGCTTTGATATATAGGACATTGATTTGATTTTTATTATTAATGCTTTGGTTCTGGGTTGTTTCCTAAAA 3600
TATCTGGGTGGCTTAAAAAAAACTCTTTAACTTGTAATAAACCCTTAACTGGCATAGGAAATGGTATCCAGAATGGAATT 3680
TTGCTACATGGGGTCTGGGTGGGGGCAAAGAGACCCAGTCAATTACATGTTTGGTACCAAGAAAGGAACCTGTCAGGGCA 3760
GTACAATGTGACTTTGAAAATATATACCGTGGGGGTAGTTTTACCCTATATCTATAAACACTGTTTGTTCCAGAATCTGT 3840
ATGATTCTATGGAGCTATTTTAAACCAATTGCAGGTCTAGA                                        3881
```

Figure 6. The nucleotide sequence of the hAT_2R gene. The shaded areas encompass exons 1-3. The underlined ATG and TAA in exon 3 denote the translation initiation and translation termination codons for the hAT_2R open reading frame. Splice donor and splice acceptor sites are underlined. The putative TATA box is also underlined. Genbank accession #U20860. (From Martin, M.M., and Elton, T.S., Biochem. Biophys. Res. Commun., in press, 1995. With permission.)

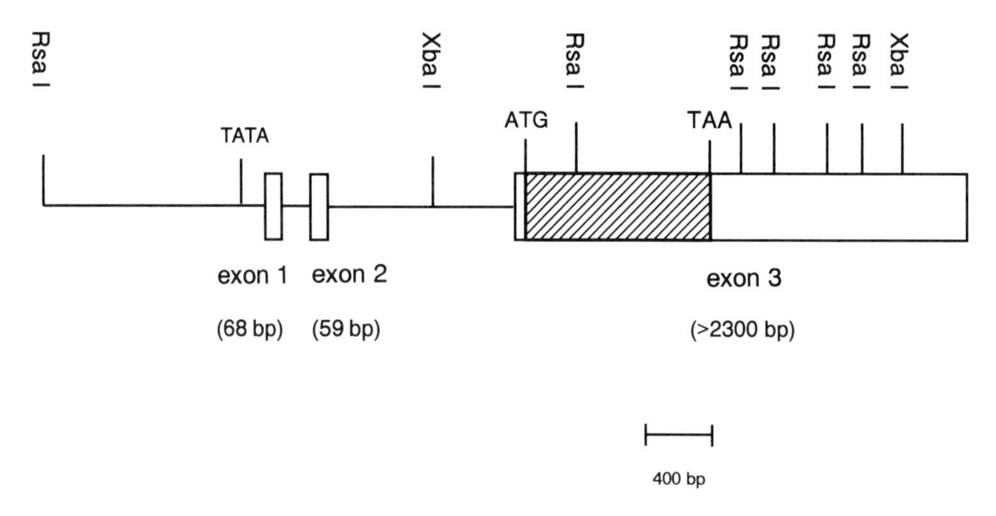

Figure 7. Schematic representation of the hAT_2R gene. Exons 1 and 2 encode for 5′ UTR mRNA sequence, while exon 3 harbors the entire uninterrupted hAT_2R open reading frame. (From Martin, M.M. and Elton, T.S., Biochem. Biophys. Res. Commun., in press, 1995. With permission.)

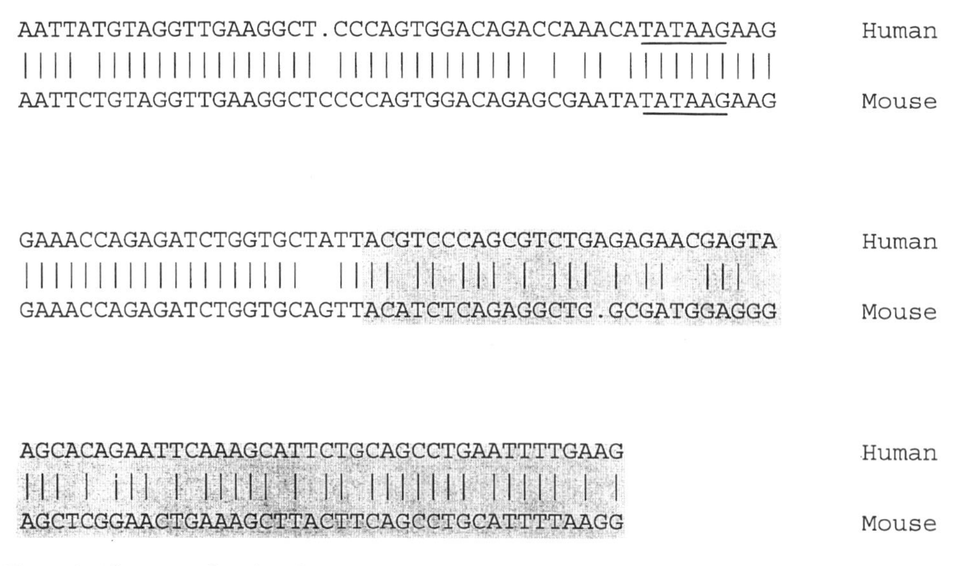

Figure 8. Alignment of nucleotide sequences surrounding the transcription initiation sites in human and mouse. The mouse AT_2R nucleotide sequences were reported by Ichiki et al. (37). The alignment was carried out using the "GAP" program in GCG. The sequence of the first exon is shaded. This exon begins at the putative transcription initiation site. The putative TATA box is underlined. (From Martin, M.M. and Elton, T.S., Biochem. Biophys. Res. Commun., in press, 1995. With permission.)

```
AACTATAGAA ACCCATCTAG TCACAAGTTT CTCCTTCCTG TGGTCTCCAT TTGGGCAATA CAATTAAGAA  -1415
                          ICSBF                      C/EBP
AAATACGTTT CCCCACATAA CAATTCTTCC AATTCTTTTC CCCTTCTTTG ATGCTTTTAT GATAGATTAT  -1345

GAATGGTCAT CAGTGATTTT CTGTAAATGC CACGTTTCCT TTTTGCCAAG ACAAAGAGTG AGACTAAGGG  -1275

ACCATCTAGT GCCTTCCTCT TGCTGGTATA GAGGCTAAAC AGCGTGAGTT CTGATATTAG TTTCCTTTGA  -1205

TTTAAATGCA AGTGTCTTCA TTTAATAGCT GTATGATCTC AGAAAAGTTA TTAAACTCTA TGAGTCCTAG  -1135

TTTCCTGATT TGTAAAGTGG GGGCTAATAT GAATAGCTTG TAATAGCACC TACCATATGC TAGGCATGAT  -1065
                        Sp1
TTTAGGCACT TTAAACATAT TAGCTCATTT AAATTTCCAT TAGCTTATTT AAAATTCCTA CTCTACGATG   -995

TAGGTAATTT ATATCATTCT CCTTTCTCAG ATGAGAAACT GAGACATGAA GAAGTAAAGT AACTTGGCCA   -925

AGGTCAGAAT TAGTAGGTGG TAGAGCTGGG TTTGTAAAGC CAGGTAGTTG GATCCCAGAG GCTATTTACT   -855
  ELP
AACCAGTGTG CTATATTGCC TCTCTATTAA ATTATTGATG ATAAATAACT CATCACAGTG TTCGGCACAT   -785
                                                        NF-E1
AACATTTATT GTCAGCTAGT ATTATCACTA CCATCACCTG TTCTTATCTT TCTTCATTTT GTCACTCTTG   -715
                        NF-E1                GRE     NF-E1
CTTCTGCTGG ATTGGTTGTG TAGTTTATTG CCCCCACTGC CACAAGAGAC AGCAGCAACA GTGATCTAGG   -645

TTGAAATAGC TAGCTAATCC CTCTTCACTT TATGTGCTTT GTATTTAAAC CACAAATATG ACAGAGACCT   -575

ATGAAAGGAG ACTAAACCTC TTTTTGGTCG TGTAAAAATT ATATTCCTGC AATCTCCAAC CCTCCAGCAA   -505

CAAGCCAACA AAAACTGCGC AAAGCAAGGC CGAAATTTTA GAAGTGGTAG ACTTCAAGTA AAATTCCTAC   -435

CATTCTTTAA TAAACAGGGG TAAAAATTTC AGAAAGTAAT GAAGGTTGTG AAAGCTGATG AAGCAAAGAC   -365

TTTGCTACTA TAATAAATTT CCTAGTCTCT ACTCACTAAT TTCTATTAGA ATGCCAATAT CAGGAAAAAA   -295
                                                                    PEA3
GGAAGAGAAA ATTCTGTAAA TAACTTACTA TATTTTTCTC CTACGAAGTA AGCCACTGAC TCAAAAATTC   -225

TCTAGCACAT TTGTGGAAAC TTCATTTTTT TTGTTTGAGA TTTATTTGAA TGAGCTGTTA TGATTGGAGA   -155
             AP-3                                                   CTF
CAGTGAGAAT TTCAGATTAA TGTTTTGCAG ACAAAAAAAA ACCTCTCTGG AAAGCTGGCA AGGGTTCATA    -85
                                                  H-APF-1
AGTCAGCCCT AGAATTATGT AGGTTGAAGG CTCCCAGTGG ACAGACCAAA CATATAAGAA GGAAACCAGA    -15
               +1                                       TATA box   PEA3
GATCTGGTGC TATTACGTCC CAGCGTCTGA                                                 bp
```

Figure 9. Sequence analysis of the 5′-flanking region of the hAT_2R gene. The sequence is numbered from the putative transcription initiation site. Consensus transcription factor recognition sites are underlined and labeled. (From Martin, M.M. and Elton, T.S. Biochem. Biophys. Res. Commun., in press, 1995. With permission.)

TISSUE DISTRIBUTION OF THE hAT_2R

As described earlier, the AT_2R mRNA is abundantly and widely expressed in rat fetal tissues (10) but just detectable in adult rodent tissues (2,5,15). To thoroughly investigate the tissue distribution of the hAT_2R mRNA, a human multiple tissue Northern blot was probed utilizing the hAT_2R cDNA as a probe. Northern results demonstrated that this gene was abundantly expressed in human adult lung (Fig. 4A, lane 4) and just detectable in adult heart (Fig. 4A, lane 1) and aorta (Fig. 4B, lane 2). Since the AT_2R is expressed in fetal rodent tissues, we investigated whether the AT_2R was also expressed in human fetal tissues (Fig. 4D). Northern results demonstrated that hAT_2R mRNA was highly expressed in human fetal kidney (Fig. 4D, lane 4) and to a lesser amount in human fetal lung (Fig. 4D, lane 2). In contrast to the human AT_2R tissue distribution, the rat AT_2R mRNA was abundantly expressed in rat brain (Fig. 4C, lane 2) and just detectable in the lung (Fig. 4C, lane 4). All

Northern blots were hybridized with a β-actin control probe to assess the relative quantity of the poly A^+ mRNA present in each lane.

CLONING OF THE HUMAN AT_2R GENE

To begin to elucidate the mechanisms that govern the developmental and tissue specific transcriptional regulation of the hAT_2R gene, this gene (i.e., exons, introns and the 5′ flanking sequence that includes the promoter region) needed to be cloned and characterized. It is now recognized that AT_1R genes are comprised of multiple exons with several exons encoding 5′ untranslated region (UTR) mRNA sequence and one uninterrupted exon encoding for the open reading frame of the receptor (33-36). Recently, it was demonstrated that the mouse AT_2R gene is also comprised of multiple exons (37). Several laboratories have isolated hAT_2R genomic clones (38,39), however, these manuscripts only describe the uninterrupted exon that encodes for the receptor. Without the hAT_2R cDNA sequence, exon/intron boundaries cannot be defined and the organization of the gene cannot be determined. Therefore, once our laboratory had isolated the hAT_2R cDNA, we could utilize this information to clone and characterize the hAT_2R gene. To accomplish this goal, a human placenta genomic DNA library was screened by standard techniques with a hAT_2R radiolabeled probe (an 850 bp Rsa I fragment that encompasses a portion of the hAT_2R coding region and some 3′ UTR sequence. Positive clones were plaque purified and analyzed by PCR and Southern blotting to determine if these genomic clones encompassed the entire hAT_2R gene. These experiments were critical since a comparison of the 5′-flanking region sequence of the published hAT_2R genomic clones (i.e., upstream from the coding region) (38,39) with the 5′ UTR of the hAT_2R cDNA (29) demonstrated that these sequences were distinct (Fig. 5) (40), suggesting that the 5′ UTR of the hAT_2R cDNA was encoded by another exon or exons located some unknown distance upstream from the coding region exon. Therefore, all of the genomic clones were subjected to Southern analysis utilizing a radiolabeled PCR probe that was specific for the 5′ UTR of the hAT_2R cDNA (29). Hybridization results using this probe demonstrated that only one hAT_2R genomic clone ($\lambda hAT_2R/3$) would hybridize, indicating that it was comprised of the exon(s) that encoded the 5′ UTR region of the hAT_2R mRNA. Therefore $\lambda hAT_2R/3$ was chosen for further characterization. By Southern blot analysis, a 2.7 kb Rsa I fragment was identified that hybridized only to the hAT_2R 5′ UTR-specific probe. Both fragments were subcloned into the pGEM-7Zf vector and sequenced (Fig. 6) (40). Sequence analysis demonstrated that the Xba I and Rsa I fragments were overlapping.

A comparison of the hAT_2R cDNA sequence (29) with the sequence obtained from the Xba I and Rsa I fragments suggests that the hAT_2R gene is comprised of at least three exons and spans only 5 kb (Fig. 6 and 7). This is in contrast to the hAT_1R gene which is comprised of at least four exons and spans greater than 60 kb (33,34). One major difference between these two genes appears to be in the sizes of their introns. The hAT_2R intron sizes are 151 bp and 1208 bp, whereas, the hAT_1R introns are all 8-10 kb or larger (33,34).

The nucleotide sequences at splice junctions (Fig. 6) of the hAT_2R gene agree well with consensus sequences for splice donor/acceptor sites (GT/AT rule) (39). Exon 1 (i.e., 68 bp, if transcription initiation begins at the "A" nucleotide designated) and exon 2 (59 bp) encode for 5′ UTR mRNA sequence and exon 3 encodes 35 bp of 5′ UTR mRNA sequence, the entire open reading frame, and the 3′ UTR mRNA sequence of the hAT_2R (Fig. 6 and 7). Unlike the hAT_1R (33,34), rat $AT_{1A}R$ (35) and rat $AT_{1B}R$ (36) mRNA transcripts, we found no evidence of alternatively spliced forms of hAT_2R mRNAs (data not shown). The open reading frame of 1089 base pairs is identical to the open reading frame of the hAT_2R cDNA shown in Figure 1.

The 140 bp nucleotide sequence of exon 1 (i.e., 68 bp) and 5′-flanking region (i.e., 72 bp) of the hAT$_2$R gene (see Fig. 6) was aligned with that of the mouse AT$_2$R gene reported by Ichiki et al. (37). This region was 82% identical between the human and mouse (Fig. 8). Since eukaryotic polymerase II transcription generally initiates at ~25-35 bp downstream from the TATA box (Fig. 8) at an "A" nucleotide (42), we assigned the nucleotide "A" 33 bp downstream (i.e., the beginning of the shaded area) from the putative TATA box to be the transcription initiation site for the hAT$_2$R gene. This nucleotide is conserved in the mouse AT$_2$R gene. To verify that this sequence (Fig. 8) represents the promoter/first exon of the hAT$_2$R gene, primer extension and RNase protection assays will be performed to localize the exact transcription start site(s). However, support for this assignment has been obtained from our laboratory by utilizing the 5′-RACE procedure.

Approximately 1500 bp of 5′-flanking region of the hAT$_2$R gene were analyzed for putative transcriptional regulatory elements (Fig. 9) by using the transcription factor database (43). In addition to the TATA box (at -33 bp), several potential binding sites for transcription factors were identified. There is one binding site for the CTF (3′ CCAATC 5′ at -163) (44) transcription factor at an inverted CCAAT-box. The CCAAT-box element seems to be related to the level of basal expression and functions in either orientation (44). There is also one putative binding site for the CCAAT and enhancer binding protein, C/EBP (5′ TTGGGCAAT 3′ at -1434) (45). There are two binding sites for the PEA3 (5′ AGGAAA 3′ at -25 and -304) transcription factor. The PEA3 motif has been shown to mediate transcriptional activation by the c-ets-1 and -2 proto-oncogenes, v-src, phorbol ester and serum components (46,47). There is also one Sp1 binding site (5′ GGGGCTAA 3′ at -1016) (47) and an AP-3 binding site (5′ TTGTGGAAA 3′ at -215) (48) and a glucocorticoid receptor binding site (5′ TGTTCT 3′ at -747) (49) present in this sequence. Additionally, the hAT$_2$R promoter region harbors a putative binding site for a hepatocyte-specific nuclear protein, H-APF-1 (5′ CTGGAAA 3′ at -108) (50) and three erythroid-cell-specific nuclear factor, NF-E1, binding sites (5′ A/C Py T/A ATC A/T Py 3′ at -743, -764 and -805) (51). There is also an interferon (IFN) consensus sequence binding protein (i.e., ICSBP) site (5′ AGTTTCTCCTTC 3′ at -1459) (51). ICSBP is a transcription factor that is expressed mainly in cells of the immune system and is preferentially induced by IFN-γ (52). Finally, the hAT$_2$R promoter region harbors a putative embryonal, long terminal repeat binding protein (ELP) site (5′ CAAGGTCA 3′ at -926) (53). ELP is a mouse homolog of drosophila FTZ-F1, which positively regulates transcription of the fushi tarazu gene in blastoderm-stage embryos of the fly. ELP is expressed in embryonal carcinoma cells (EC cells), which are derived from cells in the blastocyst stage (54) and is thought to play a role in early-stage mammalian embryogenesis (53). Thus, the promoter region of the hAT$_2$R gene harbors some very interesting putative transcription factor binding sites.

CONCLUSION

Although the function of the AT$_2$R is not known, it has been suggested that the AT$_2$R may play an important role in growth and development since the receptor is transiently expressed at high levels in the late stages of rat embryonic development (10) and in the immature rat brain (14). We have demonstrated that the hAT$_2$R is abundantly expressed in human fetal kidney and is just detectable in human fetal lung. Grone et al. (55) characterized AII receptor subtypes by *in vitro* autoradiography in fetal and adult human renal tissue utilizing [^{125}I] Sar^1,Ile^8-AII. They demonstrated that in human adult kidney, the hAT$_2$R was only present in the large preglomerular vessels. In contrast, they demonstrated that in human fetal kidney, the hAT$_2$R was present throughout the medulla and cortex. Therefore, the autoradiographic results support our Northern blot data. The presence of the hAT$_2$R in human fetal kidney strengthens the hypothesis that AII could be important for renal development. The fact that the hAT_2R promoter harbors a consensus ELP transcription factor binding site also suggests that this receptor has the proper molecular instructions to be expressed during embryogenesis.

The cloning of the hAT_2R cDNA will allow us to begin to investigate the physiological significance of this receptor. It will be of particular interest to investigate whether this receptor can inhibit protein tyrosine phosphatase activity. Tyrosine phosphorylation is most often associated with cell proliferation and differentiation. Since it has been demonstrated that the AT_2R mediates a reduction of intracellular tyrosine phosphorylation (56), it has been suggested that the AT_2 receptor may participate in the negative regulation of cell proliferation. Interestingly, it has been demonstrated that AT_2R numbers are extremely low in actively growing cells, while in confluent, quiescent cells, the expression of AT_2R sites is markedly increased (57). Additionally, serum and growth factors rapidly cause a decrease in the number of AT_2 receptor sites (57,58). Since growth factors upregulate the AT_1R (58), the function of the AT_2R might be to counteract the action of the AT_1R. With the cloning of the hAT_2R gene we can begin to elucidate the mechanisms that govern the developmental and tissue-specific transcriptional regulation of this gene. These studies will facilitate our understanding of the functional importance of this receptor.

ACKNOWLEDGMENTS

This work was supported by research grants NIH/HL 48848 and HL 48848-52.

REFERENCES

1. Peach, M.J. (1977). Physiol. Rev. *57*:313-370.
2. Catt, K.J., Mendelsohn, F.A.O., Millan, M.A. and Aguilera, G. (1984). J. Cardiol. Pharmacol. *6*:S575.
3. de Gasparo, M., Whitebread, S., Mele, M., Motani, A.S., Whitcombe, P.J., Ramjoue, H.P. and Kamber, B.J. (1990). Cardiovasc. Pharmacol. *16*:S31-S35.
4. Hausdorff, W.P., Sekura, R.D., Aguilera, G. and Catt, K.J. (1987). Endocrinology *120*:1668-1678.
5. Chiu, A.T., Herblin, W.F., McCall, D.E., Ardecky, R.J., Carini, D.J., Dunica, J.V., Pease, L.J., Wong, P.C., Wexler, R.C., Johnson, A.L. and Timmermans, P.B.M.W.M. (1989). *165*:196-203.
6. Whitebread, S., Mele, M., Kamber, B. and de Gasparo, M. (1989). Biochem. Biophys. Res. Commun. *163*:284-291.
7. Dudley, D.T., Panek, R.L., Major, T.C., Lu, G.H., Bruns, R.F., Klinkefus, B.A., Hodges, J.C. and Weishaar, R.E. (1990). Mol. Pharmacol. *38*:370-377.
8. Chang, R.S.L., Lotti, V.J., Chen, T.B. and Faust, K.A. (1990). Biochem. Biophys. Res. Commun. *171*:813-817.
9. Zemel, S., Millan, M.A. and Aguilera, G. (1989). Endocrinology *124*:1774-1780.
10. Grady, E.F., Sechi, L.A., Griffin, C.A., Schambelan, M. and Kalinyak, J.E. (1991). J. Clin. Invest. *88*:921-933.
11. Sasaki, K., Yamano, Y., Bardhan, S., Iwai, N., Murray, J.J., Hasegawa, M., Matsuda, Y. and Inagami, T. (1991). Nature *351*:230-233.
12. Murphy, T.J., Alexander, R.W., Griendling, K.K., Runge, M.S. and Bernstein, K.E. (1991). Nature *351*:233-236.
13. Tsutsumi, K. and Saavedra, J.M. (1991). Am. J. Physiol. *261*:H667-H670.
14. Millan, M.A., Jacobowitz, D.M., Aguilera, G. and Catt, K.J. (1991). Proc. Natl. Acad. Sci. USA *88*:11440-11444.
15. Pucell, A.G., Hodges, J.C., Sen, I., Bumpus, F.M. and Husain A. (1991). Endocrinology *128*:1947-1959.
16. Kambayashi, Y., Bardhan, S., Takahashi, K., Tsuzuki, S., Inui, H., Hamakubo, T. and Inagami, T. (1993). J. Biol. Chem. *268*:24543-24546.
17. Mukoyama, M., Nakajima, M., Horiuchi, M., Sasamura, H., Pratt, R.E. and Dzau, V.J. (1993). J. Biol. Chem. *268*:24539-24542.

18. Nakajima, M., Mukoyama, M., Pratt, R.E., Horiuchi, M. and Dzau, V.J. (1993). Biochem. Biophys. Res. Commun. *197*:393-399.
19. Bottari, S.P., Taylor, V., King, I.N., Bogdal, Y., Whitebread, S. and de Gasparo, M. (1991). Eur. J. Pharmacol. *207*:157-163.
20. Dudley, D.T., Hubbell, S.E. and Summerfelt, R.M. (1991). Mol. Pharmacol. *40*:360-367.
21. Leung, K.H., Roscoe, W.A., Smith, R.D., Timmermans, P.B.M.W.M. and Chiu, A.T. (1992). Eur. J. Pharmacol. *227*:63-70.
22. Jaiswal, N., Tallant, E.A., Diz, D.I., Khosla, M.C. and Ferrario, C.M. (1991). Hypertension *17*:1115-1120.
23. Buisson, B., Bottari, S.P., de Gasparo, M., Gallo, P.N. and Payet, M.D. (1992). FEBS Lett. *309*:161-164.
24. Sumners, C., Tang, W., Zelezna, B. and Raizada, M.K. (1991). Proc. Natl. Acad. Sci. USA *88*:7567-7571.
25. Bottari, S.P., King, I.N., Reichlin, S., Dahlstroem, I., Lydon, N. and de Gasparo, M. (1992). Biochem. Biophys. Res. Commun. *183*:206-211.
26. Brechler, V., Levens, N.R., de Gasparo, M. and Bottari, S.P. (1993). Regul. Peptides *44*:207-213.
27. Takahasi, K., Bardhan, S., Kambayashi, Y., Shirai, H. and Inagami, T. (1994). Biochem. Biophys. Res. Commun. *198*:60-66.
28. Bottari, S.P., de Gasparo, M., Steckelings, U.M. and Levens, N.R. (1993). Front. Neuroendo. *14*:123-171.
29. Martin, M.M., Su, B. and Elton, T.S. (1994). Biochem. Biophys. Res. Commun. *205*:645-651.
30. Kemp, B.E. and Pearson, R.B. (1990). Trends Biochem. Sci. *15*:342-346.
31. Hubbard, S.C. and Ivatt, R.J. (1981). Annu. Rev. Biochem. *50*:555-583.
32. Servant, G., Dudley, D.T., Escher, E. and Guillemette, G. (1994). Mol. Pharmacol. *45*:1112-1118.
33. Guo, D., Furata, H., Mizukoshi, M. and Inagami, T. (1994). Biochem. Biophys. Res. Commun. *200*:313-319.
34. Su, B., Martin, M.M., Beason, K.B., Miller, P.J. and Elton, T.S. (1994). Biochem. Biophys. Res. Commun. *204*:1039-1046.
35. Takeuchi, K., Alexander, R.W., Nakamura, Y., Tsujino, T. and Murphy, T.J. (1993). Circ. Res. *73*:612-621.
36. Guo, D.F. and Inagami, T. (1994). Biochim. et Biophys. Acta *1218*:91-94.
37. Ichiki, T., Herold, C.L., Kambayashi, Y., Bardhan, S. and Inagami, T. (1994). Biochim. et Biophys. Acta *1189*:247-250.
38. Tsuzuki, S., Ichiki, T., Nakakubo, H., Kitami, Y., Guo, D., Shirai, H. and Inagami, T. (1994). Biochem. Biophys. Res. Commun. *200*:1449-1454.
39. Koike, G., Horiuchi, M., Yamada, T., Szpirer, C., Jacob, H.J. and Dzau, V.J. (1994). Biochem. Biophys. Res. Commun. *203*:1842-1850.
40. Martin, M.M. and Elton, T.S. Biochem. Biophys. Res. Commun. (1995, in press).
41. Csank, C., Taylor, F.M. and Martindale, D.W. (1990). Nucleic Acids Res. *18*:5133-5141.
42. Xu, L., Thali, M. and Schaffner, W. (1991). Nucleic Acids Res. *19*:6699-6704.
43. Ghosh, D. (1990). Nucleic Acids Res. *18*:1749-1756.
44. Morgan, W.D., Williams, G.T., Morimoto, R.I., Greene, J., Kingston, R.E. and Tjian, R. (1987). Mol. Cell. Biol. *7*:1129-1138.
45. Akira, S., Isshiki, H., Sugita, T., Tanabe, O., Kinoshita, S., Nishio, Y., Nakajima, T., Hirano, T. and Koshimoto, T. (1990). EMBO J. *9*:1897-1906.
46. Wasylyk, C., Flores, P., Gutman, A. and Wasylyk, B. (1989). EMBO J. *8*:3371-3378.
47. Wasylyk, B., Wasylyk, S., Flores, P., Beque, A.I., Leprinc, D. and Stehelin, D. (1990). Nature *346*:191-193.
48. Kadonaga, J.T., Carner, K.R., Mariarz, F.R. and Tjian, R. (1987). Cell *51*:1079-1090.
49. Chiu, R., Imugawa, M., Imbra, R.J., Bockoven, J.R. and Karin, M. (1987). Nature *329*:648-651.
50. Majello, B., Arcone, R., Toniatti, C. and Ciliberto, G. (1990). EMBO J. *9*:457-465.
51. Wall, L., deBoer, E. and Grosveld, F. (1988). Genes and Develop. *2*:1089-1100.
52. Driggers, P.H., Ennist, D.L., Gleason, S.L., Mak, W.H., Marks, M.S., Levi, B.Z., Flanagan, J.R., Appella, E. and Ozato, K. (1990). Proc. Natl. Acad. Sci. USA *87*:3743-3747.
53. Tsukiyama, T., Ueda, H., Hirose S. and Niwa, O. (1992). Mol. Cell. Biol. *12*:1286-1291.
54. Tsukiyama, T., Niwa, O., and Yokoro, K. (1989). Mol. Cell. Biol. *9*:4670-4676.
55. Grone, H.J., Simon, M. and Fuchs, E. (1992). Am. J. Physiol. *262*:F326-F331.
56. Nahmias, C., Cazaubon, S.M., Briend-Sutren, M.M., Lazard, D., Villageois, P. and Strosberg, A.D. (1995). Biochem. J. *306*:87-92.
57. Dudley, D.T. and Summerfelt, R.M. (1993). Reg. Peptides *44*:199-206.
58. Kambayashi, Y., Bardhan, S. and Inagami, T. (1993). Biochem. Biophys. Res. Commun. *194*:478-482.

17

MOLECULAR AND FUNCTIONAL CHARACTERIZATION OF ANGIOTENSIN II AT2 RECEPTOR IN NEUROBLASTOMA N1E-115 CELLS

Clara Nahmias, Sylvie M. Cazaubon, Malène Sutren, Maryline Masson, Daniel Lazard, Phi Villageois, Nathalie Elbaz, and A. Donny Strosberg

ICGM and CNRS UPR 0415
22, rue Méchain 75014 Paris
France

INTRODUCTION

Since the discovery in 1989 of a second subtype (AT2) of angiotensin II (Ang II) receptors (1, 2), its functional significance has remained a matter of debate. Expression of AT2 receptors was detected in adrenals, uterus, ovary, heart, brain, and at highest levels in the developing fetus. This subtype has been associated with reduction of intracellular cGMP levels (3-5), modulation of tyrosine phosphorylation (4, 6-9), inhibition of a T-type calcium channel (10, 11), opening of potassium channels (12, 13), release of prostaglandin (14) and production of arachidonic acid (15). Recent molecular cloning of the AT2 receptor cDNA in rat (6, 16), mouse (17) and human (18, 19) revealed a structural organization typical of G-protein coupled receptors, but did not provide yet a clear understanding of transduction pathways and physiological functions associated with this subtype.

To address the question of AT2 signaling, we have undertaken a detailed pharmacological, molecular and functional analysis of this receptor subtype in the murine neuroblastoma cell line N1E-115 (9). These cells express a single molecular class of AT2 receptors and no detectable AT1 binding site. In these cells, AT2 receptors are functionally coupled to protein tyrosine dephosphorylation. The N1E-115 cell line thus constitutes a simple and well-characterized cellular model for further analyses of AT2 receptor regulation and function.

PHARMACOLOGICAL CHARACTERIZATION OF ANG II RECEPTORS IN N115 CELLS

Subtypes (AT1 and AT2) of Ang II receptors expressed in neuroblastoma N1E-115 cells were analyzed using the non-selective iodinated Ang II antagonist : ^{125}I-(Sar, Ile)-AngII,

Recent Advances in Cellular and Molecular Aspects of Angiotensin Receptors
Edited by Mohan K. Raizada et al., Plenum Press, New York, 1996

Table 1. Binding parameters of various ligands to membranes of N1E-115 cells and COS-AT2 cells. Values are from Nahmias et al. (9). n.d. stands for non determined

	N1E-115		COS-AT2	
Radioligand	Kd(pM)	Bmax(fmol/mg)	Kd(pM)	Bmax(fmol/mg)
^{125}I-(Sar,Ile)AngII	233±33	298±53	n.d.	n.d.
^{125}I-CGP 42112	91±19	321±42	75±4	125±11
Competitor	IC50 (nM)		IC50 (nM)	
(Sar,Ile)AngII	0.45±0.09		0.39±0.07	
CGP 42112	0.53±0.15		0.45±0.08	
Ang II	1.22±0.12		0.84±0.28	
PD 123319	4.60±0.21		4.74±0.55	
DUP 753	> 10,000		>10,000	

and the AT2-selective radioligand ^{125}I-CGP 42112. Scatchard analyses revealed that both ligands bound to a single class of high affinity binding sites with the same Bmax (Table I), indicating that most if not all Ang II receptors were of the AT2 subtype. The absence of the AT1 subtype was confirmed by the inefficiency of the AT1-selective ligand DUP 753 to compete for binding of ^{125}I-(Sar, Ile)-AngII (Table I). In contrast, the AT2-selective ligands CGP 42112 and PD 123319 were able to totally displace the binding of ^{125}I-(Sar, Ile)-AngII in monophasic curves, indicating the presence of a single pharmacological class of AT2 receptors, which binding parameters are presented in Table I.

Fluharty and colleagues (20, 21) previously reported expression of both AT1 and AT2 receptor subtypes in N1E-115 cells. More recently, these authors identified two subpopulations of AT2 receptors (designated as peak I and peak III, respectively) that could be separated by heparine sepharose chromatography (22). These two populations differed in their affinity for the AT2-selective ligand PD 123319, their sensitivity to DTT and to analogs of GTP, and their reactivity towards a polyclonal anti-AT2 receptor antiserum. Discrepancy between these data and ours may be explained by the divergence occuring in cell lines that were independently passaged for many years under different culture conditions. The AT2 subtype expressed in our cell line shows moderate affinity for PD 123319, increased binding in the presence of reducing reagent and insensitivity to GTPgS (9), and would thus correspond to the so-called “peak III”.

Cloning of AT2 receptor cDNA from N1E-115

For a further molecular characterization, cDNA clones were isolated from a N1E-115 cDNA library by homology screening, using a DNA probe derived by a PCR strategy on the basis of the rat AT2 receptor sequence (9). Nucleotide sequencing revealed that AT2 cDNA clones from N1E-115 belong to the same molecular subtype as those isolated from rat (16) and mouse (17) fetal tissues, as well as from rat pheochromocytoma PC12W cells (6), human myometrium (18) and human lung (19).

One remarkable feature of the mouse AT2 cDNA sequence is the presence of two in-frame initiator ATG codons that potentially generate two polypeptides of different length (Fig.1). The downstream ATG codon is more likely utilized, as its position is in agreement with Kozak’s consensus for initiation of translation (23), and corresponds to the single initiator ATG codon present in the rat (6, 16) and the human (18, 19) AT2 sequences. It cannot be excluded however that in certain conditions, translation is initiated at the upstream ATG codon, leading to expression of a polypeptide carrying 30 additional amino-terminal residues. Deletion of the cDNA sequence corresponding to these 30 amino acids did not affect the binding properties of the receptor expressed in COS cells (data not shown). Translation of these additional residues may nevertheless affect the stability or the routing of the molecule inside the cell.

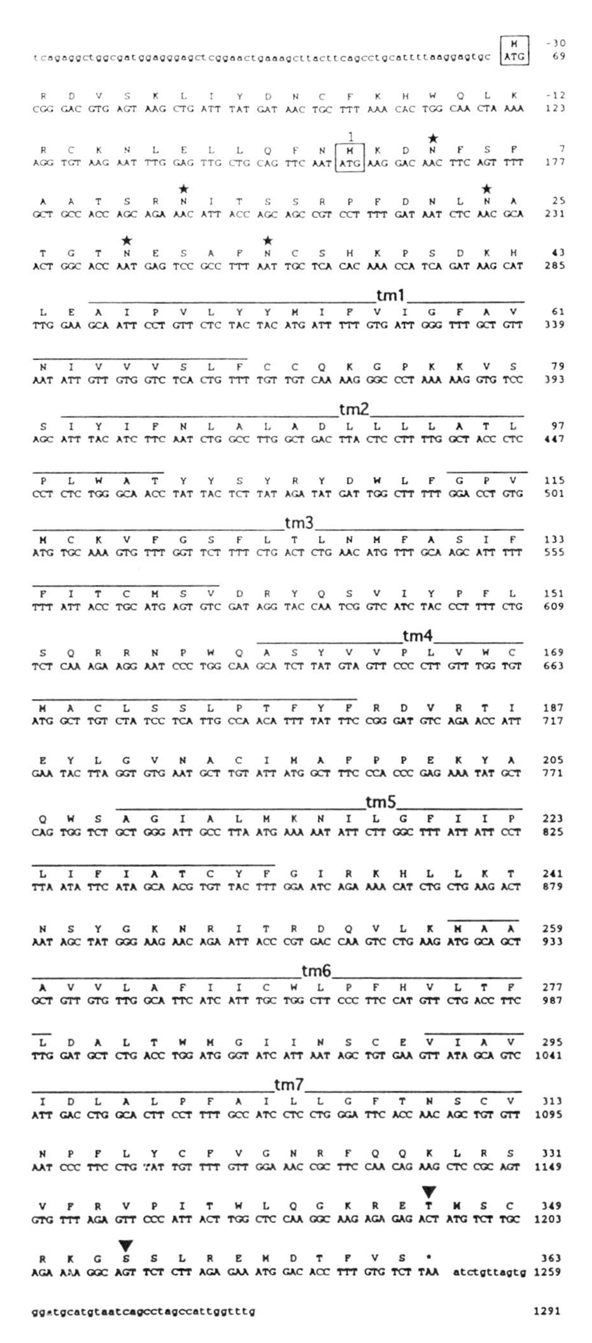

Figure 1. Nucleotide and deduced amino-acid sequences of the AT2 cDNA from N1E-115 cells. Two potential initiator ATG codons are boxed. Amino-acid number one corresponds to the downstream initiator ATG codon (position 157 of the nucleotide sequence). The seven transmembrane segments (tm1-tm7) are highlighted by solid bars. Potential N-linked glycosylation sites are indicated by stars. Putative sites for phosphorylation by protein kinase A and protein kinase C are indicated by triangles.

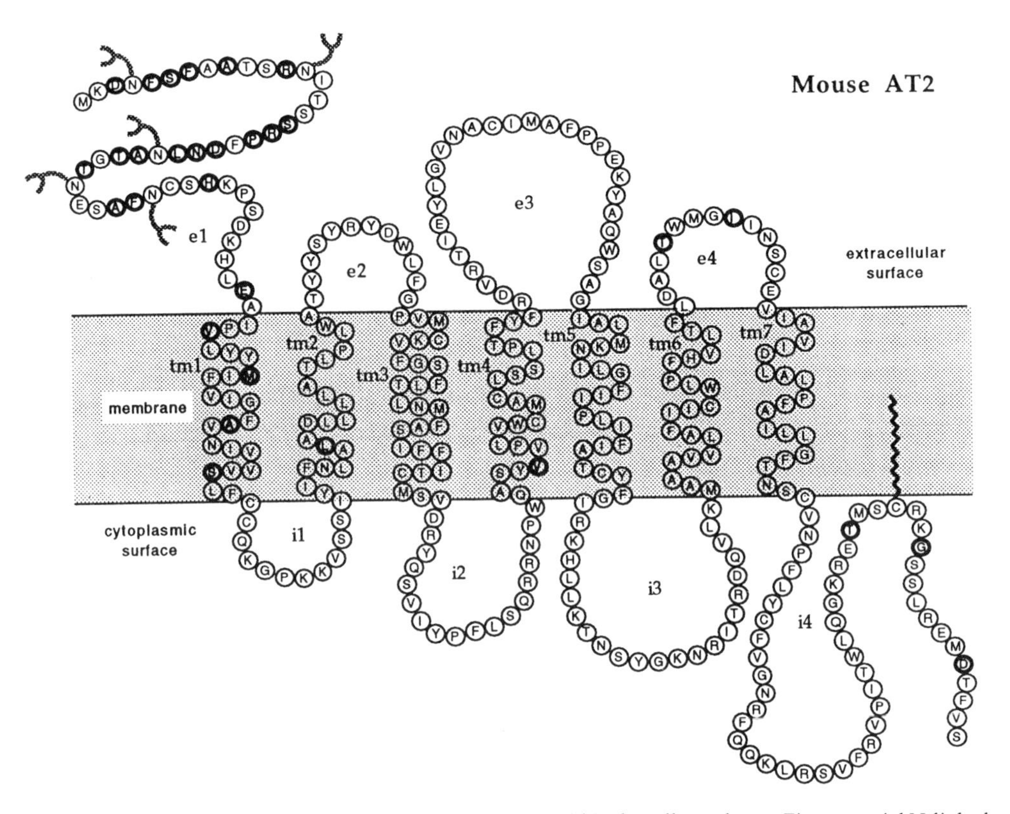

Figure 2. Putative organization of the murine AT2 receptor within the cell membrane. Five potential N-linked glycosylated sites in the amino-terminus and a putative palmitylated cysteine in the carboxy-terminus are indicated. Residues which differ between mouse and human AT2 sequences are in bold.

The deduced amino-acid sequence of the mouse AT2 polypeptide (Fig. 2) displays all the features characteristic of G-protein coupled receptors (24), including the presence of seven membrane-spanning domains, potential N-linked glycosylation sites, and consensus sites for phosphorylation by protein kinases in the carboxy terminal tail. Sequence comparison indicates high conservation (92% amino-acid sequence homology) between human and murine AT2 receptors (Fig. 2). Most divergent residues are clustered in the extracellular, N-terminal portion of the molecule. Despite these differences, five putative N-linked glycosylation sites are conserved at identical positions in the amino-terminal part of the AT2 receptor in both species.

The isolated cDNA clones were functionally expressed in COS cells, yielding pharmacological parameters identical to those of endogeneous Ang II receptors from N1E-115 cells (Table I). This confirms that N1E-115 cells express exclusively the AT2 receptor subtype, and identifies this receptor entity at the molecular level.

FUNCTIONAL STUDIES OF AT2 RECEPTORS: COUPLING TO PROTEIN TYROSINE DEPHOSPHORYLATION

The pharmacological and molecular identification of a single class of AT2 receptors in N1E-115 cells prompted us to further analyze this receptor subtype at the functional level.

The relationship between AT2 signaling and intracellular protein tyrosine phosphorylation was investigated by Western blot analysis of total cell lysates using anti-phosphotyrosine antibodies. Treatment of N1E-115 cells with Ang II led to a reduction in tyrosine phosphorylation of several endogeneous proteins, of apparent molecular masses 80, 97, 120, 150 and 180 kDa, respectively (Fig. 3). This effect was rapid and transient, showing a maximum at 5 to 10 minutes and declining at 30 minutes. This response could be blocked in the presence of an excess of the antagonist (Sar, Ile)-Ang II and was also obtained following treatment with the AT2-selective agonist CGP 42112 (9).

Preliminary studies of Chinese Hamster Ovary (CHO) cells stably expressing the mouse AT2 receptor indicated that in this cellular model, AT2 stimulation also results in the transient dephosphorylation of cellular proteins on tyrosine residues (data not shown). The same intracellular effect could not however be detected in COS cells transiently transfected with either the murine or the human AT2 receptor cDNA. This may indicate that COS cells lack one or several components of the AT2 signaling pathway. It is worth noting that other groups reporting the cloning of AT2 receptor cDNA failed to detect any intracellular effect of this receptor after both transient (16) and stable (6) overexpression in COS cells.

Our results showing a reduction of intracellular protein tyrosine phosphorylation may be explained by stimulation of a protein tyrosine phosphatase (PTP) activity, or by inhibition of a tyrosine kinase-mediated pathway. Given the complexity of intracellular cascades of kinases and phosphatases (25-27), it is indeed possible that AT2 receptors differentially affect several types of enzymes acting at different levels. In the PC12W cell line, AT2 receptors have been reported to stimulate a PTP activity by a G-protein-independent mechanism (4, 7) and to inhibit a PTP through a pertussis toxin-mediated pathway (6, 8). Progress in understanding AT2 signaling will greatly benefit from investigating molecular determinants of AT2-mediated pathway, including identification of the G-protein(s) involved, search for cellular "interaction partners" of the receptor, elucidation of PTPs/TKs activities modulated by the AT2 subtype, and identification of endogeneous phosphoproteins which are dephosphorylated following AT2 receptor activation.

The relationship between AT2 receptors and tyrosine phosphorylation suggests a link with physiological processes such as cell growth, differentiation or cellular adhesion, which are known to involve cascades of protein kinases and phosphatases (28). Possible role for AT2 receptors in these processes is also consistent with increased levels of expression of this subtype during embryonic development (29) and in pathological situations such as vascular growth (30) and tissue repair (31). In vivo studies have indeed identified AT2 receptors as

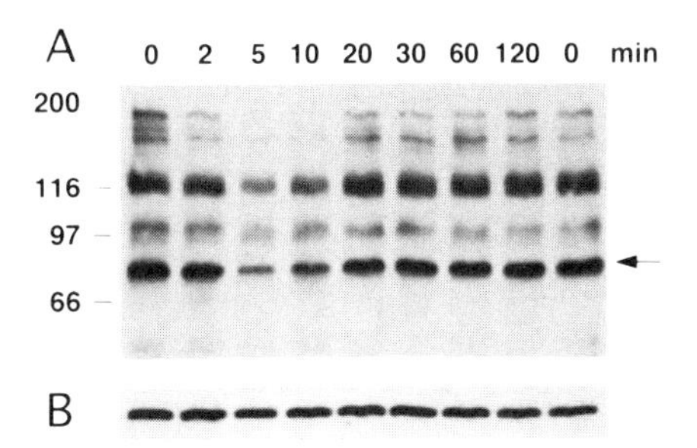

Figure 3. Ang II-induced tyrosine dephosphorylation in N1E-115 cells. (A) N1E-115 cells were incubated with Ang II for various periods of time, before lysis and subsequent immunoblotting with anti-phosphotyrosine antibodies. On the left are indicated the positions of standard molecular weight markers, and the arrow on the right shows the position of a major phosphoprotein of 80 kDa. (B) After removal of bound antibodies, the same blot was incubated with anti-ezrin antibodies for internal control of the amount of proteins in each lane.

participating in vascular neointima formation (30) and angiogenesis (32). More recently, anti-proliferative effects of AT2 receptors in primary cultures of rat endothelial coronary cells have been demonstrated (33). Other studies performed in cultured cells tend to indicate, however, inverse relationship between expression of AT2 receptors and cell proliferation (34-36).

ARE THERE MULTIPLE SUBCLASSES OF AT2 RECEPTORS?

Studies conducted in the rat brain have pointed out possible heterogeneity of AT2 receptors. Quantitative autoradiography revealed differential sensitivity of CGP 42112-binding sites to analogs of GTP (37) and to reducing reagents (38, 39). In addition, immunohistochemical staining of rat brain sections using anti-AT2 polyclonal antibodies correlated with, but was not identical to patterns of radioligand binding (40). As mentioned before, two pharmacologically, biochemically and immunologically distinct populations of AT2 receptors could be identified in N1E-115 cells (22). These observations, as well as the diverse and sometimes contradictory reports of AT2 signaling, may be due to expression of a single molecular entity in different cell environments, and/or may reflect the existence of multiple subclasses of AT2 receptors.

The AT2 receptor gene exists as a single copy in the genome of both human (18) and mouse (17). This argues against the possibility that several highly homologous AT2 genes exist, as is the case for rodent AT1a and AT1b. In addition, the coding region of the AT2 gene is devoid of introns, thus ruling out the hypothesis that alternative splicing may lead to multiple forms of receptors. It remains possible however that "AT2 receptors" in fact consist of several molecular entitites, that do not share high amino-acid sequence homology but that all exhibit high affinity for ligands such as CGP 42112 and PD 123319 which define the AT2 subtype pharmacologically.

Further characterization of AT2 receptors in a variety of cellular environments is now made possible by the development of series of new ligands (41, 42) and may help discriminate between putative AT2 receptor subtypes. The molecular basis of AT2 receptor heterogeneity will also certainly be investigated in the coming years, as more and more AT2 cDNA clones will be isolated from different tissues and cell lines.

ACKNOWLEDGMENTS

This work was supported by the Centre National de la Recherche Scientifique, the Institut National de la Santé et de la Recherche Médicale, University Paris VII, and the Ministry for Research. We are also grateful to the Association pour la Recherche sur le Cancer, the Fondation pour la Recherche Medicale and the Ligue Nationale contre le Cancer.

REFERENCES

1. Chiu, A. T. *et al.* (1989). Eur J Pharmacol *170*: 117-118.
2. Whitebread, S., Mele, M., Kamber, B. and de Gasparo, M. (1989). Biochem Biophys Res Commun *163*: 284-291.
3. Sumners, C., Tang, W., Zelezna, B. and Raizada, M. K. (1991). Proc Natl Acad Sci U S A *88*: 7567-7571.
4. Bottari, S. P., King, I. N., Reichlin, S., Dahlstroem, I., Lydon, N. and de Gasparo, M. (1992). Biochem Biophys Res Commun *183*: 206-211.
5. Brechler, V., Levens, N. R., de Gasparo, M. and Bottari, S. P. (1993). Regulatory Peptides *44*: 207-213.

6. Kambayashi, Y., Bardhan, S., Takahashi, K., Tsuzuki, S., Inui, H., Hamakubo, T. and Inagami, T. (1993). J Biol Chem *268*: 24543-24546.
7. Brechler, V., Reichlin, S., de Gasparo, M. and Bottari, S. P. (1994). Receptors and Channels *2*: 89-97.
8. Takahashi, K., Bardhan, S., Kambayashi, Y., Shirai, H. and Inagami, T. (1994). Biochem Biophys Res Commun *198*: 60-66.
9. Nahmias, C., Cazaubon, S. M., Briend-Sutren, M.-M., Lazard, D., Villageois, P. and Strosberg, A. D. (1995). Biochem J *306*: 87-92.
10. Buisson, B., Bottari, S. P., de Gasparo, M., Gallo, P. N. and Payet, M. D. (1992). FEBS lett. *309*: 161-164.
11. Buisson, B., Laflamme, L., Bottari, S. P., de Gasparo, M., Gallo-Payet, N. and Payet, M. D. (1995). J Biol Chem *270*: 1670-1674.
12. Kang, J., Posner, P. and Sumners, C. (1994). Am J Physiol *267*: C1289- C1397.
13. Sumners, C., Raizada, M. K., Kang, J., Lu, D. and Posner, P. (1994). Frontiers in Neuroendocrinology *15*: 203-230.
14. Jaiswal, N., Tallant, E. A., Diz, D. I., Khosla, M. C. and Ferrario, C. M. (1991). Hypertension *17*: 1115-1120.
15. Lokuta, A. J., Cooper, C., Gaa, S. T., Wang, H. E. and Rogers, T. B. (1994). J Biol Chem *269*: 4832-4838.
16. Mukoyama, M., Nakajima, M., Horiuchi, M., Sasamura, H., Pratt, R. E. and Dzau, V. J. (1993). J Biol Chem *268*: 24539-24542.
17. Nakajima, M., Mukoyama, M., Pratt, R. E., Horiuchi, M. and Dzau, V. J. (1993). Biochem Biophys Res Commun *197*: 393-399.
18. Lazard, D., Briend-Sutren, M.-M., Villageois, P., Mattei, M.-G., Strosberg, A. D. and Nahmias, C. (1994). Receptors and Channels *2*: 271-280.
19. Martin, M. M., Su, B. G. and Elton, T. S. (1994). Biochem Biophys Res Commun *205*: 645-651.
20. Reagan, L. P., Ye, X., Mir, R., DePalo, L. R. and Fluharty, S. J. (1990). Mol Pharmacol *38*: 878-886.
21. Reagan, L. P., Ye, X., Maretzski, C. H. and Fluharty, S. J. (1993). J Neurochem *60*: 24-31.
22. Siemens, I. R., Reagan, L. P., Yee, D. K. and Fluharty, S. J. (1994). J Neurochem *62*: 2106-2115.
23. Kozak, M. (1987). Nuc Acids Res *15*: 8125-8132.
24. Strosberg, A. D. (1991). Eur J Biochem *196*: 1-10.
25. Vogel, W., Lammers, R., Huang, J. and Ullrich, A. (1993). Science *259*: 1611-1614.
26. Sun, H. and Tonks, N. K. (1994). Trends Biochem. Sci. *19*: 480-485.
27. Hunter, T. (1995). Cell *80*: 225-236.
28. Schlessinger, J. and Ullrich, A. (1992). Neuron *9*: 383-391.
29. Grady, E. F., Sechi, L. A., Griffin, C. A., Schambelan, M. and Kalinyak, J. E. (1991). J Clin Invest *88*: 921-933.
30. Janiak, P., Pillon, A., Prost, J. F. and Vilaine, J. P. (1992). Hypertension *20*: 737-745.
31. Viswanathan, M. and Saavedra, J. M. (1992). Peptides *13*: 783-786.
32. le Noble, F. A. C., Schreurs, N., van Straaten, H. W. M., Slaaf, D. W., Smits, J. F. M. and Struyker Boudier, H. A. J. (1992). FASEB J *6*: A937 (Abstract).
33. Stoll, M., Steckelings, M., Paul, M., Bottari, S. P., Metzger, R. and Unger, T. (1995). J Clin Invest *95*: 651-657.
34. Dudley, D. T. and Summerfelt, R. M. (1993). Regul Pept *44*: 199-206.
35. Grady, E. F. and Kalinyak, J. E. (1993). Regul Pept *44*: 171-80.
36. Kambayashi, Y., Bardhan, S. and Inagami, T. (1993). Biochem Biophys Res Commun *194*: 478-82.
37. Tsutsumi, K. and Saavedra, J. M. (1992). Mol Pharmacol *41*: 290-297.
38. Tsutsumi, K., Zorad, S. and Saavedra, J. M. (1992). Eur. J. Pharmacol. *226*: 169-173.
39. Speth, R. C. (1993). Regulatory peptides *44*: 189-197.
40. Reagan, L. P., Flanagan-Cato, L. M., Yee, D. K., Ma, L.-Y., Sakai, R. R. and Fluharty, S. J. (1994). Brain Res. *662*: 45-59.
41. VanAtten, M. K., Ensinger, C. L., Chiu, A. T., McCall, D. E., Nguyen, T. T., Wexler, R. R. and Timmermans, P. B. (1993). J Med Chem *36*: 3985-3992.
42. Chang, L. L.*et al.* (1994). J Med Chem *37*: 4464-4478.

CHARACTERIZATION OF THE AT_2 RECEPTOR ON RAT OVARIAN GRANULOSA CELLS

Masami Tanaka,[1] Junji Ohnishi, Yasuhiro Ozawa, Masataka Sugimoto, Satoshi Usuki,[2] Mitsuhide Naruse,[1] Kazuo Murakami, and Hitoshi Miyazaki

Institute of Applied Biochemistry
Gene Experiment Center
University of Tsukuba, Ibaraki 305, Japan
[1] Department of Medicine
Institute of Clinical Endocrinology
Tokyo Women's Medical College
Shinjuku-ku, Tokyo 162, Japan
[2] Institute of Clinical Medicine
University of Tsukuba
Ibaraki 305, Japan

INTRODUCTION

Recent pharmacological and molecular biological studies have identified two distinct subpopulations of Ang II receptors, designated AT_1 and AT_2 (1-5). Most of the known physiological functions of Ang II are mediated by the AT_1 receptor. In contrast, the functional roles of the AT_2 receptor have not yet become apparent. The AT_1 receptor is widely distributed in adult tissues, whereas the AT_2 receptor exhibits widespread and abundant expression in fetal tissues including skin, tongue, brain, and skeletal muscles but decreases dramatically and rapidly after birth (1, 4-6). The AT_2 receptor is also present in adult tissues including ovary and skin where its expression is strictly regulated (7-9). In this study, to provide a basis for elucidating the physiological roles of local renin-angiotensin system through the AT_2 receptor in the ovary, we examined the change in the content of the rat ovarian AT_2 receptor under several conditions as well as its biochemical and physiological properties.

THE AT_2 RECEPTOR IN THE OVARY

In 1988 we identified novel Ang II receptors in the bovine ovary, corresponding to the AT_2 receptor according to the present nomenclature, based upon their sensitivity to

Recent Advances in Cellular and Molecular Aspects of Angiotensin Receptors
Edited by Mohan K. Raizada et al., Plenum Press, New York, 1996

dithiothreitol (DTT) (9). That is, DTT stimulated the Ang II binding capacity of ovarian membranes by increasing the ligand binding affinity, whereas the agent reduced the Ang II binding capacity of adrenal membranes in which the AT1 receptor predominates. In addition to Ang II receptors, there is accumulating evidence for the presence of all the components of the renin-angiotensin system in the ovary (10-14). Husain et al. (11) reported that the concentration of ovarian Ang II is 8- to 75-fold greater than that of plasma and is not reduced in bilaterally nephrectomized rats. These data suggested physiological importance of the local renin-angiotensin system in the ovary. Furthermore, autoradiographic studies by Pucell et al. (7) have demonstrated selective expression of the AT_2 receptor on granulosa cells of rat ovarian atretic follicles, which undergo apoptosis. Recent studies indicated the AT_2 receptor-mediated activation of protein tyrosine phosphatase and serine/threonine protein phosphatases, enzymes which are thought to closely correlate with the regulation of cell growth (15-17). Together these data suggested the possibility that Ang II produced in the ovary may modulate the onset and/or progression of follicle atresia involving apoptosis of granulosa cells through the AT_2 receptor. To elucidate the roles of this receptor in the ovary, it is of importance to know the regulation of the AT_2 content during differentiation and apoptosis of granulosa cells.

Here we describe the change in the AT_2 receptor content on rat ovarian granulosa cells in culture, focusing upon the effects of Ang II and follicle stimulating hormone (FSH), a differentiation factor of these cells, following biochemical and pharmacological characterization of the receptor (18, 19). We also suggest the relation of the AT_2 receptor content with apoptosis of cultured granulosa cells (19).

Biochemical and Pharmacological Properties of the AT_2 Receptor on Rat Cultured Granulosa Cells

Effect of DTT on the AT_2 Receptor. The existence of the AT_2 receptor has been suggested by several recent studies mainly based on the difference in their sensitivity to DTT and GTPγS. For example, Tsutsumi et al. (20, 21) have demonstrated that the AT_2 receptor in the inferior olive is not influenced by DTT and GTPγS, whereas the ligand binding of the receptor localized in the ventral thalamic and medial geniculate nuclei is decreased by these two reagents. In our experiment (9), the ligand binding of the AT_2 receptor on bovine ovarian membranes is increased by DTT but insensitive to GTPγS. The existence of the AT_2 receptor on rat granulosa cells in culture has been found by means of competitive binding studies using the subtype-selective antagonists such as Dup753 for the AT_1 and PD123319 for the AT_2 receptors, but correlation with the responses to DTT was not examined (7). Therefore, we examined the effect of DTT on the AT_2 receptor on rat granulosa cells in culture.

In this study, granulosa cells were isolated from ovaries of diethylstilbestrol-treated Sprague-Dawley rats according to the method of Knecht et al. (22). Briefly, immature 21-day-old rats were implanted with diethylstilbestrol for five days to stimulate granulosa cell proliferation, then granulosa cells were harvested by follicular puncture. The cells were cultured in McCoy's 5A medium under serum-free conditions with or without several reagents in a humidified 95% air, 5% CO_2 atmosphere at 37 °C.

We performed a competitive binding study in the presence and absence of DTT using ^{125}I-[Sar^1,Ile^8]Ang II as the ligand. Unlabeled [Sar^1,Ile^8]Ang II inhibited the radioligand binding in a dose-dependent manner under both conditions. However, the IC_{50} value (4×10^{-10} M) in the cells exposed to DTT was 7.5-fold lower than that (3×10^{-9} M) in the untreated cells. The agent also increased the specific binding of ^{125}I-[Sar^1,Ile^8]Ang II up to 3.0-fold over the control level measured in the absence of the competitor. These data suggested that exposing of cells to DTT increased the affinity of Ang II receptors for the ligand

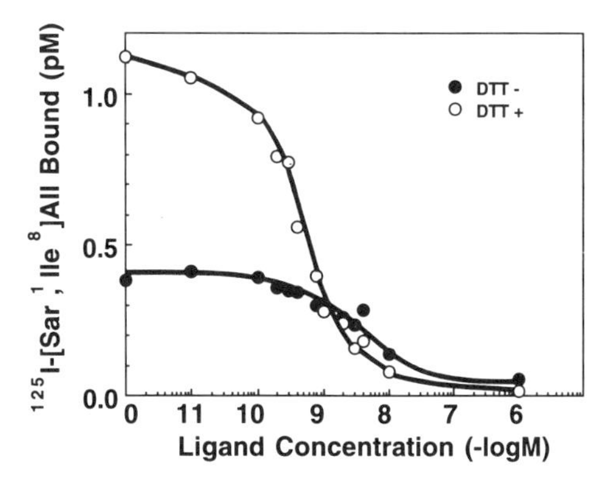

Figure 1. Competitive inhibition curves of ^{125}I-[Sar^1,Ile^8]Ang II binding to cultured ovarian granulosa cells. Cells exposed (open circle) or not (closed circle) to DTT were incubated with ^{125}I-[Sar^1 Ile^8]Ang II in the presence of the indicated concentrations of unlabeled [Sar^1 ,Ile^8]Ang II at 4 °C for 2 h. The data represent the mean of two determinations.

[Sar^1,Ile^8]Ang II. Scatchard plot analysis of ^{125}I-[Sar^1 ,Ile^8]Ang II binding, transferred from equilibrium binding study, yielded an apparent straight line indicative of one class of binding sites in cultured granulosa cells exposed to DTT or not (data not shown). Dithiothreitol caused little change in the Bmax value but markedly decreased the Kd value by about 65%; Kd = 2.0 x 10^{-9} M, Bmax = 12,600 sites/cell without DTT and Kd = 7.5 x 10^{-10} M, Bmax = 11,300 sites/cell with DTT. These Kd values were consistent with the IC_{50} values determined above with and without DTT, respectively. Thus, DTT caused a significant increase in the affinity of Ang II receptor for [Sar^1 ,Ile^8]Ang II, demonstrating that Ang II receptor in rat granulosa cells is the DTT-stimulated AT_2 receptor. This receptor must be identical with the AT_2 receptor of which the cDNA has recently been cloned from rat fetal and PC12W cell cDNA libraries (4, 5), because DTT increased the Ang II binding activity of the cloned receptor transiently expressed in COS-7 cells (4). In fact, the DNA fragment of the cloned AT_2 receptor cDNA was amplified by polymerase chain reaction using RNA extracted from cultured granulosa cells.

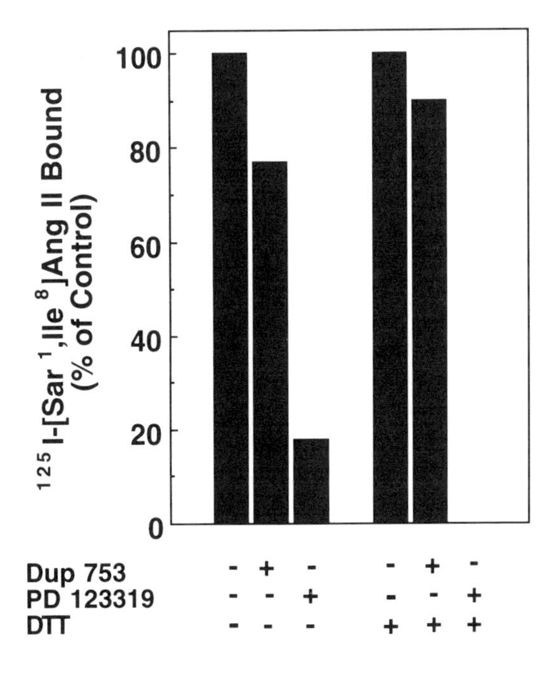

Figure 2. Effect of subtype-selective Ang II antagonists on ^{125}I-[Sar^1, Ile^8]Ang II binding to cultured ovarian granulosa cells. Cells exposed or not to DTT were incubated with ^{125}I-[Sar^1, Ile^8]Ang II (50 pM) in the presence or absence of Dup753 (100 μM) or PD123319 (100 μM). Results are given as the percent of control binding measured in the absence of the competitors. Each data represents the mean of two independent experiments each determined in duplicate.

Selectivity of Ang II Antagonists to DTT-Treated AT_2 Receptor. We next compared the binding specificity of the AT_2 receptor for the AT_1- and AT_2-selective antagonists Dup753 and PD123319, respectively, between the cells exposed to DTT or not (Fig. 2). PD123319 (100 μM) caused an 82% inhibition in ^{125}I-[Sar^1, Ile^8]Ang II binding, whereas 100 μM Dup753 decreased the binding by only 23% in the absence of DTT, again indicating the predominance of the AT_2 receptor in granulosa cells. Similar data were obtained from the DTT-treated cells. PD123319 at 100 μM completely abolished the radioligand binding but there was little inhibition by 100 μM Dup753. Therefore, the specificity of the AT_2 receptor for the two antagonists was the same, whether or not it was exposed to DTT. Together with the data from the competitive binding study, these results clearly showed that DTT increases the ligand binding activity of the AT_2 receptor without changing the selectivity for the subtype-specific antagonists.

The predominance of the AT_2 receptor in granulosa cells of atretic follicles in vivo (7) suggested the possibility that Ang II might play an important role via this type of receptor in the development of atresia. However, the involvement of a small population of the AT_1 receptor cannot be excluded based upon our data since 100 μM Dup753 induced a slight decrease in the ^{125}I-[Sar^1, Ile^8]Ang II binding in the absence of DTT. In fact, Currie et al. (23) have reported that 25% of granulosa cells prepared from rat ovarian follicles exhibit an increase in intracellular Ca^{2+} concentrations in response to Ang II, which is characteristic of the AT_1 receptor. To elucidate the roles of the AT_2 receptor in the process of atresia, specific detection of the AT_2 receptor is essential. For this purpose, DTT, which markedly decreases the ligand binding activity of the AT_1 receptor and increases that of the AT_2 receptor without affecting the antagonist selectivity, will provide a useful method.

Change in the AT_2 Receptor Content during Differentiation and Apoptosis of Rat Cultured Granulosa Cells

Effect of FSH and Ang II on the AT_2 Receptor Content. To investigate the change in the AT_2 receptor content during differentiation of cultured granulosa cells by FSH, ^{125}I-[Sar^1, Ile^8]Ang II was incubated with granulosa cells maintained in the presence or absence of FSH under serum-free conditions. To specifically detect the AT_2 receptor, cells were pretreated with DTT. This treatment is necessary because the involvement of a small population of the AT_1 receptor has previously been demonstrated in rat ovarian granulosa cells as described

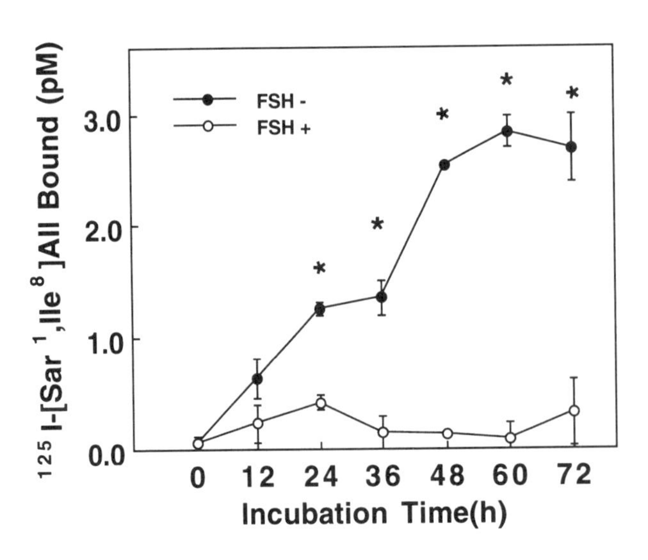

Figure 3. Effect of FSH on the AT_2 receptor content. Granulosa cells (5 x 10^5 cells/tube) were cultured in the presence (open circle) or absence (closed circle) of FSH (200 ng/ml) for indicated periods of time. Thereafter, ^{125}I-[Sar^1 ,Ile^8]Ang II (50 pM) binding assay was performed. The data represent the mean ± SD of triplicate determinations. The same results were obtained in all of three separate experiments. *, $p < 0.01$ compared to FSH-treated cells.

above. As shown in Fig. 3, FSH (200 ng/ml) did not produce any significant change in ^{125}I-[Sar^1,Ile^8]Ang II binding during 72 h of culture. In contrast, the radioligand binding was markedly increased without FSH in a time-dependent manner. These results are in agreement with the finding that the AT_2 receptor is selectively expressed on granulosa cells of atretic follicles which are not stimulated by FSH due to the lack of FSH receptors (7). Furthermore, the increase in the AT_2 content in FSH-free media is a unique characteristic of the AT_2 receptor because levels of most receptors for hormones and growth factors on granulosa cells, such as epidermal growth factor (EGF), FSH, prolactin, and leutinizing hormone (LH), are known to be up-regulated by FSH (24, 25).

To study the effect of Ang II on homologous regulation of the AT_2 receptor, the cells were cultured with Ang II under FSH-free conditions, then ^{125}I-[Sar^1,Ile^8]Ang II binding was examined. Ang II at 1 μM augmented the radioligand binding by over 2-fold compared with control cells incubated in the absence of Ang II during 72 h of culture (Fig. 4). This result is inconsistent with the data by Pucell et al. (25) that indicated Ang II-induced suppression of ^{125}I-[Sar^1,Ile^8]Ang II binding to cultured granulosa cells. Under our experimental conditions the cells were pretreated with acid buffer (50 mM glycine/150 mM NaCl, pH 3.0) to remove Ang II bound to its receptors before the binding assay, whereas Pucell et al. did not perform the acid treatment. Therefore, this discrepancy may reflect a difference in experimental conditions of ^{125}I-[Sar^1,Ile^8]Ang II binding assay.

Effect of FSH and Ang II on the AT_2 Receptor Expression. To examine whether the change in the AT_2 receptor content occurs at the mRNA level as well as at the protein level, the AT_2 receptor mRNA was measured by a competitive reverse-transcriptional polymerase chain reaction method. The expression level of the receptor mRNA in granulosa cells cultured with FSH (200 ng/ml) for 48 h was 10.0-fold lower than that in control cells incubated without FSH (data not shown). Ang II at 1 μM augmented an increase in the receptor mRNA expression by 1.8-fold after a 48-h incubation compared to that in control cells incubated without Ang II and FSH (data not shown). These effects of FSH and Ang II are comparable to those on the AT_2 receptor at the protein level.

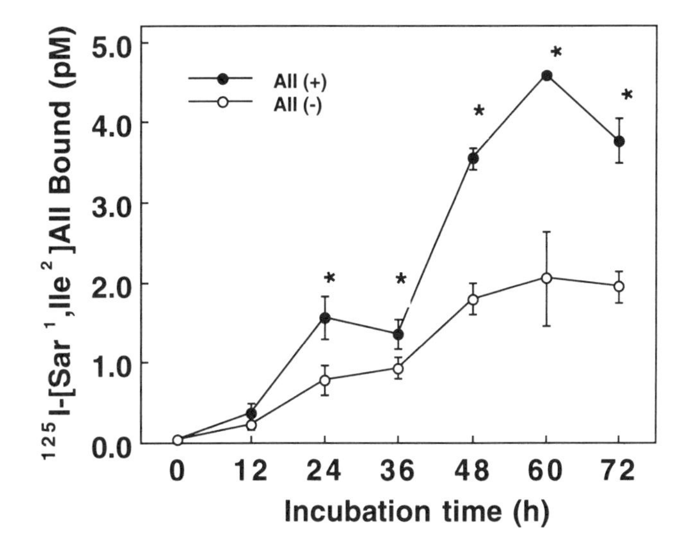

Figure 4. Effect of Ang II on the AT2 receptor content. Granulosa cells (5 x 10^5 cells/tube) were cultured in the presence (closed circle) or absence (open circle) of Ang II (100 nM) for indicated periods of time. Thereafter, ^{125}I-[Sar^1,Ile^8]Ang II (50 pM) binding assay was performed. The data represent the mean ± SD of triplicate determinations. The same results were obtained in all of three separate experiments. *, $p < 0.01$ compared to values without Ang II.

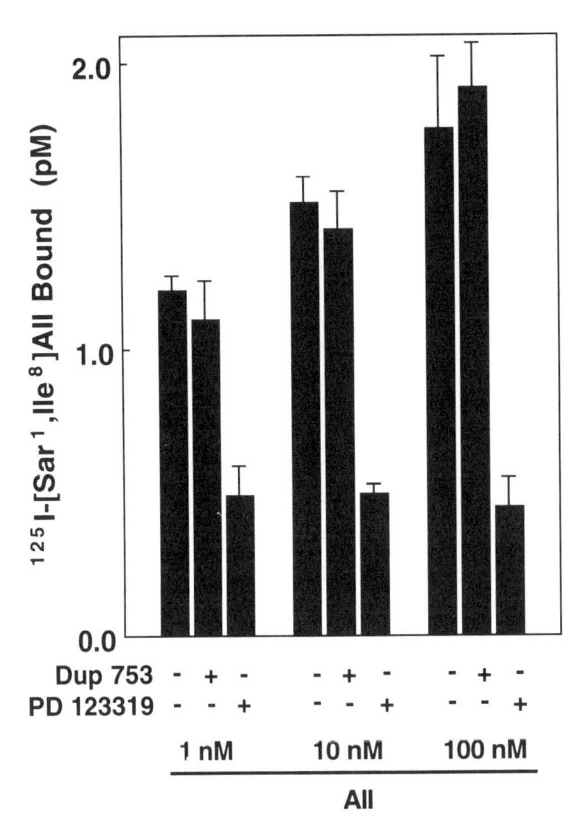

Figure 5. Effect of Ang II antagonists on Ang II-induced increase in the AT_2 receptor. Granulosa cells (5 x 10^5 cells/tube) were cultured with Ang II alone and with increasing concentrations of Ang II for 48 h in the presence and absence of the Ang II antagonists (1 μM) Dup753 or PD123319. Thereafter, ^{125}I-[Sar^1,Ile^8]Ang II (50 pM) binding assay was performed. The data represent the mean ± SD of triplicate determinations.

Which Subtype of Ang II Receptor Mediates Ang II-Induced Increase in the AT_2 Receptor Content? The effects of the respective AT_1- and AT_2-selective antagonists, Dup753 and PD123319, on Ang II-induced increase in ^{125}I-[Sar^1,Ile^8]AII binding were examined to know whether this phenomenon is mediated by the AT_1 or AT_2 receptor (Fig. 5). Ang II at 1 nM to 100 nM produced a concentration-dependent increase in the radioligand binding to granulosa cells cultured in the absence of FSH for 48 h. Dup753 at 1 μM had no effect on Ang II-induced increase in the radioligand binding, whereas PD123319 at 1 μM inhibited this Ang II action. Moreover, in contrast to that Dup753 or PD123319 alone produced no change in the AT_2 receptor content during 48 h of culture, CGP42112B, an AT_2 specific agonist, increased the receptor content up to the same level as Ang II (data not shown). These data demonstrated that Ang II-caused augmentation in the AT_2 receptor content under FSH-free conditions is mediated by the AT_2 receptor itself.

Relation of Change in the AT_2 Receptor Content with Apoptosis of Cultured Granulosa Cells. Since granulosa cells cultured in serum-free media have been shown to undergo a spontaneous onset of apoptotic cell death (27), we investigated whether internucleosomal DNA fragmentation biochemical characteristic of apoptosis occurs in granulosa cells under our experimental conditions. Expectedly, time-dependent DNA cleavage was observed in serum- and FSH-free media (Fig. 6). Conjunction with the data described above, this result shows that the AT_2 receptor content on cultured granulosa cells increases under conditions in which apoptosis of the cells is observed.

Tilly et al (27) have demonstrated that treatment of cultured granulosa cells or isolated preovulatory follicles with EGF, transforming growth factor-α (TGF-α), or basic fibroblast

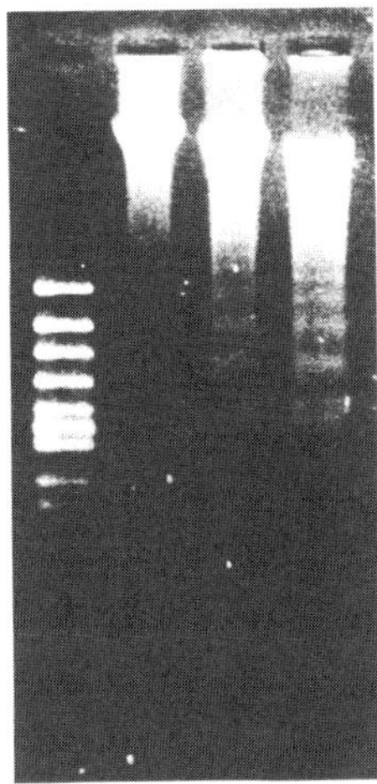

Figure 6. DNA fragmentation. Granulosa cells were cultured without FSH for indicated periods of time. Thereafter, genomic DNA was extracted from the cells, fractionated through 2.0% agarose gel (2 μg DNA/lane), then the gel was stained with ethidium bromide.

growth factor (bFGF) inhibits the spontaneous onset of apoptotic DNA fragmentation. These factors are known to exert their physiological functions through the tyrosine kinase pathway. On the other hand, Bottari et al. (15) have indicated that the AT_2 receptor activates protein tyrosine phosphatase in PC12W cells and adrenal cortex membranes. Very recently, Buisson et al. (17) have demonstrated the AT_2 receptor-mediated inhibition of the T-type Ca^{2+} channel through the protein tyrosine phosphatase pathway. Based on these data, it is of great interest to speculate that Ang II might inhibit EGF-, TGF-α-, or bFGF-induced suppression of apoptosis of granulosa cells through the AT_2 receptor.

CONCLUSION

The AT_2 receptor content was not changed during differentiation by FSH of rat ovarian granulosa cells in culture. In contrast, the receptor content was markedly increased in FSH-free media although internucleosomal DNA fragmentation characteristic of apoptosis was observed. Ang II up-regulated the AT_2 receptor content via AT_2 itself at both protein and mRNA levels. These results strongly support the hypothesis that AT_2 might modulate the initiation and/or progression of follicle atresia involving granulosa cell apoptosis. Furthermore, the demonstration of AT_2 receptor-mediated up-regulation of the AT_2 expression in this study shows that serum-free culture of granulosa cells provides a useful model to study the second messenger pathways of this receptor.

ACKNOWLEDGMENTS

This work was supported by a grant-in-aid from the Ministry of Education, Science and Culture of Japan, and a grant for a Special Research Project on Circulation Biosystem from the University of Tsukuba.

REFERENCES

1. Bottari, S. P., de Gasparo, M., Steckelings, U., and Levens, N. R. (1993). Front. Neuroendocrinol. *14*: 123-1716
2. Murphy, T. J., Alexander, R. W., Griendling, K. K., Runge, M. S., and Bernstein, K. E. (1991). Nature *351*: 233-236
3. Sasaki, K., Yamano, Y., Bardhan, S., Iwai, N., Murray, J., Hasegawa, M., Matsuda, Y., and Inagami, T. (1991). Nature *351*: 230-233
4. Mukoyama, M., Nakajima, M., Horiuchi, M., Sasamura, H., Pratt, R. E., and Dzau, V. J. (1993). J. Biol. Chem. *268*: 24539-24542
5. Kambayashi, Y., Bardhan, S., Takahashi, K., Tsuzuki, S., Inui, H., Hamakubo, T., and Inagami, T. (1993). J. Biol. Chem. *268*: 24543-24546
6. Grady, E. F., Sechi, L. A., Griffin, C. A., Schambelan, M., and Kalinyak, J. E. (1991). J. Clin. Invest. *88*: 921-933
7. Pucell, A. G., Hodges, J. C., Sen, I., Bumpus, F. M., and Husain, A. (1991). Endocrinology *128*: 1947-1959
8. Viswanathan, M. and Saavedra, J. M. (1992). Peptides *13*: 783-786
9. Miyazaki, H., Kondoh, M., Ohnishi, J., Masuda, Y., Hirose, S., and Murakami, K. (1988). Biomed. Res. *9*: 281-285
10. Glorioso, N., Atlas, S. A., Laragh, J. H., Jewelewicz, R., and Sealey, J. E. (1986). Endocrinology *133*: 1609-1616
11. Husain, A., Bumpus, F. M., de Silva, P., and Speth, R. C. (1987). Proc. Natl. Acad. Sci. USA. *84*: 2489-2493
12. Kim, S-J., Shinjo, M., Tada, M., Usuki, S., Fukamizu, A., Miyazaki, H., and Murakami, K. (1987). Biochem. Biophys. Res. Commun. *146*: 989-995
13. Palumbo, A., Jones, C., Lightman, A., Carcangiu, M. L., de Cherney, A. H., and Naftolin, F. (1989). J. Obstet. Gynecol. *160*: 8-14
14. Schultze, D., Brunswig, B., and Mukhopadhyay, A. M. (1989). Endocrinology *124*: 1389-1398
15. Bottari, S. P., King, I. N., Reichlin, S., Dahlstroem, I., Lydon, N., and de Gasparo, M. (1992) Biochem. Biophys. Res. Commun. *183*: 206-211
16. Kang, J., Posner, P., and Sumners, C. (1994) Am. J. Physiol. *267*: C1389-C1397
17. Buisson, B., Laflamme, L, Bottari, S. P., de Gasparo, M., Gallo-Payet, N., and Payet, M. D. (1995). J. Biol. Chem. *270*: 1670-1674
18. Ohnishi, J., Tanaka, M., Naruse, M., Usuki, S., Murakami, K., and Miyazaki, H. (1995) Biochim. Biophys. Acta *1192*: 286-288
19. Tanaka, M., Ohnishi, J., Ozawa, Y., Sugimoto, M., Usuki, S., Naruse, M., Murakami, K., and Miyazaki, H. (1995). Biochem. Biophys. Res. Commun. *207*: 593-598
20. Tsutsumi K. and Saavedra, J. M. (1991). Mol. Pharmacol. *41*: 290-297
21. Tsutsumi, K., Zorad, S., and Saavedra, J. M. (1992). Eur. J. Pharmacol. *226*: 169-173
22. Knecht, M., Katz M. S., and Catt, K. J. (1981). J. Biol. Chem. *256*: 34-36
23. Currie, W. D., Li, W., Baimbridge, K. G., Yuen, B. H., and Leung, P. C. (1992).Endocrinology *130*: 1837-1843
24. Hsueh, A. J. W., Adashi, E. Y., Jones, P. B. C., and Welsh, T. H., Jr. (1984). Endocr. Rev. *5*: 76-127
25. Feng, P., Knecht, M., and Catt, K. (1987). Endocrinology *120*: 1121-1126
26. Pucell, A. G., Bumpus, F. M., and Husain, A. (1988) J. Biol. Chem. *263*: 11954-11961
27. Tilly, J. L., Billig, H., Kowalski, K. I., and Hsueh, A. J. (1992). Mol. Endocrinol. *6: 1942-1950*

19

AT_2 RECEPTOR EXPRESSION IN OVARIES: A REVIEW

A. H. Nielsen, A. Hagemann, B. Avery,[1] and K. Poulsen

Department of Anatomy and Physiology and
[1]Department of Clinical Studies and Reproduction
The Royal Veterinary and Agricultural University
DK-1870 Frederiksberg C, Denmark

INTRODUCTION

All the components of the renin-angiotensin system (RAS) have been demonstrated in the mammalian ovary, indicating the existence of a functional local tissue RAS (1,2). The activity of the ovarian RAS increases during pregnancy, in relation to ovulation, and during stimulation of oocyte maturation and ovulation with gonadotropins as reflected by markedly increased prorenin concentrations in plasma (3,4,5). In accordance, Einspanier et al. (6) have demonstrated a pulsative angiotensin (Ang) II secretion in ovarian follicles and corpora lutea of freely moving pigs with the highest average secretion in the periovulatory period by using implanted microdialysis systems. Several studies have suggested effects of the ovarian RAS on oocyte maturation, ovulation, steroidogenesis and angiogenesis, but the functional roles of the ovarian RAS are not fully clarified. Since the effects of the ovarian RAS are mediated by Ang II receptors, the expression of these receptors and their intracellular messenger systems are of major interest. This paper reviews some of the data on the expression and functional roles of Ang II receptors in ovaries.

ANGIOTENSIN II RECEPTORS IN OVARIAN FOLLICLES

The first identification of Ang II receptors in the ovaries was made by Speth et al. (7) in rats. By using autoradiography they observed that the Ang II receptors were primarily localized in the follicles with the highest density in the granulosa cell layer. The granulosa cells were later demonstrated to express exclusively the AT_2 receptor by using subtype-selective nonpeptide Ang II receptor antagonists (8). In bovine ovaries, Brunswig-Spickenheier and Mukhopadhyay (9) showed that Ang II receptors were present only on thecal cells. The bovine follicular Ang II receptors (Fig. 1) were demonstrated to be almost exclusively AT_2 receptors (9,10). In pubertal monkey ovaries the Ang II receptors were also predominantly expressed on thecal cells (11). The receptor subtype was not determined. Except for

Recent Advances in Cellular and Molecular Aspects of Angiotensin Receptors
Edited by Mohan K. Raizada et al., Plenum Press, New York, 1996

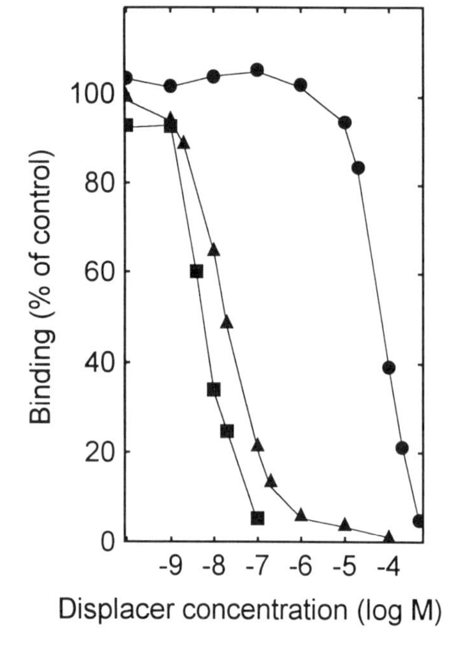

Figure 1. Displacement of the specific binding of ^{125}I-[Sar^1-Ile^5-Ile^8]-Ang II by [Val^5]-Ang II (■), PD 123319 (▲) and Losartan (●) in membrane fractions from bovine ovarian follicles, indicating that the Ang II receptors are almost exclusively AT_2 receptors. Modified with permission from (10).

the finding that no Ang II receptors were found on cultured human luteinized granulosa cells (12), little is known about the localization and subtype of Ang II receptors in adult human ovaries.

Variation of Angiotensin II Receptor Expression in Ovarian Follicles

As discussed above the distribution of Ang II receptors on thecal or granulosa cells differs between species. The density of Ang II receptors and concentrations of active renin and prorenin also vary between species. In the follicular wall tissue of porcine follicles (Table 1), the Ang II receptor density and active renin concentration were about 35-fold and 15-fold lower, respectively, than in bovine follicles (13). The cellular localization and subtype of the Ang II receptors in the porcine follicular wall tissue were not determined (13). Relatively high concentrations of prorenin were found in bovine follicular wall tissue, but no prorenin could be detected in pigs (13). These findings indicate the existence of species differences in the physiological roles of the ovarian RAS.

The Ang II receptor expression depends on the developmental stage of the ovarian follicle. In rats Ang II receptor expression was found in secondary follicles but not in primordial or primary follicles (11). A marked postovulatory increase in ovarian Ang II receptors was demonstrated in rats (14) with the highest expression in atretic follicles (15). Cycle-related changes of Ang II receptor expression in rat ovaries were also found by Lightman et al. (16). The highest expression was in diestrus. In cattle, the Ang II expression correlated positively with the follicular size and negatively with the active renin concentration in the follicular wall tissue (10). No relations were found with indicators of atresia such as the E_2/P_4 ratio and prorenin (10), suggesting that a high expression of Ang II receptors is not confined to atretic follicles in cattle as shown in rats (15).

In cultured bovine thecal cells, Brunswig-Spickenheier and Mukhopadhyay (9) demonstrated an up-regulation of Ang II receptor expression by luteinizing hormone in a dose-dependent fashion via a cAMP-dependent mechanism. In a rat granulosa cell culture,

Table 1. Ang II receptor densities, and active renin and prorenin concentrations in porcine and bovine ovarian follicular wall tissue. Values are given as mean with range in parentheses, except the data marked with (*), which are given as median and range. b.d.: below the detection limit. GU: Goldblatt units. Modified with permission from (13).

	Pigs	Cattle
Receptor density (fmol/mg protein)	47 (19 - 97; n=13)	1640 (90 - 5990; n=34)
Active renin intissue (GU/kg)	1.32 (0.40 - 3.43; n=23)	18.0$^{(*)}$ (2.1 - 107; n=101)
Prorenin in tissue (GU/kg)	b.d. (n=23)	11.7$^{(*)}$ (<2.6 - 142; n=101)

Pucell et al. (17) demonstrated a decrease of Ang II receptor expression by follicle-stimulating hormone, Ang II and testosterone. The decrease of Ang II receptor expression by Ang II may reflect receptor down-regulation. In accordance the Ang II receptor density and the active renin concentration correlated negatively in the wall tissue of bovine follicles (10). The Ang II receptor expression was not affected by progesterone and estradiol in cultured rat granulosa cells (17) in accordance with the finding that the Ang II receptor expression did not correlate with the estradiol or progesterone concentrations in bovine follicular fluid (10).

ANGIOTENSIN II RECEPTORS IN CORPORA LUTEA

Some studies could not demonstrate Ang II receptors in corpora lutea from rats (7) and cattle (9). Other studies found Ang II receptors in the rhesus monkey corpus luteum (11) and low levels of Ang II receptors in some rat corpora lutea (15). The subtype and the binding characteristics of the Ang II receptors were not determined in these studies. In bovine corpora

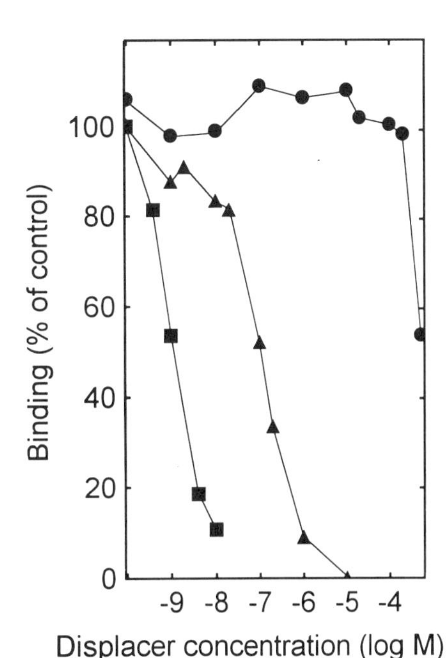

Figure 2. Displacement of the specific binding of ^{125}I-[Sar^1-Ile^5-Ile^8]-Ang II by unlabelled [Sar^1-Ile^5-Ile^8]-Ang II ( ), PD 123319 ( ) and losartan ( ) in membrane fractions from bovine corpora lutea, indicating that the Ang II receptors are almost exclusively AT_2 receptors. Reproduced with permission from (13).

lutea (Fig. 2), Nielsen et al. (13) found AT_2 receptors with a density about 80-fold lower than in the bovine follicular wall tissue. In porcine corpora lutea, the Ang II receptor density was very low and the receptor subtype could not be determined (13). There are no data on the variation of the Ang II receptor density with the developmental stage of the corpus luteum.

PHYSIOLOGICAL ASPECTS OF OVARIAN ANGIOTENSIN II RECEPTORS

The Ang II receptors in the ovarian follicles and corpora lutea are almost exclusively AT_2 receptors. However, The presence of low levels of AT_1 receptors can not be excluded with the methods previously used for identification of receptor subtypes. Therefore, Ang II might have an effect in the ovary mediated by both AT_1 and AT_2 receptors. Ang II may influence steroidogenesis, oocyte maturation and ovulation in the ovarian follicle, and angiogenesis and steroidogenesis in the corpus luteum. The Ang II receptor antagonist saralasin was demonstrated to inhibit gonadotropin-induced ovulation in-vivo in rats (18). However, the reproducibility of these data was later debated (19). Ovulation and oocyte maturation were since then shown to be directly stimulated by Ang II in perfused rabbit ovaries (20,21,22). Ovulation induced by gonadotropins was inhibited by saralasin (21,22,23), indicating a modulating effect of Ang II. Interestingly, angiotensin converting enzyme (ACE) inhibitors did not affect ovulation in rat ovaries in-vivo and in-vitro (24,25,26). In accordance, treatment of women for arterial hypertension with ACE inhibitors has not been reported to reduce fertility. This may be caused by a low accessibility of ACE inhibitors in the follicular wall tissue or by the formation of Ang II by enzymatical pathways not involving ACE. Alternatively, inhibition of ACE may increase the concentration of bradykinin in the follicles. Bradykinin has previously has been shown to be obligatory for ovulation in perfused rabbit ovaries (26). Estrogen but not progesterone secretion was stimulated by Ang II in incubated rat ovaries (14) and perfused rabbit ovaries (22). In pigs, Einspanier et al. (6) could not demonstrate any effect of Ang II on estradiol or progesterone secretion in ovarian follicles. In accordance with the absence of effect on ovulation, ACE inhibitors did not affect estradiol and progesterone secretion in the perfused rat ovary (25). No functional studies of the ovarian RAS using specific Ang II receptor antagonists have been published. Such studies are essential for the further clarification of the physiological roles of the ovarian RAS.

ACKNOWLEDGMENTS

This work was supported by the Danish Agricultural and Veterinary Research Council, the Research Fund of the Danish Medical Association, the Nordic Insulin Fund and the NOVO Fund.

REFERENCES

1. Lumbers, E.R. (1993) in The renin-angiotensin system (eds. J.I.S. Robertson and M.G. Nicholls), Gower Medical Publishing, London, pp 46.1-46.12
2. Morris, R.S. and Paulson, R.J. (1994). Fertil. Steril. *62*:1105-1114
3. Sealey, J.E., Wilson, M., Morganti, A.A., Zervoudakis, I. and Laragh, J.H. (1982). Clin. Exp. Hypertens. A *4*:2373-2384

4. Sealey, J.E., Atlas, S.A., Glorioso, N., Manapat, H. and Laragh, J.H. (1985). Proc. Natl. Acad. Sci. U.S.A. *82*:8705-8709
5. Itskovitz, J., Sealey, J.E., Glorioso, N. and Rosenwaks, Z. (1987). Proc. Natl. Acad. Sci. U.S.A. *84*:7285-7289
6. Einspanier, A., Jarry, H., Pitzel, L., Holtz, W. and Wuttke. W. (1991). Endocrinology *129*:3403-3409
7. Speth, R.C., Bumpus, F.M. and Husain, A. (1986). Eur. J. Pharmacol. *130*:351-352
8. Pucell, A.G., Hodges, J.C., Sen, I., Bumpus, F.M. and Husain, A. (1991). Endocrinology *128*:1947-1959
9. Brunswigs-Spickenheier, B. and Mukhopadhyay, A.K. (1992). Endocrinology *131*:1445-1452
10. Nielsen, A.H., Hagemann, A., Svenstrup, B., Nielsen, J. and Poulsen, K. (1994). Clin. Exp. Pharmacol. Physiol. *21*:463-469
11. Aguilera, G., Millan, M.A. and Harwood, J.P. (1989). Am. J. Hypertens. *2*:395-402
12. Rainey, W.E., Bird, J.M., Byrd, W. and Carr, B.R. (1993). Fertil. Steril. *59*:143-147
13. Nielsen, A.H., Hagemann, A., Avery, B. and Poulsen, K. (1995). Exp. Clin. Endocrinol. in press
14. Pucell, A.G., Bumpus, F.M. and Husain, A. (1987). J. Biol. Chem. *262*:7076-7080
15. Daud, A.I., Bumpus, F.M. and Husain, A. (1988). Endocrinology *122*:2727-2734
16. Lightman, A., Jones, C.L., MacLusky, N.J., Palumbo, A., DeCherney, A.H. and Naftolin, F. (1988). Am. J. Obstet. Gynecol. *159*:526-530
17. Pucell, A.G., Bumpus, F.M. and Husain, A. (1988). J. Biol. Chem. *263*:11954-11961
18. Pellicer, A., Palumbo, A., DeCherney, A.H. and Naftolin, F. (1988). Science *240*:1660-1661
19. Daud, A.I., Bumpus, F.M. and Husain, A. (1989). Science *245*:870-871
20. Kuo, T.C., Endo, K., Dharmarajan, A.M., Miyazaki, T., Atlas, S.J. and Wallach, E.E. (1991). J. Reproduc. Fertil. 92:469-474
21. Yoshimura, Y., Karube, M., Koyama, N., Shiokawa, S., Nanno, T. and Nakamura, Y. (1992). F.E.B.S. Lett. *307*:305-308
22. Yoshimura, Y., Karube, M., Oda, T., Koyama, N., Shiokawa, S., Akiba, M., Yoshinaga, A. and Nakamura, Y. (1993). Endocrinology *133*:1609-1616
23. Peterson, C.M., Zhu, C., Mukaida, T., Butler, T.A., Woessner, J.F.J. and LeMaire, W.J. (1993). Am. J. Obstet. Gynecol. *168*:242-245
24. Daud, A.I., Bumpus, F.M. and Husain, A. (1990). Endocrinology *126*:2927-2935
25. Peterson, C.M., Morioka, N., Zhu, C., Ryan, J.W. and LeMaire, W.J. (1993). Reprod. Toxicol. *7*:131-135
26. Yoshimura, Y., Koyama, N., Karube, M., Oda, T., Akiba, M., Yoshinaga, A., Shiokawa, S., Jinno, M. and Nakamura, Y. (1994). J. Clin. Invest. *93:180-187*

20

HETEROGENEITY OF RAT ANGIOTENSIN II AT_2 RECEPTOR

G. M. Ciuffo,[1] O. Johren, G. Egidy, F. M. J. Heemskerk, and J. M. Saavedra

Section on Pharmacology
National Institute of Mental Health, NIH
Bldg.10, Room 2D:45
9000 Rockville Pike
Bethesda, Maryland 20892
[1] Facultad de Química, Bioquímica y Farmacia
Universidad Nacional de San Luis
Chacabuco y Pedernera
5700 San Luis, Argentina

INTRODUCTION

Angiotensin II (Ang II) plays a key role in cardiovascular homeostasis in developing animals as well as in adults by regulating blood pressure, fluid electrolite balance and renal hemodynamics(1-5).

The discovery of new peptidic and non peptidic ligands for Ang II receptors demonstrated the existence of two main Ang II receptor subtypes, (1-4) defined by their pharmacological characteristics. AT_1 receptors are specifically blocked by losartan (2,4,6) and AT_2 receptors are selectively displaced by CGP42112 or PD123177 (3,6,7). CGP42112 is a modified peptide, derived from the 4-8 core of Ang II, with a high affinity and selectivity for Ang II AT_2 receptors (about 10^{-9} M) (7-10) and a very low affinity for Ang II AT_1 receptors (> 100 nM) (7).

AT_1 receptors are present in many different tissues and are responsible for most of the well-known actions of Ang II. Little is known about the role of the AT_2 subtype. In the adult, AT_2 receptors are present in myometrium, a few brain nucleus, adrenal and brain arteries (1-4, 11). The possible involvement of the AT_2 subtype in growth and development has been supported by the high expression of these receptors early during development (12-18). AT_2 receptors are very highly expressed in the fetus (12-15,18) and most of them disappear soon after birth.

Molecular cloning techniques have made possible the elucidation of the molecular structure of the receptor subtypes. Both AT_1 and AT_2 receptors have sequences which suggest the presence of seven transmembrane domains, characteristic of G-protein coupled receptors

Recent Advances in Cellular and Molecular Aspects of Angiotensin Receptors
Edited by Mohan K. Raizada et al., Plenum Press, New York, 1996

(19-22). AT_2 receptors have been recently cloned from different sources (21-22), which express full homologous sequences. However, the reported sequence from fetal rat AT_2 receptors (22) has only a partial homology (30%) with the sequence of AT_1 receptors (20). Although the cloned AT_2 subtype belongs to the seven transmembrane G-coupled superfamily, no G-protein related second messengers had so far been reported. Recently, several authors (23-25) have reported the effect of Ang II on tyrosine phosphorylation of different proteins mediated by AT_2 receptor subtype.

In brain and a neuroblastoma cell line, heterogeneity of AT_2 receptors have been previously proposed, based on different G-protein coupling properties (10,26). Tsutsumi et al. first reported heterogeneity of Ang II AT_2 receptors in the rat brain (10), where different brain nuclei showed a difference in their sensitivity to GTPγS and pertusis toxin pretreatment. No G-protein coupling has been demonstrated in peripheral tissues.

Polyacrylamide gel electrophoresis of AT_2 receptors from various tissues from different species has shown bands with molecular weights ranging from 68 to 113 kDa. This size heterogeneity was attributed to differences in glycosilation of the same receptor (27,28).

We provide additional evidence, based on structure-activity studies using peptide analogs of CGP42112, and on molecular biology techniques, in support of AT_2 receptor heterogeneity.

ANG II RECEPTOR SUBTYPES IN KIDNEY DEVELOPMENT

The rat kidney is immature at birth, allowing all stages of glomerular development to be present in a single kidney. The least mature glomeruli are found near the renal capsule, and the most mature near the medulla.

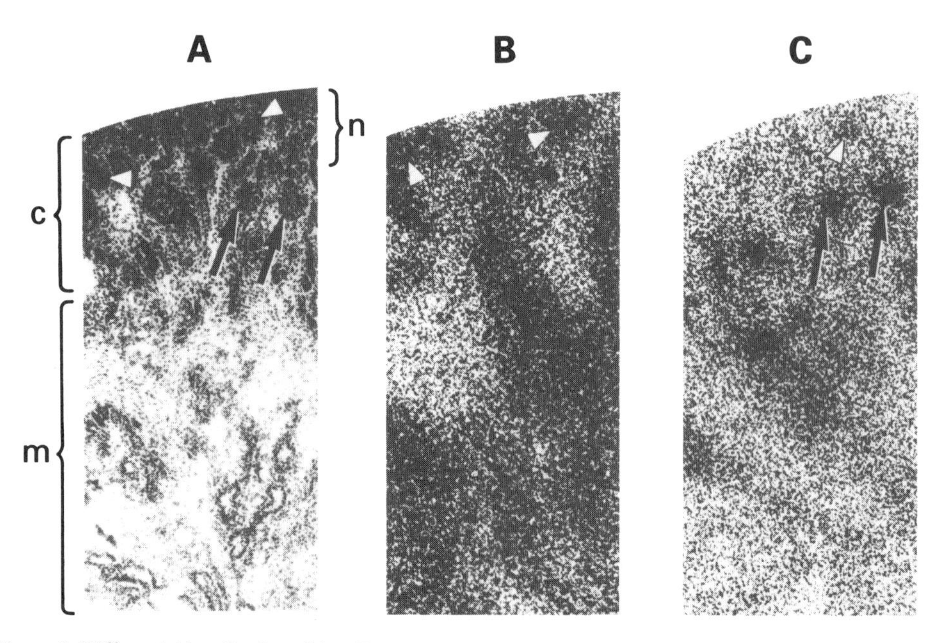

Figure 1. Differential localization of Ang II receptor subtypes in the developing kidney: c cortex, m medulla, n nephrogenic zone. A: section stained with hematoxylin-eosin; B: binding of [^{125}I]Sar^{1}Ang II in the presence of losartan (AT_2 receptors). C: binding in presence of CGP 42112 (AT_1 receptors).

Autoradiographic studies showed that Ang II receptor subtypes are developmentally regulated (5,14,16). In the adult, only AT_1 receptors are present, and they are associated only with fully developed glomeruli, and with the inner stripe of the outer medulla (14,16).

In the developing kidney, AT_2 receptors are associated with early stages during glomeruli development (S-shaped and comma-shaped bodies) in the nephrogenic zone of the cortex and the immature medulla of fetal and newborn rat kidney (Figure 1). These AT_2 receptors disappear one week after birth (14). Based on this association and the timing of expression of Ang II receptor subtypes, we proposed that AT_2 receptors could be involved in kidney organogenesis (14).

HETEROGENEITY WITHIN THE AT_2 RECEPTOR SUBPOPULATIONS

Evidence Based on Structure-Activity Studies

[^{125}I]CGP42112, the selective AT_2 competitor, recognizes AT_2 receptors in both fetal and newborn kidney membranes. Unlabeled CGP42112 selectively competes with [^{125}I]CGP42112 binding in both preparations, with an affinity ten times lower in fetal membranes than in newborn kidney membranes (see Table 1, Figure 2). Ang II, PD123177 and PD123319, also competed in both preparations, the last two compounds showed an IC_{50} in the range of 0.1 μM (see Table 1), as reported for other tissues (7,8). Thus, the affinity profiles clearly confirmed that both preparations contain AT_2 receptors, as classically defined.

To further characterize the structural requirements of the AT_2 receptors, we developed a number of CGP42112 analogs, designed on the basis of CGP42112 structure and previous structure-activity studies on Ang II receptors (29-31). Their affinity for newborn kidney and fetal AT_2 receptors was tested in membrane preparations, by competition with [^{125}I]CGP42112 (0.5 nM) as described earlier.

In analog III, the CBZ group of CGP42112 was replaced with an acetyl group, to keep a similar charge distribution as on the parent compound. Analog III exhibited an affinity very close to that of CGP42112 for fetal membranes (IC_{50} 7.7 nM) and newborn kidney

Table 1. Comparative table with IC_{50} values from competition curves obtained for the analogs tested.

Name	Formula	newborn kidney IC_{50}[a] (M)	fetus IC_{50}[a] (M)
CGP42112	Nic-Tyr-Lys-His-Pro-IleOH-Arg-CBZ	$3.0\ 10^{-10}$	1.9 10-9
Ang II	Asp-Arg-Val-Tyr-Ile-His-Pro-Phe	$5.72\ 10^{-9}$	$7.07\ 10^{-9}$
PD123177	non-peptidic	$2.37\ 10^{-6}$	$6.4\ 10^{-7}$
PD123319	non-peptidic	$8.6\ 10^{-8}$	$1.25\ 10^{-7}$
III	Nic-Tyr-Lys-His-Pro-IleOH-Arg-Ac	$2.16\ 10^{-9}$	7.7 10-9
IV	Nic-Tyr-Lys-His-Pro-IleOH	$3.675\ 10^{-9}$	$2.0\ 10^{-7}$
V	Tyr-Lys-His-Pro-IleOH	n.d	$7.25\ 10^{-5}$
VI	Nic-Tyr-Lys-His-Ala-His	$7.05\ 10^{-6}$	$4.23\ 10^{-7}$

[a] IC_{50} values are the mean of two independent experiments obtained on competition curves by using 6-9 increasing concentrations. n.d.: under identical conditions, no displacement.

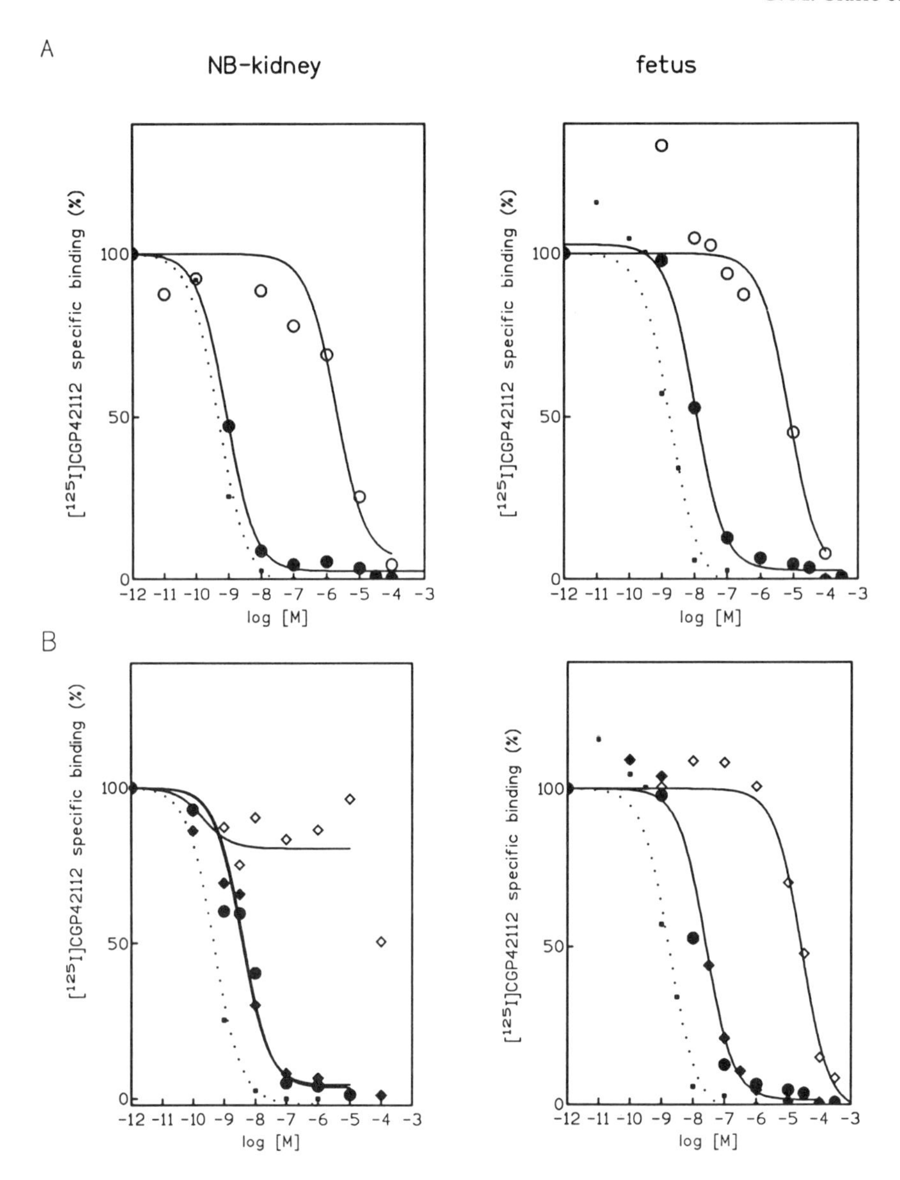

Figure 2. Competition curves performed on newborn kidney (left panels) and fetal membranes (right panels). Binding was performed by using 0.5 nM [^{125}I]CGP42112 and increasing concentrations of the compounds tested. Each point represent the mean of duplicate determinations and data are representative of 2-3 separate experiments.

membranes (2.16 nM) (Table 1, Figure 2A). This suggested that the CBZ group is not a critical element for recognition at AT_2 receptors.

The side chain of CGP 42112 was deleted in analog IV, giving a linear peptide with a sequence that contains most of the elements common to the Ang II sequence. The affinity of analog IV for the AT_2 receptor on fetal membranes (200 nM) is 100 times lower than analog III and CGP42112. On the other hand, the affinity of analog IV in newborn kidney membranes is very close to that of analog III (3.7 nM) (Table 1, Figure 2B). This suggests that the Arg residue is more important for the fetal than for the newborn kidney AT_2 receptor. Such striking affinity change may represent a significant difference in the AT_2 receptor structure between the kidney and fetal receptors.

When the Nic residue of the N-terminal portion of analog IV was removed, we obtained a compound which containing only the backbone of the main chain of CGP42112 (see Table 1) (analog V). The resulting compound has both free N- and C-terminal groups, being a linear pentapeptide containing the Tyr and His-Pro of the original Ang II sequence. Analog V does not compete for the AT_2 receptors from newborn rat kidney, but it still shows some affinity for the AT_2 receptors from fetal membranes (see Figure 2, panel B). However, the affinity of analog V on fetal membranes is 1000 times lower than that of analog IV and 10^4 times less than that of CGP42112 (Figure 2, Table 1).

In order to verify the importance of the spatial disposition of the COOH terminal portion, we designed analog VI, where Pro-Ile was replaced by Ala-His (Table 1). Analog VI has a free carboxyl terminal group, but the amino-terminal residue remains blocked by the nicotinic (Nic) residue. Surprisingly, the analog retained a high affinity for AT_2 receptors from fetal membranes, only 100 times lower than CGP42112 and comparable to that of the PD123177 and PD123319 (Figure 2A, Table 1). However, in newborn kidney membranes the analog exhibited a 10 times lower affinity when compared with fetal membranes, several orders of magnitude lower than the affinity of CGP42112.

PANEL A. Comparison of competition curves for CGP42112 (...) and analogs III (closed circles) and VI (open circles).

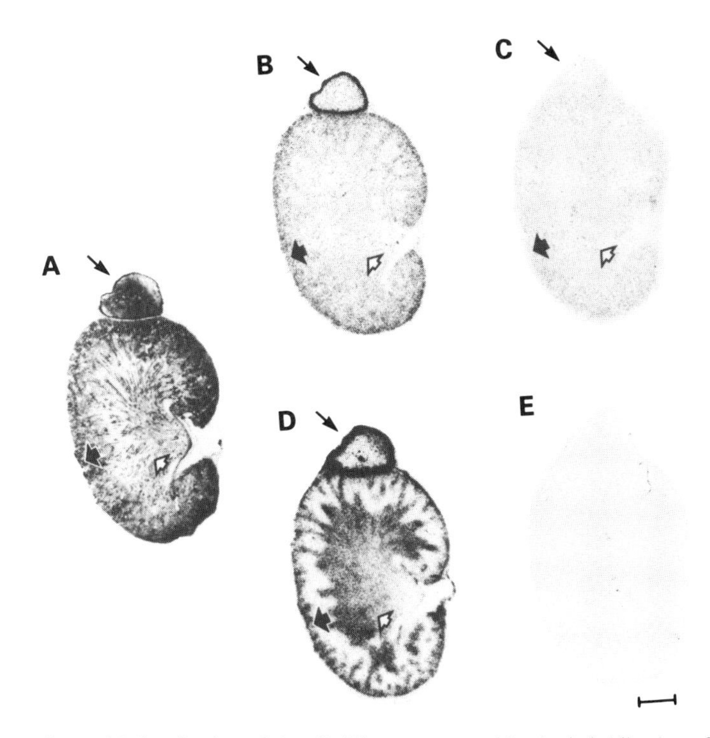

Figure 3. Autoradiographic localization of Ang II AT_2 receptors, and in situ hybridization of AT_2 mRNA in newborn rat kidney. A. Hematoxylin-eosin staining. B. Consecutive section showing in situ hybridization with the antisense probe. C. Consecutive section with in situ hybridization with the sense probe. D-E. Autoradiogram of [^{125}I]CGP42112 labeling of AT_2 receptors. D: total binding; E: consecutive section with non specific binding determined in the presence of 1μM CGP42112. Arrows point to the zona glomerulosa in adrenal gland; black arrow heads: nephrogenic zone in the kidney; white arrow heads: kidney medulla.

PANEL B. Comparison of competition curves for CGP42112 (...) and analogs IV (closed diamonds) and V (open diamonds).

Evidence Based on mRNAs Expression

A. Localization of AT_2 Receptors in Newborn Rat Kidney. In order to compare the expression of AT_2 binding with that of the AT_2 mRNA, autoradiography and in situ hybridization were performed on consecutive sections. AT_2 receptors were selectively labeled with $[^{125}I]$CGP42112 by autoradiography. In the newborn kidney, $[^{125}I]$CGP42112 selectively labeled the nephrogenic zone and the medulla (Figure 3) and this binding was displaced by unlabeled Ang II (results not shown). This was in agreement with our previous results using $[^{125}I]Sar^1$Ang II (14).

For in situ hybridization, a selective $[^{35}S]$-labeled antisense riboprobe of 371 bp XbaI/BglI fragment in the 3′ end no coding region, with no homology for the AT_1 subtype was used. Radiolabeled sense probes were used to asses the background signals, and this showed extremely low background. Sections were incubated with sense and antisense riboprobes for 16-18 h at 55 °C and, after washing, were dried and exposed to ^{3}H-Hyperfilm (Amersham).

In the newborn kidney, the antisense probe specifically labeled the nephrogenic zone, but not the kidney medulla (n: 4 individual experiments) (Figure 3B and C). As a positive control, the adrenal zona glomerulosa was clearly labeled by the AT_2 riboprobe (Figure 3).

The mismatch between AT_2 binding and AT_2 mRNA distribution in the newborn kidney supports the hypothesis of AT_2 receptor heterogeneity.

B. PCR Amplification. Analysis of the mRNAs was performed to better characterize the dissimilarities observed on the pharmacological properties of the two sources of AT_2 receptors. Primers designed for the 5′ and 3′ end including the full coding region of the

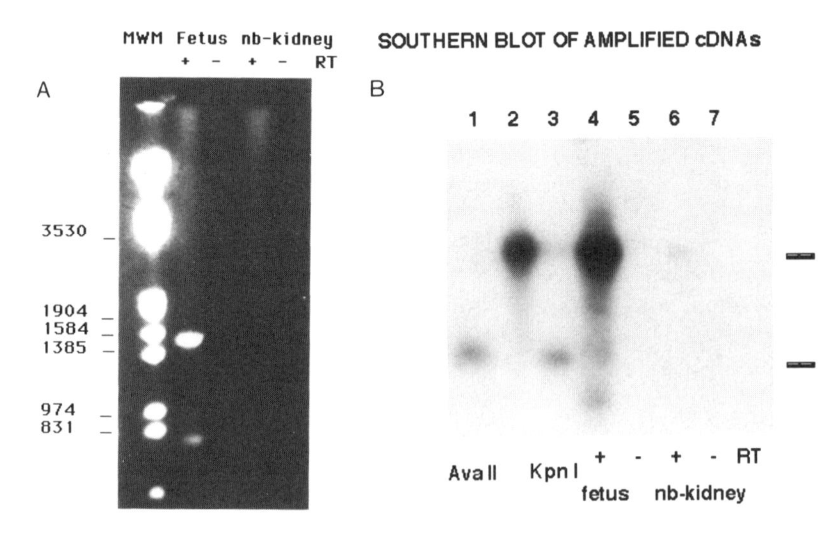

Figure 4. PCR amplification of cDNAs in fetal and newborn kidney. A. Ethidium bromide stained agarose gel of the fetal and newborn kidney amplified products. B. Autoradiogram of the Southern blot tested with a probe including the full coding region. Lane 1: 1534 bp band (fetus) after Ava II digestion. Lane 2: one third of the excised 1534 bp band. Lane 3: 1534 bp band (fetus) after Kpn I digestion. Lane 4: PCR products from fetus. Lane 6: PCR products from newborn kidney. Lanes 5 and 7: controls without reversed transcriptase (RT).

cloned AT_2 receptor (17,18) were used to amplify cDNA from fetus and newborn kidney. A clear band of the expected molecular weight (1534 bp, see Figure 4A) was amplified in fetus while a very poor signal was observed in the newborn kidney preparation. To rule out contamination of genomic DNA in our mRNA preparations, we subjected RNAs from the tissues used on the study to PCR omitting the reverse transcriptase step: no PCR product was detected on the agarose gel (see Figure 4A and Figure 4B, lanes 5 and 7). GAPDH amplification of both cDNA samples was comparable (data not shown). The identity of the DNA generated by PCR was confirmed by enzyme digestion and Southern blot analysis (Figure 4B). The Ava II and Kpn I restriction digestion pattern of the 1534 bp band generated fragments of the expected size (Figure 4B, lanes 1 and 3). The amplified products and the digested fragments were studied by Southern blot performed by hybridization to a 1000 bp ^{32}P-labeled Xba fragment of the AT_2 (from a cloned obtained from Dr. T. Inagami). Our results confirm that the amplification products, the excised band and its digested fragments corresponded to AT_2 receptors.

DISCUSSION

The heterogeneity of Ang II receptors, suggested from early structure-activity studies, has been confirmed with the synthesis of specific competitors, clarification of physiological differences and the cloning of different receptor subtypes (AT_1 and AT_2). Further heterogeneity, based on a number of biochemical differences, has been proposed within the AT_2 subtype (10).

We first report that both membrane preparations from newborn kidney and fetus showed a characteristic AT_2 affinity profile when tested with the classical AT_2 competitors. In addition, we report further differences on AT_2 receptors from newborn rat kidney and rat fetus based on new structure-activity studies and mRNA analysis.

Using CGP42112 as the parent coumpound, we studied the importance of different residues on the CGP 42112 molecule for binding to AT_2 receptors in rat newborn kidney and fetal membrane preparations.

Removal of the CBZ group (analog III) present in the side chain of CGP42112, did not affect significantly the binding affinity to either fetal or newborn kidney receptors, suggesting that the CBZ group is not important for binding to AT_2 receptors.

On the other hand, the Arg residue located on the side chain of CGP 42112 allowed us to differentiate between the two AT_2 receptors. The affinity of analog IV for newborn kidney receptors was two orders of magnitude higher than that for AT_2 receptorsi in fetal membranes.

The presence of the Nic residue was also important for selective affinity, since analog V completely lost its affinity for newborn kidney AT_2 receptors, while retaining a low affinity for fetal membrane AT_2 receptors. Surprinsingly, analog VI which retains only two residues from the original Ang II C-terminal sequence, had an affinity close to that of PD123177 and PD123319. This suggests that the increase in the mobility of the COOH residue, consequence of the introduction of the Ala residue, is not critical for binding.

The affinity profile of the compounds tested is therefore different for the two different sources of AT_2 receptors studied, as follows:

newborn kidney membranes

CGP42112 > Ang II ≈ analog III > PD123177 ≈ PD123319 ≈ analog IV ≈ analog VI > analog V

fetal membranes

CGP42112 > Ang II ≈ analog III ≈ analog IV > PD123319 > PD123177 ≈ analog VI >> analog V

A difference of one order of magnitude in ligand affinity for newborn kidney and fetal AT_2 receptors was observed for CGP42112, PD123177, PD123319 and analog VI. In addition, analog IV and analog V showed an even larger difference in affinity between the two AT_2 receptor subtypes.

Comparative studies of the mRNA and receptor expression confirmed the previous observations. AT_2 receptors were selectively labeled by [^{125}I]CGP42112 in the nephrogenic zone and medulla of the newborn kidney, in agreement with our previous report (14). However, AT_2 mRNA, localized by in situ hybridization with a selective probe for AT_2 receptors, was recognized only in the nephrogenic zone. The distribution pattern of binding and mRNA expression observed by autoradiography and in situ hybridization are clearly different, since the medulla, intensively labeled by [^{125}I]CGP42112, is not recognized by the antisense probe.

The discrepancy mentioned above was confirmed by PCR analysis. PCR amplification of cDNAs from newborn kidney and fetal membranes showed a strong band of the expected size on the fetal cDNA and a very low signal for newborn kidney cDNA. The identity of the PCR products was confirmed by restriction enzyme digestion and Southern blot analysis of both the PCR products and the restriction fragments.

We are reporting that, based on structure-activity differences, different receptor expression and different mRNA recognition with homologous probes, AT_2 receptors from rat fetus and rat newborn kidney may be different. In addition, the affinity requirements for AT_2 receptors with respect to the Ang II molecule are less stringent that those of the AT_1 receptor. This raises the intriguing possibility that maybe the endogenous ligand for AT_2 receptors could be a peptide different than Ang II.

REFERENCES

1. Saavedra, J.M.(1992). Endodocrine Rev. *13*: 329-380.
2. Timmermans, P.B.M.W.M, Wong, P.C., Chiu, A.T., Herblin, W.F., Benfield, P., Carini, D.J., Lee. R.J., Wexler, R.R, Saye, J.M. and Smith, R.D.(1993). Pharmacol. Rev. *45*: 205-251.
3. Bottari, S.P., De Gasparo, M., Steckelings, U.M., and Levens, N.R.(1993). Front. in Neuroendocrinol. *14*: 123-171.
4. Bumpus, F.M., Catt, K.J., Chiu, A.T., De Gasparo, M., Goodfriend, T., Husain, A., Peach, M.J., Taylor, Jr. D.G., and Timmermans, P.B.M.W.M. (1991). Hypertension *17*: 720-721.
5. Tufro-McReddie, A., Johns, D.W., Geary, K.M., Dagli, H., Everett, A.D., Chevalier, R.L., Carey, R.M., and Ariel Gomez, R. (1994). Am. J. Physiol. *266*: F911-F918.
6. Chiu, A.T., Herblin, W.F., McCall, D.E., Ardecky, R.J., Carini, D.J., Duncia, J.V., Pease, L.J., Wong, P.C., Wexler, R.R., Johnson, A.L., and Timmermans, P.B.M.W.M. (1989). Biochem. Biophys. Res. Commun. *165*: 196-203.
7. Whitebread, S., Mele, M., Kamber, B., and De Gasparo, M. (1989). Biochem. Biophys. Res. Commun. *163*: 284-291.
8. Heemskerk, F.M.J., Zorad, S., Seltzer, A., and Saavedra, J.M. (1993). NeuroReport *4*: 103-105,.
9. Speth, R. C. (1993). Regul. Pept. *44*: 189-197.
10. Tsutsumi, K. and Saavedra, J.M. (1992). Molecular Pharmacology *41*: 290-297.
11. Tsutsumi, K. and Saavedra, J.M. (1991). Am. J. Physiol. *261*: H667-H670.
12. Tsutsumi, K., Strömberg, C., Viswanathan, M., and Saavedra, J.M. (1991). Endocrinology *129*: 1075-1082.
13. Viswanathan, M., Tsutsumi, K., Correa, F.M.A. and Saavedra, J.M. (1991). Biochem. Biophys. Res. Commun. *179*, 1361-1367.
14. Ciuffo, G.M., Viswanathan, M. Seltzer, A.M., Tsutsumi, K. and Saavedra. J.M. (1993). Am. J. Physiol. *265*: F264-271.

15. Grady, E.F., Sechi, L.A., Griffin, C.A., Schambelan. M. and Kalinyak J.E.(1991) J. Clin. Invest. *88*:921-933.
16. Aguilera G., Kapur S., Feuillan P., Sunar-Akbasak B., Bathia A.J. (1994). Kidney Int. *46*: 973-979.
17. Gröne H-J, Simon, M. and Fuchs, E. (1992) Am. J. Physiol. *262*: F326-F331.
18. Sechi L.A., Grady, E.F., Griffin, C.A., Kalinyak, J.E. and Schambelan, M. (1992). Am. J. Physiol. *262*: F236-F240.
19. Murphy T.J., Alexander R.W., Griendling K.K., Runge M.S., Bernstein K.E. (1991). Nature *351*:233-236.
20. Sasaki, K., Yamano, Y., Bardhan, S., Iwai, N., Murray, J.J., Hasegawa, M., Matsuda, Y., and Inagami, T. (1991) Nature *351*: 230-233.
21. Kambayashi Y., Bardhan S., Takahashi K., Tsuzuki S., Inui H., Hamakubo T., Inagami T. (1993). J. Biol. Chem. *268*: 24543-24546.
22. Mukoyama M., Nakajima M., Horiuchi M., Sasamura H., Pratt R.E., Dzau V.J. (1993). J. Biol. Chem. *268*: 24539-24542.
23. Brechler, V., Levens, N.R., De Gasparo, M., and Bottari, S. In: Receptors and Channels, Vol.2, pp79-87. Harwood Acad. Pub.,U.S.A. 1994.
24. Takahasi, K., Bardham, S., Kambayashi, Y., Shirai, H., and Inagami, T. (1994). Biochem. Biophys. Res. Commun. *198*, 60-66.
25. Nahmias, C., Cazaubon, S.M., Briend-Sutren, M.M., Lazard, D., Villageois, P. and Strosberg, A.D. (1995). Biochem. J. *306*: 87-92.
26. Siemens, I.R, Yee, D.K., Reagan, L.P., Fluharty, S.J.(1994) J. Neurochem. *62*: 257-264.
27. Lazard, D., Villageois, P., Briend-Sutren, M.M., Cavaille, F., Bottari, S., Strosberg, A.D., and Nahmias, C. (1994). Eur. J. Biochem. *220*: 919-926.
28. Servant, G., Dudley, D.T., Escher, E., and Guillemette, G. (1994). Mol. Pharmacol. *45*: 1112-1118.
29. Matsoukas, J.M., Hondrelis, J., Keramida, M., Mavromoustakos, T., Makriyannis, A., Yamdagni, R., Wu, Q., and Moore, G.J. (1994). J. Biol. Chem. *269*: 5303-5312.
30. Moore G.J., Goghari M.H., and Franklin K.J. (1993). Int. J. Peptide Protein Res. *42*: 445-449.
31. Nikiforovich G.V. and Marshall G.R. (1993).Biochem. Biophys. Res. Commun. *195*: 222-228.

21

HETEROGENEITY OF ANGIOTENSIN TYPE 2 (AT_2) RECEPTORS

L. P. Reagan, D. K. Yee, P. F. He, and S. J. Fluharty

Departments of Animal Biology Pharmacology
and Institute of Neurological Sciences
University of Pennsylvania
Philadelphia, Pennsylvania

INTRODUCTION

Angiotensin II (Ang II) is one of the most important hormones involved in the regulation of body fluid and cardiovascular homeostasis (1). The rate limiting step in the synthesis of Ang II is the release of the enzyme renin from the juxtaglomerular cells of the kidney into the blood. Several physiological and pathophysiological conditions induce renin release from the secretory granules of the juxtaglomerular cells of the kidney (*c.f.* (2)). Once released into circulation, renin exhibits high specificity for the plasma α-globulin angiotensinogen. Angiotensinogen is continuously synthesized and released from the liver into circulation. While several hormones, including Ang II itself, may regulate the synthesis of angiotensinogen, the concentration of angiotensinogen itself is not rate-limiting in the synthesis of Ang II. Renin cleaves the 10 N-terminal amino acids from angiotensinogen to produce the biologically inactive precursor, Ang I. Subsequently, the decapeptide Ang I serves as a substrate for the dipeptidyl carboxypeptidase angiotensin converting enzyme (ACE), an indiscriminate enzyme which cleaves dipeptides from substrates with dissimilar sequences. The potent vasodilator bradykinin also serves as a substrate for ACE. One of the determinants of ACE substrate selectivity is that the penultimate amino acid must be a proline, which explains why ACE does not inactivate Ang II. Once produced, Ang II is known to exert its diverse effects by acting at membrane bound receptors present on several peripheral target tissues, such as kidney, adrenal cortex, heart and vascular smooth muscle (1). In addition to its many peripheral effects, all of the components of the RAS have also been localized in the central nervous system (3). Moreover, it is now recognized that many of the physiological, endocrinological and behavioral actions of Ang II are mediated by specific receptors present in the brain (4).

The diversity of actions mediated by Ang II was among the first pieces of evidence leading to the suggestion of multiple receptor subtypes for Ang II. The unequivocal demonstration of such multiple receptor subtypes, however, came with the development of subtype-specific antagonists (5), which are now referred to as Angiotensin Type 1 (AT_1) and

Recent Advances in Cellular and Molecular Aspects of Angiotensin Receptors
Edited by Mohan K. Raizada et al., Plenum Press, New York, 1996

Angiotensin Type 2 (AT_2) receptors. The existence of at least two receptor subtypes for Ang II has initiated the search for a greater understanding of these receptors at the cellular and molecular levels. While the development of subtype selective antagonists has facilitated the analysis of Ang II receptor subtypes in the periphery, much less is known about the pharmacology, biochemistry and functional properties of central Ang II receptors. This relative paucity of information is due to the fact that central Ang II receptors are found in very discrete neuronal populations in very low densities. In order to circumvent these potential problems, numerous laboratories have adopted an alternative approach to investigate the properties of central Ang II receptors, namely, the use of clonal neuronal cell lines. In this regard, our laboratory has chosen to utilize the murine neuroblastoma N1E-115 cell line in order to study the pharmacological, functional and biochemical properties of neuronal Ang II receptors.

MURINE NEUROBLASTOMA N1E-115 CELLS AS A MODEL SYSTEM

The initial characterization of Ang II receptors expressed on N1E-115 cells, using the high affinity antagonist ^{125}I-[Sarc^1, Ile^8]-angiotensin II, revealed that these binding sites were saturable and of high affinity (6). The rank order potency of ^{125}I-[Sarc^1, Ile^8]-angiotensin II binding activity, as well as the association and dissociation kinetics, were in close agreement with studies performed on Ang II receptors expressed in several peripheral tissues. *In vitro* differentiation of N1E-115 cells increased the density of Ang II receptors nearly 20 fold without affecting the affinity of these binding sites (7). Use of the subtype selective antagonists in radioligand binding studies revealed that differentiated N1E-115 cells expressed both AT_1 and AT_2 receptor subtypes (8). In addition, increased expression of the AT_2 receptor subtype was responsible for the proliferation of Ang II binding sites as a result of differentiation, while the density of AT_1 sites was unaffected. Therefore, differentiated N1E-115 cells provided an excellent model system in which to study the pharmacological, biochemical, and functional properties of neuronal Ang II receptors.

Biochemical analysis of Ang II receptor subtypes expressed on N1E-115 cells began with the successful solubilization of functional Ang II receptors using the detergent 3-[(3-cholamidopropyl)dimethylammonio]-1-propanesulfonate (CHAPS) (9). Scatchard analysis of ^{125}I-[Sar^1, Ile^8]-angiotensin II binding activity in solubilized N1E-115 membranes predicted a B_{max} of 120 fmols/mg protein and a K_D of approximately 1 nM, which is in good agreement with Ang II receptors expressed on differentiated N1E-115 cells. Pharmacologic analysis of CHAPS solubilized N1E-115 cell membranes revealed the exclusive solubilization of the AT_2 receptor subtype (9). Displacement of ^{125}I-Ang II binding activity with the AT_2 selective peptide antagonist CGP42112A was monophasic and of high affinity, while the AT_1 receptor antagonist losartan was ineffective at displacing ^{125}I-Ang II binding activity. However, the AT_2 nonpeptide antagonist PD123319 exhibited a biphasic competition curve, suggesting heterogeneity among the AT_2 receptor sites present in solubilized N1E-115 membranes. In an attempt to further analyze these two pharmacologically distinct AT_2 binding sites, solubilized N1E-115 membranes were subjected to heparin-Sepharose chromatography in an effort to separate these two potentially unique receptor populations. Heparin-Sepharose chromatography of solubilized N1E-115 membranes produced two major peaks of ^{125}I-Ang II binding activity, referred to as Peak I and Peak III (10). Scatchard analysis of Peak III material predicted a B_{max} of 32 fmols/mg protein and a K_D of approximately 1.3 nM, while Peak I material possessed a B_{max} of 214 fmols/mg protein with a similar K_D. These results indicated that Peak I material represents over 80% of the AT_2 binding activity originally present in solubilized N1E-115 membranes. Despite representing the large majority of ^{125}I-Ang II binding activity present in solubilized N1E-115 cells, Peak I material accounts for less than 7% of the total protein applied to the heparin-Sepharose column. Therefore, Peak I material represents a significant enrichment of AT_2 binding activity from N1E-115 cells. At this point, our laboratory decided to use this partially-purified population of AT_2 receptors as an immunogen in the development of protein directed polyclonal antisera.

DEVELOPMENT OF AT_2-SELECTIVE ANTISERA

The use of a partially purified immunogen raises the possibility that, in addition to producing AT_2-directed antisera, antibodies were also produced against non AT_2 receptor protein present in Peak I. Therefore, we needed to develop a strategy in order to selectively purify AT_2 directed antisera from the crude serum. Purification of AT_2-directed antisera was accomplished by applying crude antisera to a protein column of Peak I material depleted of AT_2 receptor protein. Antibodies directed against non-AT_2 receptor proteins present in the

I. Affinity Purification of AT_2 Receptors

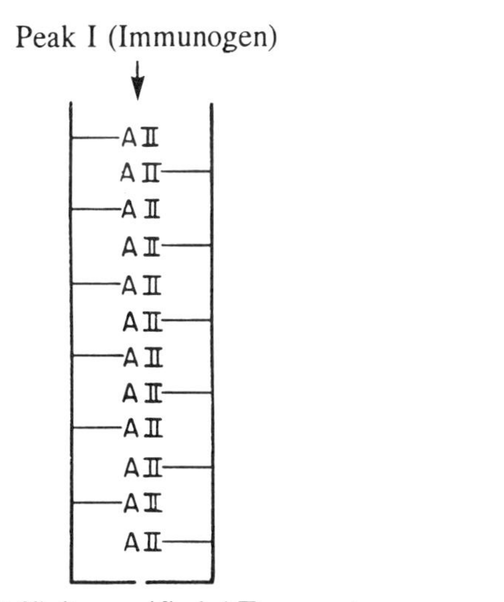

1. Affinity purified AT_2 receptors
2. Peak I material devoid of AT_2 receptors [PI-AT2]

II. Purification of AT_2 Antisera

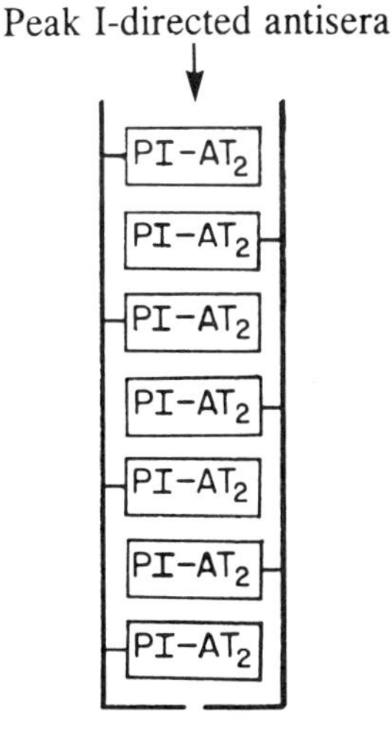

1. Antisera fractions
2. TBS wash fractions
3. $MgCl_2$ wash fractions

Figure 1. Purification of AT_2-directed antisera from crude Peak I antisera. Peak I material which was not retained by the Ang II affinity column were concentrated and the absence of AT_2 receptors was verified by ^{125}I-Ang II binding assays. A protein column of Peak I material devoid of AT_2 binding activity was prepared and crude antisera was applied to this column. AT_2-directed antisera flows through this column while non-AT_2 antisera immunodetect their target proteins present in the Peak I protein column and are retained.

immunogen would be retained by this column, while AT_2 directed antibodies would flow through this column.

The success of this purification strategy was tested in both immunoblot and immunoprecipitation analysis. This method therefore leads to the purification of the AT_2 directed antisera, which could now be used to analyze the biochemical properties of the AT_2 receptor, as well as to localize AT_2 receptors in the brain via immunohistochemical techniques.

Characterization of the AT_2 antisera began with immunoblot and immunoprecipitation studies (11). These antisera were able to immunodetect proteins with the approximate molecular weights of 66 and 110 kDa in solubilized N1E-115 cells, which is in good agreement with several other independent estimates of the molecular weights of AT_2 receptors expressed in N1E-115 cells. In addition, these antisera were able to immunodetect affinity purified AT_2 receptors. These antisera were also able to immunoprecipitate AT_2 receptor binding activity from solubilized N1E-115 cells, as well as immunoprecipitate AT_2 receptors which had been crosslinked with ^{125}I-Ang II. Collectively, these results demonstrated the specificity of these antisera for the AT_2 receptor subtype. These AT_2-selective antisera were then used to localize AT_2 receptor populations in rat brain via immunohistochemical techniques.

While the results of this immunohistochemical study were largely consistent with previous autoradiographic and radioligand binding studies, there were some notable differences in the distribution of immunoreactivity (12). In addition, these studies confirmed the

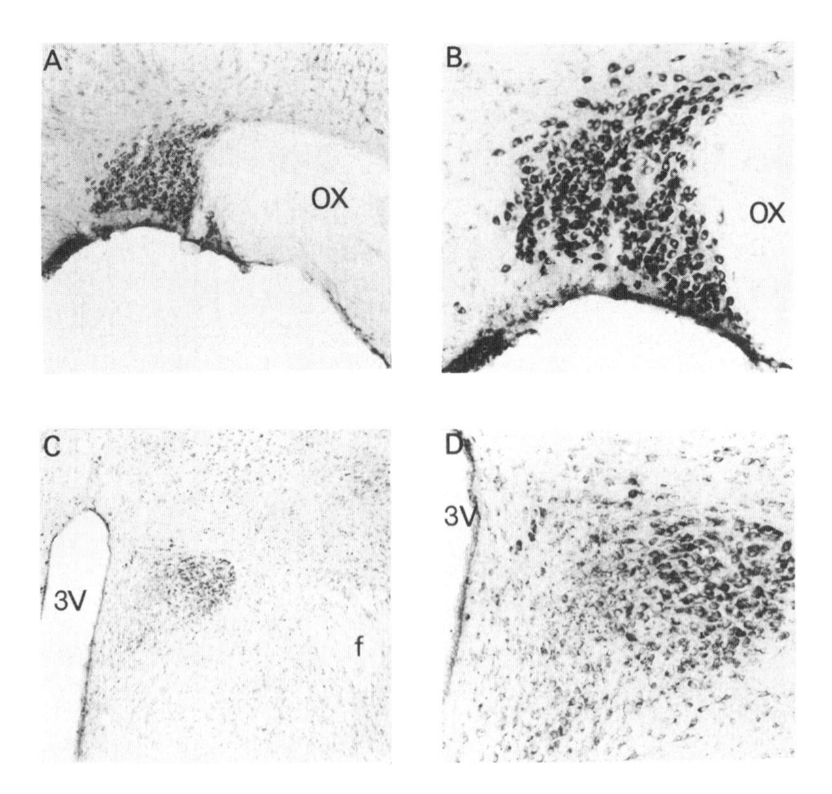

Figure 2. Light micrographs of coronal rat brain sections demonstrating AT_2 receptor immunoreactivity in selected hypothalamic nuclei. Panels A,B: Supraoptic nucleus. Panels C,D: Paraventricular nucleus, with staining associated with the lateral magnocellular neurons. [ox = optic chiasm; f = fornix; 3V = third ventricle].

presence of AT_2 receptors in regions which the expression of AT_2 had previously been equivocal.

For example, AT_2 receptor immunoreactivity was associated with the reticular thalamic nucleus and the bed nucleus of the accessory olfactory tract (BAOT), two regions which had previously been shown to express AT_2 receptors (13-15). AT_2 receptor immunoreactivity was also seen in the supraoptic nucleus (SON) and the paraventricular hypothalamic nucleus (PVN). Higher power magnification of AT_2 receptor immunoreactivity in the SON and the PVN demonstrated that the staining appears to be associated with the lateral magnocellular neurons, which are known to be vasopressin and oxytocin containing neurons.

One of the central effects of Ang II is the release of vasopressin and oxytocin from the lateral magnocellular neurons of the paraventricular hypothalamic nucleus and the supraoptic nucleus. AT_2 receptors expressed in the SON and the PVN may therefore play a role in Ang II-induced release of vasopressin and oxytocin from these regions. In support of this hypothesis, Phillips and colleagues have recently shown that Ang II-induced release of vasopressin is mediated by both AT_1 and AT_2 receptors (16). The present study demonstrates that AT_2 receptor immunoreactivity in the PVN is associated with magnocellular neurons and the SON, as has previously been demonstrated for the AT_1 subtype (17). These results raise the possibility that both AT_1 and AT_2 receptors may be expressed on the same neurons and therefore may play a role in Ang II mediated vasopressin release, as has been suggested by the work of Phillips and colleagues.

AT_2 receptor immunoreactivity was also seen in the cerebellum, which is in good agreement with previous quantitative autoradiographic studies (13,14,18). However, in the immunohistochemical studies, AT_2-receptor immunoreactivity was shown to be associated specifically with the Purkinje cell layer and the deep cerebellar nuclei. Previous autoradiographic studies have suggested that AT_2 receptors are expressed in the molecular layer of the cerebellum (13,14,18). These results demonstrate the greater cellular resolution that is possible using immunohistochemical techniques.

AT_2 receptor immunoreactivity was particularly dense in the locus coeruleus, a region which has previously been shown in quantitative autoradiographic studies to express a high density of AT_2 receptors (14,15). However, AT_2 receptor immunoreactivity was not seen in the inferior olive, a region which is also known to express a high density of AT_2 receptors (14). In fact, evidence continues to accumulate in support of heterogeneity within the AT_2 receptor subtype. The initial evidence was provided by Saaverdra and colleagues (21), who demonstrated that AT_2 receptors in the locus coeruleus are sensitive to the actions of GTP and pertussis toxin, while AT_2 receptors expressed in the inferior olive are insensitive. In addition, electrophysiological studies have predicted heterogeneity between AT_2 receptors in the inferior olive and the locus coeruleus. For example, Ang II has been shown to increase the firing rate of inferior olive neurons, an effect which is mediated through the AT_2 subtype (19). Conversely, AT_2 receptors in the locus coeruleus have been shown to depress glutamate-induced depolarizations and excitatory post synaptic potentials (20). The results of this immunohistochemical study, therefore, may represent a more selective survey of AT_2 receptor expression in rat brain and strengthens the hypothesis of AT_2 receptor heterogeneity.

HETEROGENEITY OF AT_2 RECEPTORS EXPRESSED IN N1E-115 CELLS

At this same time in our laboratory, we were more fully characterizing the Peak I and Peak III populations of AT_2 receptors from solubilized N1E-115 cells isolated by heparin-Sepharose chromatography (10). Pharmacologic studies using the AT_2-selective nonpeptide antagonist PD123319 was the first result that predicted heterogeneity of AT_2 receptors present in crude solubilized N1E-115 membranes. In order to address this issue, competition studies were performed in Peak I and Peak III receptors using both selective and nonselective Ang II ligands. While both Peak I and Peak III AT_2 receptors demonstrated similar pharmacologies using the nonselective Ang II peptidic compounds, interesting differences were seen with the AT_2 selective antagonists. For example, the selective peptide antagonist CGP42112A and the nonpeptide antagonist PD123319 were equally effective at competing for ^{125}I-Ang II binding activity in Peak I material. However, when similar competition studies were performed in Peak III material, CGP42112A was 30 to 40 fold more selective than was PD123319 in displacing ^{125}I-Ang II binding activity. This decreased affinity of PD123319 for AT_2 receptors present in Peak III material likely is the explanation for the biphasic competition curves generated by this selective AT_2 receptor antagonist in crude, CHAPS solubilized N1E-115 membranes.

Activation of a G protein coupled receptors leads to the disassociation of heterotrimeric subunits of G proteins, an effect which can be measured as a shift in agonist binding activity from a high to a low affinity state. In this regard, the effects of the nonhydrolyzable analog of GTP, GTPγS, was tested in radioligand binding analysis of Peak I and Peak III material, in order to determine if these AT_2 receptor populations were coupled to G proteins. The results demonstrated that specific binding activity of Peak I type AT_2 receptors was reduced by approximately 40% in the presence of 100 μM GTPγS, while Peak III binding activity was unaffected. These results suggested that AT_2 receptors which are present in Peak I material are coupled to G proteins, while the Peak III population of AT_2 receptors are not.

Previous studies have demonstrated that ^{125}I-Ang II binding activity may be modulated by sulfhydryl reducing agents, such as dithiothreitol (DTT). Radioligand binding assays performed in the presence of DTT with Peak I and Peak III material again produced dissimilar results within these AT_2 receptor populations. In particular, increasing concentration of DTT increases the specific ^{125}I-Ang II binding activity of Peak III AT_2 receptors, while that of Peak I receptors decreased.

In an effort to determine whether AT_2 receptors present in Peaks I and III are immunologically distinct, Peak I and Peak III proteins were screened by immunoblotting using antisera raised against the Peak I population of AT_2 receptors. A four fold excess of Peak III material was used in immunoblot analysis to compensate for the greater ^{125}I-Ang II binding activity present in Peak I material when compared to Peak III material. Peak I exhibited two major immunoreactive bands of 110 and 66 kDa, consistent with previous observations (9,22). Conversely, no immunoreactive bands were detected in Peak III,

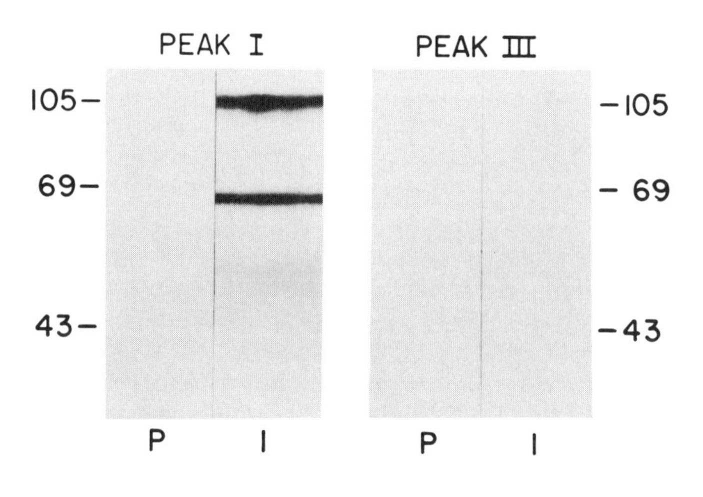

Figure 3. Immunoblot analysis of AT_2 receptors from Peaks I and III derived heparin-Sepharose chromatography.

suggesting that this AT_2 receptor population is immunologically distinct from AT_2 receptors present in Peak I.

To further verify that ^{125}I-Ang II binding sites in Peak I and Peak III were immunologically distinct, immunoprecipitation assays were performed in Peak I and Peak III with the AT_2-directed antisera. These antisera were able to immunoprecipitate AT_2 receptors present in Peak I material, as evidenced by the appearance of immune specific ^{125}I-Ang II binding activity in the precipitated pellet. Preimmune serum was unable to immunoprecipitate significant binding activity. In contrast to Peak I, the AT_2 antisera were ineffective at immunoprecipitating ^{125}I-Ang II binding activity in Peak III.

CLONING OF AT_2 RECEPTOR FROM N1E-115 CELLS

Because the neuroblastoma N1E-115 cell line is a good model for the study of AngII receptors, our laboratory actively pursued the cloning of these receptors from these cells. Previously, we have successfully used homology-based polymerase chain reaction (PCR) to obtain an AT_1 receptor cDNA from these cells. Subsequent sequence analysis revealed that this clone is highly homologous (98%) to murine AT_{1A} receptors. When expressed in COS-1 cells, these receptors exhibited all of the characteristics of other AT_1 receptors; they possessed the expected high affinity for the AT_1 antagonist losartan and not for AT_2 specific compounds, demonstrated GTP-sensitivity of agonist binding, and mediated AngII-induced Ca^{2+} mobilization. Most recently, two groups have reported the cloning of an AT_2 receptor, first from a rat fetal cDNA library and PC12w cells (23,24), and later from mice (25). Each of these cDNAs encode for a 363 amino acid protein that apparently conforms to the seven transmembrane domain structural motif of G-protein coupled receptors. In view of these recent developments, we also began using PCR employing primers from the most conserved regions of these clones to isolate a cDNA from our N1E-115 library. This approach has been successful, as we have been able to isolated an AT_2 clone from an N1E-115 cell library that is identical with other cloned AT_2 receptors.

Pharmacological, biochemical and immunological analysis of this AT_2 receptor clone isolated from N1E-115 cells revealed that it has the identical characteristics or properties of the Peak III receptor. The AT_2 receptor isolated from N1E-115 cells and transfected in COS cells exhibited a similar pharmacology as the Peak III AT_2 receptor, in that CGP42112A was more effective at displacing ^{125}I-Ang II binding activity than was PD123319. These results are not unexpected in that the AT_2 receptor subtype that was originally cloned from whole rat fetus displays the same pharmacology. Interestingly, AT_2 receptors expressed in rat fetal tissues exhibit the identical pharmacology. The AT_2 receptor clone was also shown to be insensitive to the actions of GTPγS, as well as demonstrated increased binding activity in the presence of DTT, identical to the Peak III AT_2 receptor. Lastly, immunoblot analysis of membranes prepared from COS-1 cells transfected with the AT_2 receptor cDNA from N1E-115 cells did not demonstrate any immune-specific bands with the AT_2-directed antisera. Comparisons of the binding properties of the cloned AT_2 receptor with those in N1E-115 cells suggest that the cloned receptor most closely resembles Peak III receptors. Both receptors possess higher affinity for the peptidic antagonist CGP42112A than the non-peptide PD123319. In addition, both receptors failed to demonstrate guanine nucleotide regulation of agonist binding and displayed increased ligand binding in the presence of dithiothreitol. Furthermore, AT_2 specific antisera that were raised against Peak I receptors and were unable to immunodetect either Peak III AT_2 receptors or the cloned receptor when transfected into COS-1 cells.

Successful cloning of one of the AT_2 receptors from N1E-115 cells has also greatly facilitated our studies of its regulation in these cells. By using both Northern blot analysis

Table 1. Characteristics of Angiotensin AT_2 receptors expressed in N1E-115 cells

	PEAK I	PEAK III	AT_2 RECEPTOR CLONE
Pharmacology	CGP = PD	CGP > PD	CGP > PD
GTP Sensitivity	+	-	-
DTT Sensitivity	decreased	increased	increased
Immunoreactivity	+	-	-

and the ribonuclease protection assay (RPA), specific oligonucleotide probes have been designed in order to examine the regulation of its mRNA in response to a variety of cell treatments. For example, as measured by radioligand binding assays, we have previously demonstrated that the induction of cell differentiation by using dimethyl sulfoxide and low serum media, caused a steady upregulation of AT_2 receptors in N1E-115 cells. Both the Northern blot analysis as well as the RPA have corroborated our previous observations; AT_2 receptor mRNA levels are also increased upon differentiation.

SUMMARY

Evidence continues to accumulate that strengthens the proposal of heterogeneity within both the AT_1 and the AT_2 receptor subtypes. Pharmacologic, biochemical and immunological studies of AT_2 receptors expressed in N1E-115 cells strengthen the hypothesis of AT_2 receptor heterogeneity. However, it is important to reassess these studies, especially in terms of how these results correlate with other reports of AT_2 receptor heterogeneity. For example, AT_2 receptor immunoreactivity was absent in some neuronal regions which have previously been proposed to express the AT_2 receptor subtype. In particular, AT_2 receptor staining was not seen in the inferior olive, a region which is known to express a high density of AT_2 receptors. Upon first examination, these results were somewhat troubling. However, when compared with earlier reports, these results should not have been unexpected. For instance, Tsutsumi and Saaverdra previously have shown that AT_2 receptors in the locus coeruleus are sensitive to the actions of guanine nucleotides, while AT_2 receptors in the inferior olive are insensitive (21). These antisera were raised against a population of AT_2 receptors which are sensitive to GTPγS and therefore, the lack of AT_2 receptor staining in the inferior olive, as well as the presence of AT_2 receptor immunoreactivity in the locus coeruleus, confirms and extends these earlier reports. In addition the AT_2 receptors expressed in the locus coeruleus have been shown to be functionally distinct from AT_2 receptors in the inferior olive. In this regard, Ang II has been shown to depress glutamate-induced EPSPs in the locus coeruleus, an effect which is mediated through the AT_2 receptor (19). Conversely, AT_2 receptors have been shown to increase the firing rate of neurons in the inferior olive (20). Collectively, these results would predict that staining should be absent in the inferior olive using these AT_2-directed antisera. Indeed, in view of these earlier physiological and pharmacological studies, the presence of AT_2 receptor immunoreactivity in the inferior olive would have been surprising.

The most convincing example of AT_2 receptor heterogeneity is the characterization of AT_2 receptors present in N1E-115 cells. Separation of solubilized N1E-115 membranes by heparin-Sepharose chromatography generates two populations of AT_2 receptors which are pharmacologically and biochemically distinct. In particular, CGP42112A was approximately 2 orders of magnitude more selective for Peak III AT_2 receptors than was PD123319. Binding activity of Peak I and Peak III AT_2 receptor populations also differed in their responses to GTPγS and DTT treatment. Lastly, the AT_2-directed antisera, raised against the Peak I population of AT_2 receptors, were not able to immunodetect the Peak III population of AT_2 receptors in immunoblot analysis, or immunoprecipiatate AT_2 binding activity from Peak III material.

Pharmacological, biochemical and immunological analysis of the AT_2 receptor clone isolated from N1E-115 cells revealed that it has the identical characteristics or properties of the Peak III receptor. The AT_2 receptor isolated from N1E-115 cells exhibited a similar pharmacology as the Peak III AT_2 receptor, in that CGP42112A was more effective at displacing ^{125}I-Ang II binding activity than was PD123319. The AT_2 receptor clone was also shown to be insensitive to the actions of GTPγS, as well as demonstrated increased binding activity in the presence of DTT, identical to the Peak III AT_2 receptor. Lastly, immunoblot analysis of membranes prepared from COS-1 cells transfected with the AT_2 receptor cDNA from N1E-115 cells did not demonstrate any immune-specific bands with the AT_2-directed antisera. Characterization of an AT_2 receptor cDNA isolated from N1E-115 cells reveals that this clone is identical to the Peak III type of AT_2 receptor. In conjunction with the analysis of AT_2 receptors expressed in N1E-115 cells, studies utilizing the AT_2-directed antisera have been able to identify and localize an Angiotensin Type 2 receptor subtype which is unique from the AT_2 receptor clone isolated from rat fetus and PC12w cells. One of the challenges that remains is to utilize these antisera to isolate this unique AT_2 receptor, by molecular cloning or immunoaffinity protein purification, which will permit for a greater understanding of the cellular, physiological and behavioral actions of Ang II in the brain.

REFERENCES

1. Peach,M.J. (1977) Physiol. Rev., *57*, 313-370.
2. Zehr,J.E., Kurz,K.D., Seymour,A.A. and Schultz,H.D. (1980) Adv. Exp. Med. Biol., *130*, 135-170.
3. Sumners,C., Myers,L.M., Kalberg,C.J. and Raizada,M.K. (1990) Prog. Neurobiol., *34*, 355-385.
4. Wayner,M.J., Armstrong,D.L., Polan-Curtain,J.L. and Denny,J.B. (1993) Peptides, *14*, 441-444.
5. Bumpus,F.M., Catt,K.J., Chiu,A.T., DeGasparo,M., Goodfriend,T., Husain,A., Peach,M.J., Taylor,D.G Jr. and Timmermans,P.B. (1991) Hypertension., *17*, 720-721.
6. Fluharty,S.J. and Reagan,L.P. (1989) J. Neurochem., *52*, 1393-1400.
7. Reagan,L.P., Ye,X.H., Mir,R., DePalo,L.R. and Fluharty,S.J. (1990) Mol Pharmacol., *38*, 878-886.
8. Ades,A.M., Slogoff,F. and Fluharty,S.J. (1991) Soc. Neurosci. Abstr., *17*, 809.(Abstract)
9. Siemens,I.R., Adler,H.J., Addya,K., Mah,S.J. and Fluharty,S.J. (1991) Mol. Pharmacol., *40*, 717-726.
10. Siemens,I.R., Reagan,L.P., Yee,D.K. and Fluharty,S.J. (1994) J. Neurochem., *62*, 2106-2115.
11. Reagan,L.P., Theveniau,M., Yang,X.-D., Siemens,I.R., Yee,D.K., Reisine,T. and Fluharty,S.J. (1993) Proc. Natl. Acad. Sci. U. S. A., *90*, 7956-7960.
12. Reagan,L.P., Flanagan-Cato,L.M., Yee,D.K., Ma,L-Y., Sakai,R.R. and Fluharty,S.J. (1994) Brain. Res., *662*, 45-59.
13. Aldred,G.P., Chai,S.Y., Song,K., Zhuo,J., MacGregor,D.P. and Mendelsohn,F.A.O. (1993) Regul. Pept., *44*, 119-130.
14. Song,K., Allen,A.M., Paxinos,G. and Mendelsohn,F.A.O. (1992) J. Comp. Neurol., *316*, 467-484.
15. Tsutsumi,K. and Saavedra,J.M. (1991) Am. J. Physiol., *261*, R209-R216.
16. Hogarty,D.C., Speakman,E.A., Puig,V. and Phillips,M.I. (1992) Brain Res., *586*, 289-294.
17. Phillips,M.I., Shen,L., Richards,E.M. and Raizada,M.K. (1993) Regul. Pept., *44*, 95-107.
18. Allen,A.M., Chai,S.Y., Clevers,J., McKinley,M.J., Paxinos,G. and Mendelsohn,F.A.O. (1988) J. Comp. Neurol., *269*, 249-264.
19. Xiong,H. and Marshall,K.C. (1994) Neuroscience, *62*, 163-175.
20. Ambuhl,P., Felix,D., Imboden,H., Khosla,M.C. and Ferrario,C.M. (1992) Regul. Pept., *41*, 19-26.
21. Tsutsumi,K. and Saavedra,J.M. (1992) Mol. Pharmacol., *41*, 290-297.
22. Siemens,I.R., Yee,D.K., Reagan,L.P. and Fluharty,S.J. (1994) J. Neurochem., *62*, 257-264.

23. Kambayashi,Y., Bardhan,S., Takahashi,K., Tsuzuki,S., Inui,H., Hamakubo,T. and Inagami,T. (1993) J. Biol. Chem., *268*, 24543-24546.
24. Mukoyama,M., Nakajima,M., Horiuchi,M., Sasamura,H., Pratt,R.E. and Dzau,V.J. (1993) J. Biol. Chem., *268*, 24539-24542.
25. Sasamura,H., Hein,L., Krieger,J.E., Pratt,R.E., Kobilka,B.K. and Dzau,V.J. (1992) Biochem. Biophys. Res. Commun., *185*, 253-259.

22

ANGIOTENSIN II STIMULATES PROTEIN PHOSPHATASE 2A ACTIVITY IN CULTURED NEURONAL CELLS VIA TYPE 2 RECEPTORS IN A PERTUSSIS TOXIN SENSITIVE FASHION

Xian-Cheng Huang, Colin Sumners, and Elaine M. Richards

Department of Physiology, College of Medicine
University of Florida
Gainesville, Florida

SUMMARY

Recent studies have suggested a role for an inhibitory G protein (G_i) and protein phosphatase 2A (PP2A) in the angiotensin II (Ang II) type 2 (AT_2) receptor mediated stimulation of neuronal K^+ currents. In the present study we have directly analyzed the effects of Ang II on PP2A activity in neurons cultured from newborn rat hypothalamus and brainstem. Ang II elicited time (30 min - 24 h)- and concentration (10 nM - 1 μM)-dependent increases in PP2A activity in these cells. This effect of Ang II involved AT_2 receptors, since it was inhibited by the AT_2 receptor selective ligand PD123319 (1 μM), but not by the Ang II type 1 receptor antagonist losartan (1 μM). Furthermore, the stimulatory effects of Ang II on PP2A activity were inhibited by pretreatment of cultures with pertussis toxin (PTX) (200 ng/ml; 24 h) indicating the involvement of an inhibitory G-protein; and by cycloheximide (CHX) (1 μg/ml; 30 min) indicating a requirement for protein synthesis. These effects of Ang II appear to be via activation of PP2A, since Western Blot analyses revealed no effects of this peptide on the protein levels of the catalytic subunit of PP2A in cultured neurons. In summary, these data suggest that PP2A is a key component of the intracellular pathways coupled to neuronal AT_2 receptors.

INTRODUCTION

It is now evident that both major subtypes of Ang II receptors are present in rat brain (1, 2, 3). Ang II type 1 (AT_1) receptors in the brain mediate all of the previously characterized effects of centrally injected Ang II (4, 5, 6, 7). In contrast, the physiological functions of CNS Ang II type 2 (AT_2) receptors are unknown (8). However, the high level of expression of AT_2 receptors in many neonatal tissues, including brain, suggests that these sites have a

Recent Advances in Cellular and Molecular Aspects of Angiotensin Receptors
Edited by Mohan K. Raizada et al., Plenum Press, New York, 1996

role in development and/or differentiation (9, 10, 11). Cloning studies have predicted that the AT_2 receptor belongs to the class of guanine nucleotide (G) binding-protein receptors (12, 13). Specifically, these studies suggested that the AT_2 receptor is coupled to a pertussis toxin (PTX)-sensitive inhibitory G protein (12). Furthermore, one of the studies indicated that activation of AT_2 receptors, both on pheochromocytoma PC12W cells and on transfected COS-7 cells, resulted in an inhibition of phosphotyrosine phoshatase (PTPase) activity (12). Despite these results, the signal transduction pathway(s) coupled to AT_2 receptors is not clear (8).

Recent studies from our laboratory have begun to shed some light on the intracellular coupling of AT_2 receptors in neuronal cells. We have utilized neurons cultured from the hypothalamus and brainstem of neonatal rats to investigate the properties of AT_2 receptors (for review see 8). These neonatal cultured neurons express AT_2 receptors which are similar, from both molecular and pharmacological standpoints, to the cloned AT_2 receptors and the AT_2 receptors present in rat brain (14, 8). We determined that AT_2 receptor activation stimulates the delayed rectifier K+ current (I_k) in these cultured neurons (15). This activation is pertussis toxin-sensitive, involving a G_i protein, (16) and this is consistent with the results obtained from cloned AT_2 receptors (12). Further, the stimulatory action of Ang II on I_k appears to involve activation of protein (serine/threonine) phosphatase 2A (PP2A), since it is inhibited by low concentrations (1-10 nM) of okadaic acid and by anti-PP2A antibodies (16). In the present study we tested the effect of Ang II on PP2A activity. Our data clearly show that Ang II elicits a concentration- and time-dependent stimulation of PP2A activity, an effect mediated by AT_2 receptors, and involving a PTX-sensitive G-protein. However, it took 30 minutes and a protein synthetic event following AT_2 receptor stimulation before significant PP2A activation occurred. Since PP2A interferes with the intracellular pathways activated by growth factors this ties in nicely with the possibility of AT_2 receptors being involved in neuronal cell growth and differentiation, but clearly differs from the immediate effect of AT_2 receptors on I_k, (15, 16). Thus it appears that PP2A may either be involved at two separate steps following AT_2 receptor activation, or that AT_2 receptors may activate two different pathways both having a PP2A component.

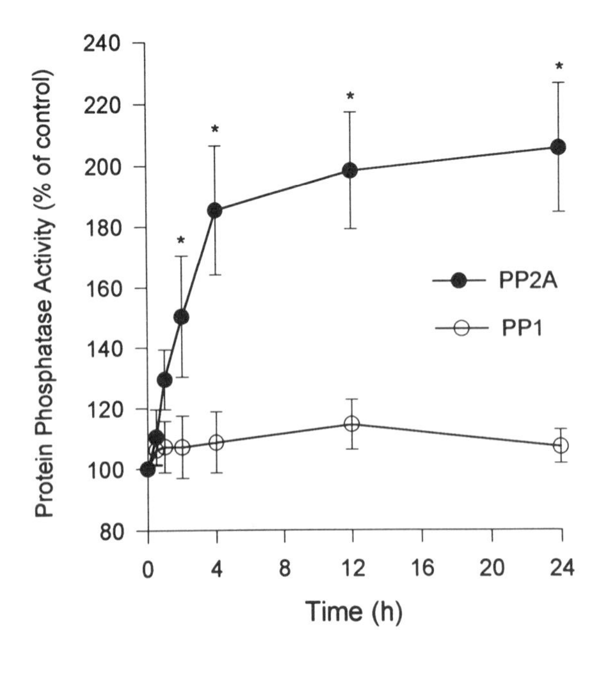

Figure 1. Effects of Ang II on okadaic acid-sensitive serine/threonine phosphatase activities in neuronal cultures as a function of incubation time. Neuronal cultures were incubated with 100 nM Ang II (A) for the indicated times, followed by analysis of PP1 (○) and PP2A (●) activities. Data are means ± SEM from 4 experiments, and are presented as a percentage of control PP1 and PP2A activities. Control enzyme activities (DMEM treatment; 100%) are plotted on the y-axis. Control PP1 and PP2A activities were 1.37 ± 0.2 and 2.05 ± 0.43 nmol/min/mg protein respectively.

RESULTS

Phosphatase activity due to PP1 and PP2A was measured using a kit supplied by Gibco, BRL, Gaithersburg, MD. The assay conditions used with this kit did not measure phosphatase activity due to PP2B or PP2C. In preliminary experiments the linear range of the dephosphorylation reaction (the index of phosphatase activity) was determined by measuring the release of ^{32}Pi from ^{32}P-labeled phosphorylase a with increasing amounts of cell extract. Cell extracts were prepared according to (17) and were not enriched for any particular cell fraction. All subsequent experiments were performed at 0.2-1.0 μg of cell extract protein which was in the linear range. A titration curve of phosphatase activity against okadaic acid concentration revealed a plateau at 1-10 nM. Thus, all subsequent experiments were carried out under the assumption that PP2A was inhibited by 3 nM okadaic acid whereas PP1 was not. Non-specific phosphatase activity was that remaining in the presence of 5 μM okadaic acid. PP2A activity was total phosphatase activity minus the activity remaining in the presence of 3 nM okadaic acid. PP1 was the difference between the activities remaining in the presence of 3 nM and 5 μM okadaic acid.

Incubation of neuronal cultures with Ang II (100 nM) elicited a significant increase in PP2A activity as a function of incubation time (30 min - 24 h) (Fig. 1). PP2A activity increased rapidly during the first 4 h and remained elevated for 24 h. The activity of PP1 was not significantly altered by the Ang II treatment (Fig. 1).

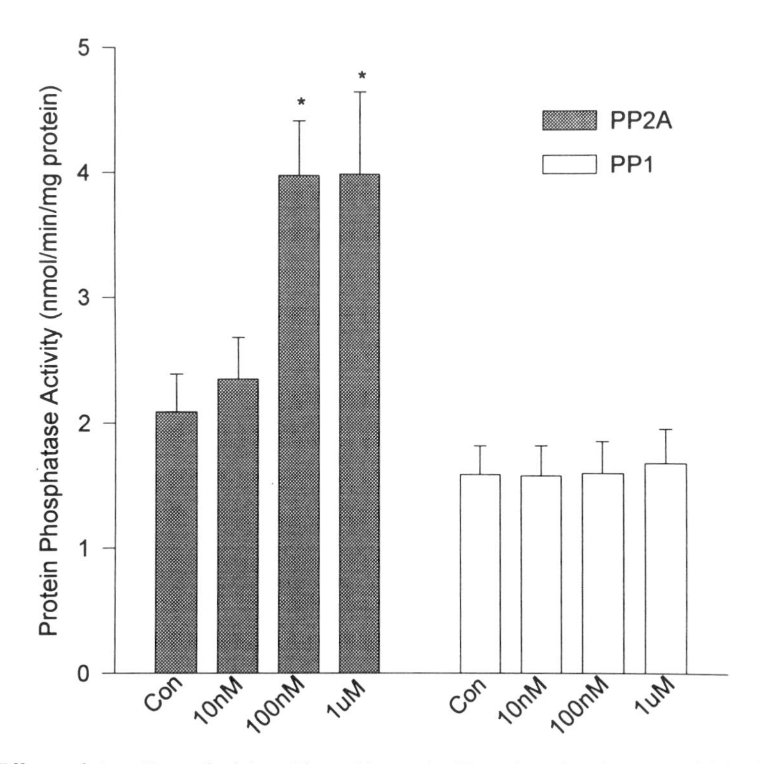

Figure 2. Effects of Ang II on okadaic acid-sensitive serine/threonine phosphatase activities in neuronal cultures as a function of concentration. Neuronal cultures were incubated with either control (DMEM), 10 nM, 100 nM or 1 μM Ang II for 24 h at 37°C. Following these treatments, PPI and PP2A activities were measured as described earlier. Data are means ± SEM from 4 independent experiments. *, significantly different from controls.

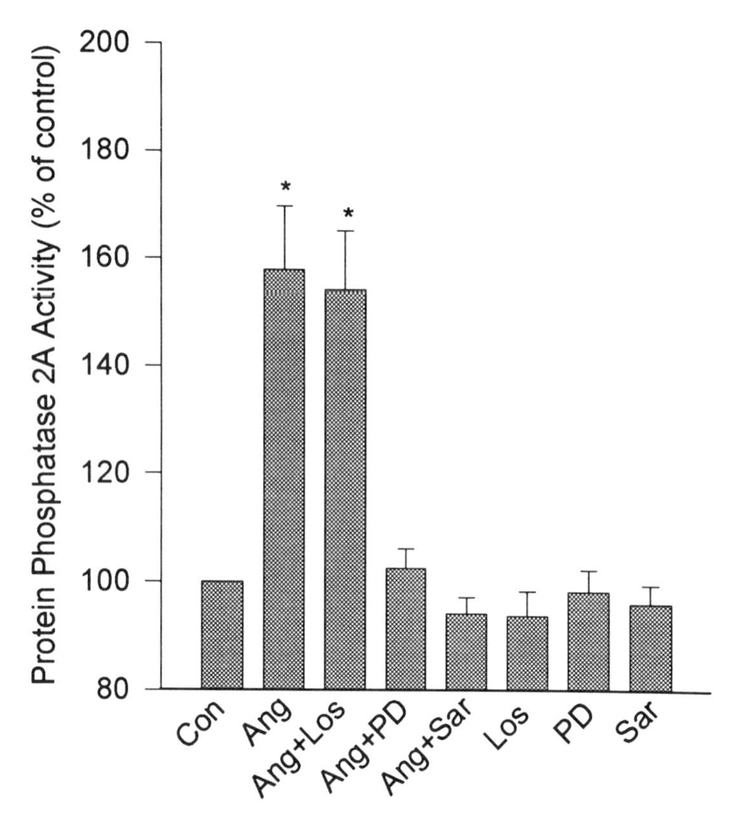

Figure 3. Effects of Ang II receptor blockers on Ang II-stimulated PP2A activity. Neuronal cultures were treated as follows. Con (DMEM 10 min, then DMEM 24 h); Ang (DMEM 10 min, then 100 nM Ang II 24 h); Ang + Los (1 μM Losartan 10 min, then 100 nM Ang II 24 h); Ang + PD (1 μM PD123319 10 min, then 100 nM Ang II 24 h); Ang + Sar (1 μM Sar^1Ile^8-Ang II 10 min, then 100 nM Ang II 24 h); Los (1 μM Losartan 10 min, then DMEM 24 h); PD (1 μM PD123319 10 min, then DMEM 24 h); Sar (1 μM Sar^1Ile^8-Ang II 10 min, then DMEM 24 h). Following these incubations, PP2A activity was analyzed as detailed above. Data are means ± SEM from 4 experiments, and are presented as a percentage of control PP2A activity. Control PP2A activity (100%) was 2.04 ± 0.22 nmol/min/mg protein.

The stimulatory effects of Ang II on PP2A activity were clearly concentration-dependent, as seen in Fig 2. Maximal effects were obtained with 100 nM Ang II.

Figure 2 also shows that PP1 activity was not altered by different concentrations of Ang II. The stimulation of PP2A activity by Ang II (100 nM) was completely abolished by pretreatment of the neuronal cultures with the selective AT_2 receptor ligand PD123319 (1 μM, 10 min) (Fig. 3).

Similarly, pretreatment with the non-selective Ang II receptor antagonist Sar^1-Ile^8-Ang II (1 μM, 10 min) also inhibited the stimulation of PP2A by Ang II (Fig. 3). By contrast, the AT_1 receptor antagonist losartan (1 μM, 10 min) did not alter the stimulatory effects of Ang II on PP2A (Fig. 3). These data suggest that the effect of Ang II on PP2A activity are mediated by AT_2 receptors. Incubation of neuronal cultures with the antagonists alone did not alter PP2A activity (Fig. 3).

Our previous studies suggested that the AT_2 receptor-mediated stimulatory effects of Ang II on neuronal K^+ currents involved both PP2A and an inhibitory guanine nucleotide regulatory protein (G_i) (16). To test whether the AT_2 receptor-mediated stimulation of PP2A

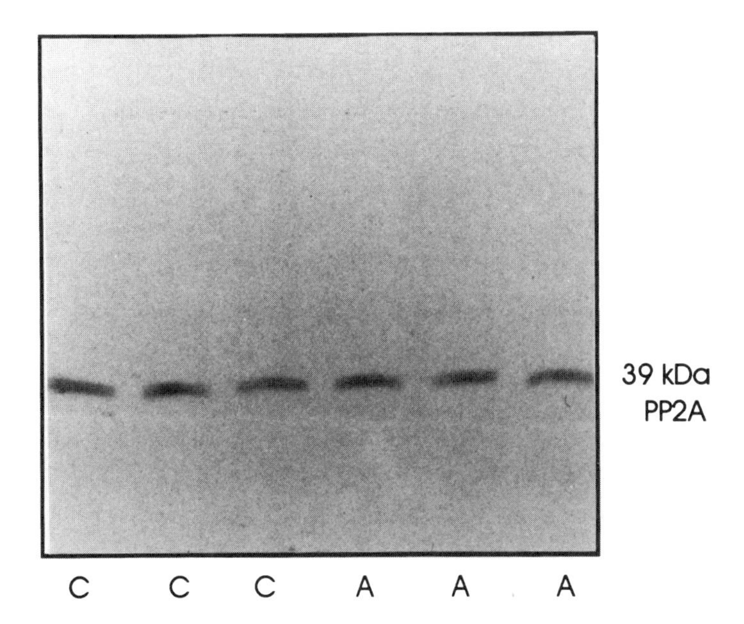

Figure 4. Effects of Ang II on PP2A protein expression. Neuronal cultures were treated with DMEM (U), 100 nM Ang II (A) for 18 h at 37°C. After isolation, 25 μg of proteins from each treatment group were subjected to SDS-polyacrylamide gel electrophoresis followed by immunoblot analysis using anti-PP2A catalytic subunit antibody. The protein recognized by anti-PP2A had a molecular mass of ~39 kDa and appears as single band. Shown here is a representative Western Blot which was repeated 4 times with similar results.

activity by Ang II involves G_i, we pretreated neuronal cultures with PTX (200 ng/ml, 24 h), which inhibits this G-protein and also G_o (18). PTX treatment completely abolished the stimulation of PP2A activity by Ang II (100 nM). PTX alone did not alter PP2A activity. The PP2A activity values (means ± SEM, for n=4 experiments) for control; PTX alone; Ang II; and Ang II + PTX were 1.7012 ± 0.03; 1.72 ± 0.03; 3.20 ± 0.14; 1.70 ± 0.10 nmol/min/mg protein, respectively. The stimulation of PP2A by Ang II persists over a 24 h period, and this may indicate that there is an increase in the synthesis of PP2A protein, or synthesis of a protein which activates PP2A. Indeed, the stimulatory effects of Ang II on PP2A are abolished by pretreatment with the protein synthesis inhibitor cycloheximide (1 μg/ml, 30 min); values of PP2A activity were control; CHX; AngII; Ang II + CHX: 1.70 ± 0.03; 1.75 ± 0.05; 3.20 ± 0.13; 1.68 ± 0.09 nmol/min/mg protein respectively. Data are means ± SEM from 4 seperate experiments. However, Western Blot analyses using a specific antibody revealed no effects of Ang II (100 nM, 18 or 24h) on the levels of the catalytic subunit of PP2A protein (Fig. 4).

As expected, there were also no effect of Ang II on PP1 protein levels, (data not shown). Thus, it is unlikely that the increase in PP2A activity includes an increase in synthesis of the catalytic subunit of this enzyme.

DISCUSSION

The studies presented here demonstrate that activation of neuronal AT_2 receptors, by Ang II elicits a stimulation of PP2A activity in these cells. Our data suggest that this serine/threonine phosphatase is a key component in the intracellular pathways affected by stimulation of neuronal AT_2 receptors. By contrast, PP1 activity in neurons was not altered by Ang II, showing that the effect of the peptide is specific for PP2A. In previous studies we determined that Ang II, via AT_2 receptors, causes a stimulation of neuronal I_K (15) through a process involving PP2A (16). The present data support this idea by showing that Ang II stimulates PP2A. However, the time course of the stimulatory response to Ang II on PP2A activity is not consistent with the previously observed involvement of this enzyme in Ang II-stimulated neuronal I_K. The latter effect occurs within a few minutes (16), while in the present study at least 30 minutes incubation with Ang II was required to elicit a significant stimulation of PP2A activity (Fig. 1). When considering that only a small percentage (5-10%) of the total cell population on each dish contain AT_2 receptors (8), one possible explanation for the discrepancy is that the assay utilized in the present work is not sensitive enough to detect small changes in PP2A activity which occur within a few minutes in a small number of cells. Another explanation could be that the short-term activation of PP2A, involved in Ang II-stimulated I_K, involves a form of the enzyme which utilizes a different natural substrate compared with that used in the present study ie. phosphorylase a is not a sensitive indicator of the PP2A activity in neuronal cells associated with increases in I_K as it is too different from the endogenous substrate of the enzyme but is adequate for that involved in the longer term response. We are currently testing phosphohistones to determine whether they are a better substrate.

The stimulatory effects of Ang II, via AT_2 receptors, on neuronal I_K also involves an inhibitory G-protein (G_i) (16). The involvement of an inhibitory G-protein in the Ang II-stimulated increase in neuronal PP2A activity is indicated because this effect is blocked by PTX pretreatment. Taken together, these findings provide good evidence that the signal transduction pathways coupled to neuronal AT_2 receptors involve an inhibitory G-protein and PP2A. While the involvement of PP2A in AT_2 receptor mediated effects is a novel idea, the involvement of this enzyme in the intracellular processes mediated by other G-protein coupled receptors has been previously reported. For example, the stimulation of calcium-activated K^+ channels by somatostatin in GH_4C_1 pituitary cells involves activation of PP2A (19). In addition, in pancreatic acinar cells cholecystokinin activates a phosphatase which is predominately PP2A (20).

Even though the present studies are indicative of an AT_2 receptor-mediated activation of PP2A in cultured neurons, the mechanisms involved in this activation are not understood. *In vitro* studies indicate that phosphorylation of PP2A results in inactivation of this enzyme. For example, tyrosine phosphorylation of the PP2A catalytic subunit in 10T 1/2 fibroblasts results in inactivation of the enzyme (21). Furthermore, *in vitro* phosphorylation of the PP2A catalytic subunit at a threonine residue, elicited by autophosphorylation-activated serine/threonine kinase, inactivates PP2A (22). It should be noted, however, that the mechanisms present *in vivo* for regulation of PP2A are not known (23). Nevertheless, the above *in vitro* results may imply that activation of PP2A would result from dephosphorylation of either tyrosine or threonine residues on the catalytic subunit. However, there is no evidence that activation of AT_2 receptors stimulates the activity of a PTPase in our cultured neurons. Rather, studies from PC12W pheochromocytoma cells and COS-7 cells transfected with the AT_2 receptor indicate that Ang II elicits a reduction in PTPase activity (12). Further studies on the regulation of PP2A in neurons are required before the mechanisms by which Ang II alters PP2A activity can be understood.

It is evident from our results that the longer term increase in neuronal PP2A activity elicited by AT_2 receptor activation requires a protein synthetic event, since it is abolished by cycloheximide. Western Blot experiments revealed no changes in the levels of PP2A catalytic subunit protein in response to Ang II. From these studies the implication is that Ang II increases the synthesis of a protein (perhaps a PTPase?) which acts to increase PP2A activity.

Another possibility is that Ang II may stimulate the synthesis of one of the regulatory (B) subunits of PP2A (24, 23), which in turn activates the enzyme.

A further question which arises from the present study concerns the role of increased PP2A activity in response to AT_2 receptor stimulation. It is now apparent that PP2A can inhibit the activity of certain protein kinases which are intermediates in the intracellular signalling cascade coupled to growth factor receptors. Namely, PP2A is able to inactivate MAP kinase and MAP kinase kinase (MEK) (24, 25). By stimulating PP2A, Ang II may be able to inhibit MAP kinase kinase and MAP kinase activity and disrupt the actions of growth factors on cell growth and differentiation. Thus, in neonate neurons where AT_2 receptors are predominant (8), Ang II may act as an inhibitory regulator of neuronal differentiation via a pathway which includes activation of PP2A and inhibition of MAP kinase. Studies are ongoing to assess the role of AT_2 receptors in neuronal differentiation in neonates.

ACKNOWLEDGMENTS

The authors thank Jennifer Brock for typing the manuscript, and Tammy Gault for preparation of neuronal cultures. This work was supported by NIH grant NS-19441.

REFERENCES

1. Rowe B.P., Grove K.L., Saylor D.L., and Speth R.C. (1990). Eur. J. Pharmacol. *186*: 339-342.
2. Tsutsumi K., and Saavedra J.M. (1991). J. Neurochem. *56*: 348-351.
3. Song K., Allen A., Paxinos, and Mendelsohn F.A.O. (1992). J. Comp. Neurol. *316*: 467-490.
4. Koepke J.P., Bovy P.R., McMahon E.G., Olins G.M., Reitz D.B., Salles K.S., Schuh J.R., Trapani A.J., and Blaine E.D. (1990). Hypertension *15*: 841-847.
5. Fregly M.J. and Rowland N.E. (1991). Brain Res. Bull. *27*: 97-100.
6. Hogarty D.C., Speakman E.A., Puig V., and Phillips M.I. (1992). Brain Res. *586*: 389-395.
7. Qadri F., Culman J., Veltmar A., Maas K., Rascher W., and Unger T. (1993). J. Pharmacol. Exp. Therap. *267*: 567-574.
8. Sumners C., Raizada M.K., Kang J., Lu D., and Posner P. (1994). Front. Neuroendocrinol. *15*: 203-230.
9. Grady E.F., Sechi L.A., Griffin C.A., Shambelan M., and Kalinyak J.E. (1991). J. Clin. Invest. *88*: 921-933.
10. Millan M.A., Kiss A., and Aguilera G. (1991). Peptides *12*: 712-737.
11. Tsutsumi K., and Saavedra J.M. (1991). Am. J. Physiol. *261*: R209-R216.
12. Kambayashi Y., Bardhan S., Takahashi K., Tsuzuki S., Inui H., Hambuko T., and Inagami T. (1993). J. Biol. Chem. *268*: 24543-24546.
13. Mukoyama M., Nakajima M., Horiuchi M., Sasamura H., Pratt R.E. and Dzau V.J. (1993). J. Biol. Chem. *268*: 24539-24542.
14. Sumners C., Tang W., Zelezna B., and Raizada M.K. (1991). Proc. Natl. Acad. Sci. USA *88*: 7567-7571.
15. Kang J., Sumners C., and Posner P. (1993). Am. J. Physiol. *265*: C607-C616.
16. Kang J., Posner P., and Sumners C. (1994). Am. J. Physiol. *267*: C1389-C1397.
17. Begum N., Robinson L.J., Draznin B., and Heidenreich K.A. (1993). Endocrinol. *133*: 2085-2090.
18. Spiegel A.M., Shenker A., and Weinstein L.S. (1992). Endocrine Rev. *13*: 536-565.
19. White R.E., Schonbrunn A., and Armstrong D.L. (1991). Nature (Lond.) *351*: 570-573.
20. Lutz M.P., Gates L.K., Pinon D.I., Shenolikar S., and Miller L.J. (1993). J. Biol. Chem. *268*: 12136-12142.
21. Chen J., Parsons S., and Brautigan D.C. (1994). J. Biol. Chem. *269*: 7957-7962.
22. Guo H., Reddy S.A.G., and Dahmuni Z. (1993). J. Biol. Chem. *268*: 11193-11198.
23. Shenolikar S. (1994). Ann. Rev. Cell Biol. *10*: 55-86.
24. Mumby M.C., and Walter G. (1993). Physiol. Rev. *73*: 673-699.
25. Matsuda, S., Kosako, H., Takenaka, K., Moriyama, K., Sakai, H., Akiyama,T., Gotoh, Y., and Nishida, E. (1992). Embo. J. *11*:973-982.

23

FUNCTIONAL ASPECTS OF ANGIOTENSIN TYPE 2 RECEPTOR

Masatsugu Horiuchi

Cardiovascular Medicine
Falk Cardiovascular Research Center
Stanford University School of Medicine
Stanford, California

Angiotensin II (AngII) exerts various actions in its diverse target tissues controlling vascular tone, hormone secretion, tissue growth, and neuronal activities. Extensive pharmacological evidence indicates that most of the known effects of AngII in adult tissues are mediated by the seven-transmembrane G-protein coupled AngII type 1 (AT_{1}) receptor. Recently, a second receptor subtype known as type 2 (AT_{2}) receptor has been described and cloned by us and others (1, 2). However, little is known about the regulation and physiological function(s) of this novel receptor. The intracellular signal transduction mechanism after the activation of AT_2 receptor is not also well defined. The AT_2 receptor is abundantly and widely expressed in fetal tissues and immature brain, but present only in at low levels in adult tissues such as adrenal gland, specific brain regions, uterine myometrium, and atretic ovarian follicles (3-5). The highly abundant expression of this receptor during embryonic and neonatal growth and quick disappearance after birth has led us to the suggestion that this receptor may be involved in growth, development and/or differentiation.

CLONING OF AT_2 RECEPTOR

Recent cloning of the AT_1 receptor from adrenal glomerulosa and vascular smooth muscle cells has revealed that it belongs to a seven-transmembrane, G protein-coupled receptor family (7, 8). Subsequent genomic analyses and homology cloning demonstrated that the AT_1 receptor comprises two isoforms, named AT_{1a} and AT_{1b}, with a striking similarity to each other in amino acid structure, pharmacological specificity and signal transduction mechanism (9, 10). To date, extensive pharmacological evidence indicates that most of the known effects of Ang II in adult tissues are attributable to the AT_1 receptor, through its ability to activate the phosphoinositide/calcium pathway, or to inhibit adenylate cyclase activity (7-10). In contrast, much less is known about the structure and function of the AT_2 receptor. When examined in cells with endogenous AT_2 receptor expression, this receptor does not appear to involve any known classical intracellular signaling pathways, nor seen to be linked to classical G proteins (6, 10, 11).

Recent Advances in Cellular and Molecular Aspects of Angiotensin Receptors
Edited by Mohan K. Raizada et al., Plenum Press, New York, 1996

Using the expression cloning strategy in COS-7 cells from a rat fetus expression library, we have reported the successful cloning of the AT_2 receptor which fulfills all the criteria of the AT_2 receptor such as ligand binding specificity, effect of dithiothreitol (DTT) on binding characteristics, lack of the effect of guanylnucleotide analogs on binding states, lack of phosphoinositide or calcium signaling, and tissue distribution and developmental pattern of receptor mRNA expression (1). The AT_2 receptor cDNA encodes a unique 363 amino acid protein with an apparent molecular weight of 41,330. Hydropathy analysis indicates that it belongs to a seven transmembrane receptor family. It is 34% identical in sequence to AT_1 receptor and it shares limited homology with receptors for somatostatin and bradykinin. Northern blot analysis confirms its abundant expression in the fetus and reduced expression in selective adult tissues such as the ovary, uterus, adrenal gland, and brain. Consequently, we have cloned mouse and human AT_2 receptor (13, 14) and also reported that the human AT_2 receptor is mapped to the X chromosome. Interestingly, the X chromosome of the rat contains a gene, *Bp3*, that cosegregates with hypertension as identified by genetic mapping of the cross between the stroke prone spontaneously hypertensive rat (SHRSP) and the Wistar-Kyoto rat (WKY) (15, 16). By synteny, the genes on the X chromosome are highly conserved between human and rat. Indeed, in this study we also mapped the rat AT_2 receptor gene to the X chromosome. Although the function of the AT_2 receptor is still unknown, it is possible that AT_2 receptor may play an important role in cardiovascular development and regulation, and may be a candidate gene for *Bp3*.

ANTIGROWTH EFFECT OF AT_2 RECEPTOR ON VASCULAR SMOOTH MUSCLE CELL GROWTH

In the past, the major limitations to the understanding of the biology and function(s) of the AT_2 receptor were the lack of information on its structure, the unavailability of molecular probes such as cDNA, and the absence of cell models for studies of its function and regulation. The successful cloning of the cDNA encoding the AT_2 receptor is a major advancement in this field and should provide the opportunity to answer many questions about the AT_2 receptor such as: "What is its function(s)?, What is its signal transduction mechanism?, Is this receptor G protein coupled?, What are the important structure-function relationships of this receptor?, and What regulates the expression of this receptor?"

Currently, the "usual" approach to discover the function of a novel gene is to use gene knock-out experiments (loss of function) or over-expression (gain of function) via transgenic techniques. These methods, while useful in certain situations, have several disadvantages. For example, transgenic/knock-out technology is time consuming and costly, the effect of the over-expressed transgene is exerted throughout development, it is impossible to target the transgenic expression to only a local site, and it is difficult to exclude the potential contribution of the systemic effect of transgene expression. If knockout of the targeted gene is lethal, it is impossible to test the specific functions by transgenic or gene targeting techniques. In those cases, *in vivo* gene transfer approach may be ideal. Thus, local gene transfer approach may be an effective method for studying the function of the AT_2 receptor. The information gained in this approach will be complementary to that obtained by transgenic technology.

A technologic advancement in the studies of AT_2 receptor function has come with the advent of *in vivo* gene transfer techniques which have recently been used by our laboratories to examine *in vivo* function. Many methods for the introduction of DNA *in vivo* have been proposed including viral-mediated gene transfer using retrovirus or adenovirus, liposome mediated gene transfer with cationic liposome or direct injection of DNA. We have recently developed a highly efficient and simple method of DNA transfer using Hemagglutinating Virus of Japan (HVJ), a form of Sendai virus. The HVJ method involves the complexing of the DNA with a neutral liposome mixture which is then coated with UV inactivated HVJ (17). The HVJ coating assists in the cellular uptake of the complex. A further modification, complexing the DNA with high mobility group 1 protein (HMG-1) increases the transport of the DNA to the nucleus. This technique possesses many ideal properties for *in vivo* gene transfer such as efficiency, safety, simplicity, brevity of incubation time, and no limitations of inserted DNA size. This method is many fold more efficient than the standard liposome method. The HVJ method has been successfully employed for gene transfer *in vivo* into various tissues including and vascular wall as well as numerous cells in culture (17-20).

The AT_2 receptor is widely and abundantly expressed in fetal tissues (e.g., aorta) and immature brain. Initially, the abundant expression of the AT_2 receptor during neonatal and embryonic growth led to the speculation that this receptor might be involved in growth. Ang II is known to be a trophic factor for many cell types in vivo as well as in vitro. However, all of the demonstrated growth promoting effects of Ang II have been shown to be inhibitable with DuP753, the AT_1 receptor antagonist. It must be remembered that many other processes occur during development such as differentiation and apoptosis (programmed cell death). Thus, it is entirely conceivable that Ang II via the AT_2 receptor may be involved in these events.

The adult vessel expresses little, if any, of the AT_2 receptor. Similar to the expression pattern seen in other tissues, the AT_2 receptor is highly expressed in neonatal and embryonic vessels (21). We hypothesize that the AT_2 receptor modulates cellular growth responses to AT_1 receptor or other stimuli either by an antiproliferative effect or by way of its actions on migration and/or apoptosis. Interestingly, following vascular injury, we have shown that the AT_2 receptor is re-expressed, constituting 10-20 % of the total specific Ang II binding sites (22). On the other hand, CGP42112A, a putative AT_2 agonist, has been reported to block neointimal development (23). The model of the injured rat carotid artery is being used for several reasons. First, this is a simple model of *in vivo* smooth muscle proliferation and migration which will permit the examination of the "antigrowth" effects of the AT_2 receptor *in vivo*. Furthermore, it has been shown in many species that this process involves the participation of the AT_1 receptor which has been shown to activate the production of autocrine growth factors and to contribute to the proliferation and migration of the vascular smooth muscle cell (VSMC). Thus, this model provides an ideal situation to study the mitogenic/antimitogenic actions induced by the AT_1 vs. the AT_2 receptor. Transfection and over-expression of the AT_2 receptor under this condition will elucidate the actions of this receptor on VSMC function.

A control vector or an AT_2 receptor expression vector was transfected into the balloon injured rat carotid artery and we examined the effect of AT_2 receptor (24). The AT_2 receptor mRNA was highly expressed in the AT_2 receptor transfected injured vessels but only at low levels in the control-vector transfected injured vessels and was not detectable in the untransfected uninjured vessels. Scatchard analysis of Ang II binding sites was consistent with the RT/PCR data. Analysis of membranes prepared from control vector transfected injured carotid arteries four days after transfection (prior to the development of the neointima) demonstrated a single class of binding sites for [^{125}I]-Sar 1, Ile 8 Ang II. No specific binding of [^{125}I]-CGP42112A was observed, consistent with the low levels of endogenous AT_2 mRNA found in these vessels. In contrast, injured vessels transfected with the AT_2 receptor expression vector contained both AT_1 and AT_2 binding sites. Specific binding sites for [^{125}I]-CGP42112A, the AT_2 receptor binding ligand, were also observed four days after AT_2 transfection, showing that one-fourth of the total Ang II receptors was of the AT_2 subtype.

Localization of the Ang II binding sites was determined by *in vitro* autoradiography at 4 days after transfection. Ang II binding localized to the media of control vector-transfected could be competed effectively by DuP753, an AT_1 receptor antagonist, but not significantly with PD123319, an AT_2 receptor antagonist. In contrast, vessels transfected with the AT_2 receptor expression vector contained detectable levels of both AT_2 and AT_1 binding sites. DuP753 only partially blocked the Ang II binding in the AT_2 transfected vessels while both PD123319 and DuP753 were necessary to block completely the Ang II binding. Quantitative analysis of the autoradiographic data using an image analyzer showed that the density of AT_2 specific binding increased significantly in the AT_2 receptor transfected vessel.

Fourteen days following injury, thick neointima developed in the control vector transfected vessels. Transfection of the AT_2 receptor expression vector resulted in a 70% reduction in neointimal area compared to vessels transfected with the control vector. This inhibitory effect on neointimal formation was blocked by treatment with the AT_2 receptor antagonist, PD123319. Treatment with PD123319 did not affect the development of the neointimal lesion in control vector transfected injured vessels. We examined the rates of DNA synthesis (as assessed by BrdUrd incorporation) 4 days after injury. In uninjured vessels, less than 1% of the cells of the media stained positively for BrdUrd while in the control transfected injured vessel, 20.2% of the medial cells stained positively. AT_2 receptor transfection significantly decreased BrdUrd incorporation to 12.6%, suggesting that one of

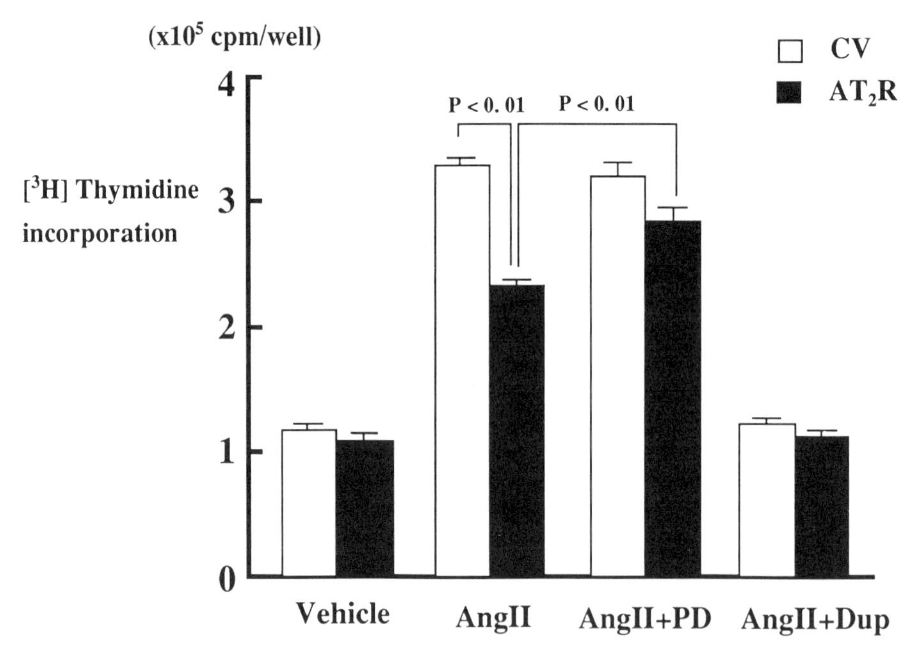

Figure 1. Effect of the AT_2 receptor expression on the incorporation of [^{3}H]thymidine in cultured VSMC. VSMC were seeded onto 24 well culture dishes (Costar Corp., Cambridge, MA) and maintained with Waymouth's medium with 5% calf serum. After suconfulence, cells were washed with three times BSS containing 2 mM $CaCl_2$. Then, 0.2ml of HVJ liposomes (0.5 mg of lipids and 3 μg of encapsulated DNA) was added to the wells. The cells were incubated at 4°C for 5 min and then at 37°C for 30 min, after changing to fresh medium with 5% calf serum, they were incubated overnight in a CO_2 incubator. After the cells were made quiescent by placing them for 48 h in a DSF medium, Ang II (3 x 10^{-7}M), PD123319(10^{-5}M) and Dup753 (10^{-5}) were added in the medium. Relative rates of DNA synthesis were assessed by determination of [^{3}H]thymidine incorporation into TCA-precipitable material. Quiescent VSMC cells grown in the dishes were pulsed for 24 h (12 h after stimulation) with [^{3}H]thymidine, washed twice with cold PBS, twice with 10% (wt/vol) cold TCA, and incubated with 10% TCA at 4°C for 30 min. Cells were rinsed in ethanol (95%) and dissolved in 0.25N NaOH at 4°C for 3 h, neutralized, the radioactivity determined by liquid scintillation spectrometry. Data shows mean±s.e.m.; N=6.

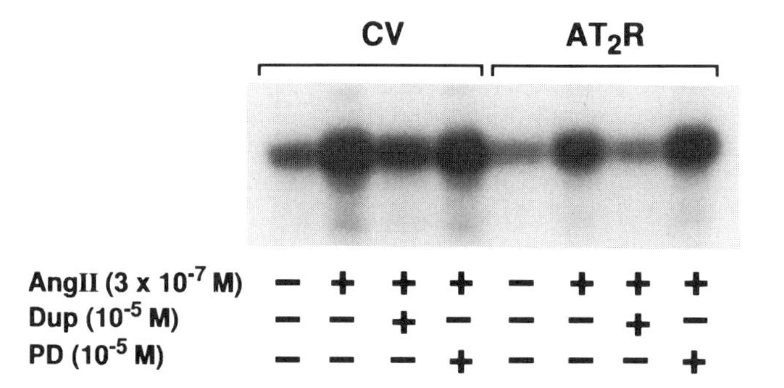

Figure 2. Effect of AT_2 receptor on MAP kinase activation. Vascular smooth muscle cells were transfected with a control vector or the AT_2 receptor expression vector as in Figure 1. MAPK activity was assayed by its ability to phosphorylate myelin basic protein (MBP). After stimulation Ang II for 15 min, the cells were lysed in lysis buffer (25 mM Tris/HCl pH 7.5, 25 mM NaCl, 0.5 mM EGTA, 10 mM NaF, 20 mM β-glycerophosphate, 1 mM vanadate (Na_3VO_4, Sigma) and 1 mM PMSF) and centrifuged. After immunoprecipitation of supernatant fraction with 5 μg of anti-rat MAPK antibody (UBI), the samples were washed and centrifuged three times. The immunoprecipitates were incubated with reaction buffer (25 mM Tris/HCl, pH 7.5, 10 mM $MgCl_2$, 1 mM DTT, 2 mM protein kinase inhibitor (Sigma), 0.5 mM EGTA, 1 mg/ml of MBP and 40 mM [γ-^{32}P]ATP (Amersham)) for 10 min. The reaction was terminated by adding Laemli buffer and the sample was run on 14 % SDS/PAGE and autoradiographed. In the control transfected cells, Ang II induces MAP kinase activity while in the cells expressing the AT_2 receptor transgene, Ang II has little effect on MAPkinase activation. In these cells, blockade of the AT_2 receptor with PD123319 allows the Ang II induced increase in MAP kinase activity.

the actions of the AT_2 receptor is the inhibition of progression of the cells into the S phase. The development of a neointimal lesion in this model can be broadly divided into three phases; replication of medial smooth muscle cells, migration of the medial smooth muscle cells into the intima and further proliferation in the intima. Our data suggest that overexpression of the AT_2 receptor transgene in the medial smooth muscle cells in the first and early phase inhibits the subsequent later development of the neointima at least in part by a reduction in the rates of DNA synthesis in the medial cells during the first phase.

Cultured rat aortic VSMC were transfected with the AT_2 receptor expression vector or control vector. AT_2 receptor binding was detected in the AT_2 receptor expression vector transfected cells but not the non-transfected cells (24). In confluent, quiescent cells, transfected with the control vector, Ang II ($3X10^{-7}$M) increased the cell number and [^{3}H]thymidine uptake (Figure 1). This increase was abolished by Dup753 but not by PD123319, demonstrating that the AT_1 receptor mediated the Ang II stimulated proliferation of these VSMC. In contrast, in cells transfected with AT_2 receptor, Ang II treatment had little effect on cell number and [^{3}H]thymidine uptake compared to the control vector transfected cells. However, in AT_2 receptor-expressing cells, treatment with Ang II plus PD123319 for three days increased the accumulation of cells to a level comparable to that observed in control vector transfected cells, suggesting that the antiproliferative effect was mediated by the AT_2 receptor.

We next examined the potential intracellular signaling pathway which may be affected by the actions of the AT_2 receptor (24). We hypothesize that the AT_2 receptor may antagonize the MAP kinase pathway since the growth actions of the AT_1 receptor are mediated in part by this signal mechanism (25, 26). Treatment of the control vector transfected cells with Ang II resulted in an AT_1 receptor mediated increase in MAP kinase activity. DuP753 but not PD123319 blocked the effects of Ang II (Figure 2). Conversely, in cells transfected with the AT_2 receptor expression vector, the AT_1 mediated increase in MAP kinase activity was greatly attenuated as compared to the control vector transfected cells.

Furthermore, in the presence of PD123319, Ang II stimulated MAP kinase activity in the AT_2 receptor expression vector transfected cells to a value indistinguishable from the Ang II treated control cells. Taken together, these results demonstrate that the two Ang II receptor subtypes exert opposing effects on MAP kinase activity and suggest that antigrowth action of AT_2 receptor is mediated by its inhibiting effect on MAP kinase. Furthermore, very recently, Stoll *et al.* (27) reported that the antiproliferative actions of the AT_2 receptor offset the growth promoting effects mediated by AT_1 receptor, supporting that AT_2 receptor exerts its antiproliferative effect by counteracting the AT_1 receptor in cardiovascular system.

MOLECULAR MECHANISM OF GROWTH DEPENDENT EXPRESSION OF ANGIOTENSIN II TYPE 2 RECEPTOR

To understand the molecular mechanism of the developmental and growth regulation of AT_2 receptor expression, we cloned mouse AT_2 receptor gene, analyzed its structure and examined the promoter activity (28). We then employed R3T3 cells, a mouse fibroblast cell line, as model to study the AT_2 receptor promoter activity since these cells express the only AT_2 subtype binding sites . Furthermore, the expression of AT_2 receptor sites in these cells appear to be modulated by the growth state of the cells. Specifically, the expression of AT_2 binding sites is very low in actively growing cells, but markedly increased in confluent, quiescent cells (29, 30). These characteristics of this cell line provides us an excellent model for studying the growth regulation of the AT_2 receptor.

We have sequenced entire exon regions both in the sense and antisense directions. Sequence comparison between genomic DNA and cDNA reveals that there are two introns in the 5′ untranslated region. Coding region in the third exon is not interrupted by any intron and there is no intron in the 3′ untranslated region. Promoter/luciferase reporter deletion analysis of AT_2 receptor in R3T3 cells showed that putative negative regulatory region located between the positions -453 and -225 which play an important role in the transcriptional control of AT_2 receptor gene expression along with the cell growth. Interestingly, as shown in Figure 3, we found in this region the several putative consensus sequences such as two repeats (5′ AAAGAGAAAGAGAA 3′) of interferon regulatory factor (IRF) binding sequence (31) at the position -283 and this hexamer motif (5′ AAATGA3′) at the position -258. We also found protein kinase C-responsive element (AP1) (32) at positions -340, and octamer binding sequence for POU domain family of transcriptional factors, (33) at a position -394.

Our result also suggests the existence of an additional mechanism for positive regulation of the AT_2 receptor gene expression in these cells from growing to confluent state. If positive regulatory mechanism is necessary to fully express AT_2 receptor, these putative positive regulatory elements will have to be located between the positions -224 and +52. In

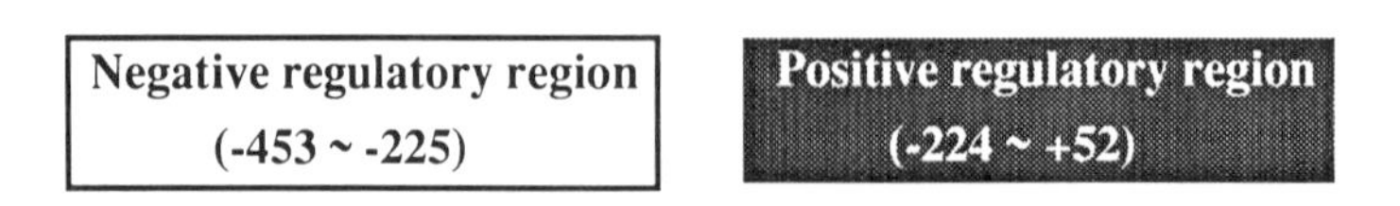

Figure 3. Putative consensus DNA elements observed in mouse AT_2 receptor 5′ flanking region.

this region, the following putative consensus sequences are present as shown in Figure 3, AP-1 (34) at position -75, AP-2 enhancer element (34) at a position -41 bp, c-myc binding sequence at positions -86 (35) and the POU domain family of transcriptional factors binding motif, octamer motif (36) at position -169. Furthermore it is assumed that specific nuclear trans-acting protein that can recognize these positive regulatory elements are highly expressed in the confluent R3T3 cells compared to the growing R3T3 cells.

If these sequences in fact act as regulatory DNA elements in the AT_2 receptor promoter, then the AT_2 receptor expression may be regulated by proto-oncogene proteins such as Fos, Jun family proteins and c-myc in addition to IRF family proteins such as IRF-1, IRF-2, interferon consensus sequence binding protein (ICSBP) which are recently reported to modulate cell proliferation and transformation (37-39). Based on these observations, we propose that AT_2 receptor is one of the target genes, whose expression is regulated by transcriptional factors that are related to the growth and differentiation such as IRFs, POU domain family proteins, thereby supporting a role for this receptor in these processes.

ACKNOWLEDGMENTS

This work was supported by NIH grants HL46631, HL35252, HL35610, HL48638, HL07708, and the American Heart Association Bugher Foundation Center for Molecular Biology in the Cardiovascular System, and by a grant from Ciba-Geigy, Basel, Switzerland. We gratefully acknowledge Dr. M. de Gasparo for CGP42112A and Dr. D. Dudley for R3T3 cells.

REFERENCES

1. Mukoyama, M., Nakajima, M., Horiuchi, M., Sasamura, H., Pratt, R. E., and Dzau, V.J. (1993) J. Biol. Chem., *268*:24539-24542
2. Kambayashi, Y., Bardhan, S., Takahashi, K., Tsuzuki, S., Inui, H., Hamakubo, T., and Inagami T. (1993) J. Biol. Chem. 268:24543-24546
3. Grady, E. F., Sechi, L. A., Griffin, C. A., Schambelan, M. and Kalinyak, J. E. (1991) J. Clin. Invest. *88*:921-933
4. Tsutsumi, K. and Saavedra, J. M. (1991) Am. J. Physiol. *261*:H667-H670.
5. Millan, M. A., Jacobowitz, D. M., Aguilera, G. and Catt, K. J. (1991) Proc. Natl. Acad. Sci. U.S.A. *88*:11440-11444
6. Pucell, A. G., Hodges, J. C., Sen, I., Bumpus, F. M. and Husain, A. (1991) Endocrinology *128*:1947-1959
7. Murphy, T.J., Alexander, R.W., Griendling, K.K., Runge, M.S. and Bernstein, K.E.
8. Sasaki, K., Yamano, Y., Bardhan, S., Iwai, N., Murray, J.J., Hasegawa, M., Matsuda, Y. and Inagami, T. (1991) Nature *351*:230-233
9. Sasamura, H., Hein, L., Krieger, J.E., Pratt, R.E., Kobilka, B.K. and Dzau, V.J. (1992) Biochem. Biophys. Res. Commun *185*:253-259
10. Iwai, N. and Inagami, T. (1992) FEBS Lett., *298*:257-260
11. Tsutsumi, K., Stromberg, C., Viswanathan, M. and Saavedra, J.M. (1991) Endocrinology *129*:1075-1082
12. Dudley, D.T., Hubbell, S.E. and Summerfelt, R.M. (1991) Mol. Pharmacol. *40*:360-67
13. Nakajima, M., Mukoyama, M., Pratt, R.E., Horiuchi, M. and Dzau, V.J. (1993) Biochem. Biophys. Res. Commun. *197*:393-399
14. Koike, G., Horiuchi, M., Yamada, T., Szpirer, C., Jacob, H.J. and Dzau, V.J. (1994) Biochem. Biophys. Res. Commun. *203*:1842-1850
15. Jacob, H. J., Lindpaintner, K., Lincoln, S. E., Kusumi, K., Bunker, R. K., Mao, Y. -P., Ganten, D., Dzau, V. J., and Lander, E. S. (1991) Cell *67*:213-224.
16. Hilbert, P., Lindpaintner, K., Beckmann, J. S., Serikawa, T., Soubrier, F., Dubay, C., Cartwright, P., De Gouyon, B., Julier, C., Takahashi, S., Vincent, M., Ganten, D., Georges, M., and Lathrop, G. M. (1991) Nature *353*,:521-529.
17. Kaneda, Y., Iwai, K. and Uchida, Y. (1989) Science *243*:375-378

18. Kato, K. Nakanishi, M., Kaneda, Y., Uchida,T. and Okada, Y. (1991) J. Biol. Chem. *266*:3361-3364
19. Morishita, R., Gibbons. G.H., Ellison, K.E., Nakajima, M., Zhang, L., Kaneda, Y., Ogihara, T. and Dzau, V.J. (1993) Proc. Natn. Acad. Sci. U.S.A. *90*:8474-8478
20. Morishita, R., Gibbons, G.H., Kaneda, Y., Ogihara, T. & Dzau, V.J. (1993) J. Clin. Invest. *91*:2580-2585
21. Viswanathan, M. Tsutsumi, K., Correa, F.M. and Saavedra, J.M. (1991) Biochem. Biophys. Res. Commun. *179*:1361-1367
22. Pratt, R.E., Wang, D., Hein, L. and Dzau, V.J. (1992) Hypertension *20*:432
23. Janiak, P., Pillon, A., Prost, J.-F. and Vilaine, J.-P. (1992) Hypertension *20*:737-745
24. Nakajima, M., Hutchinson, H, Fujinaga, M., Hayashida, W., Morishita, R., Zhang, L., Horiuchi, M., Pratt, R.E. and Dzau, V.J. (submitted)
25. Duff, J.L., Marrero, M.B., Paxton, W.G., Charles, C.H., Lau, L.F., Bernstein, K.E. and Berk, B.C. (1993) J. Biol. Chem. *268*:26037-40
26. Booz, G.W., Dostal, D.E., Singer, H.A. and Baker, K.M. (1993) Amer. J. Physiol. *267*:C1308-18
27. Stoll, M., Steckelings, M., Paul, M., Bottari, S.P., Metzger, R. and Unger, T. (1995) J. Clin. Invest. 95:651-657
28. Horiuchi, M., Koike, G., Yamada, T., Mukoyama, M., Nakajima, M. and Dzau, V.J. (submitted)
29. Dudley, D.T., Hubbell, S.E. and Summerfelt, R.M. (1991) Mol. Pharmacol. *40*:360-367
30. Dudley, D.T. and Summerfelt, R.M. (1993) Regulatory Peptides *44*:199-206
31. Fujita, T., Sakakibara, J., Sudu Y., Miyamoto M., Kimura, Y. and T. Taniguchi, T. (1988). EMBO J. *7*:3397-3405
32. Angel, P., Imagawa, M., Chiu, R., Stein, B., Imbra, R.J., Rahmsdorf, H.J., Jonat, C. , Herrlich, P. and M. Karin. (1987) Cell *50*: 847-861.
33. Herr, W., Sturm, R.M., Clerc, R.G., Corcoran, L.M., Baltyimore, D., Sharp, P.A., Ingraham, H.A., Rosenfeld, M.G., Finney, M., Ruvkun, G. and Horvitz, H.R. (1988) Genes Dev. *2*:1513-1516.
34. Imagawa, M., Chiu, R. and Karin, M. (1987) Cell *51*:251-260.
35. Blackwell, T.K., Kretzner, L., Blackwood, R.N., Eisenman, R.N. and Weintraub, H. (1990) Science *250*:1149-1151
36. Ruvkin, G., and Finney, M. (1991) Cell *64*:475-478
37. Ledwith, B.J., Manam, S., Kranak, A.R., Nichols, W.W. and Bradley, M.O. (1990) Mol. Cell Biol. *10*:1545-1555
38. Matsuyama, T., Kimura, T., Kitagawa, M., Preffer, K., Kawakami, T., Watanabe, N., Kundig, T.M., Amakawa R., Kishihara, K., Wakeham, A., Potter J., Furlonger, C.L., Narendran, A., Suzuki, H., Ohashi, P.S., Paige, C.J., Taniguchi, T. and Mak, T.W. (1993) Cell *75*:83-97
39. Nakabeppu, Y., Oda, S. and Sekiguchi, M. (1993) Mol. Cell Biol. *13*:4157-4166

24

ANGIOTENSIN RECEPTOR HETEROGENEITY IN THE DORSAL MEDULLA OBLONGATA AS DEFINED BY ANGIOTENSIN-(1-7)

Debra I. Diz and Carlos M. Ferrario

Hypertension Center
Bowman Gray School of Medicine
Winston-Salem, North Carolina 27157-1032

INTRODUCTION

Our research efforts over the past few years addressed the anatomical origin and pharmacological characteristics of the saturable, high affinity angiotensin (Ang) II binding sites found in the dorsomedial medulla and the intra- and extracranial segments of the vagus nerves. We related these findings to the production of hemodynamic responses following direct application of the peptide at physiologically relevant doses at sites within the dorsomedial medulla oblongata. Work also began to determine whether other endogenous peptides in the angiotensin family, particularly the des-Phe^8-Ang II or Ang-(1-7), may selectively act at angiotensin receptors in various tissues. Studies addressing the functional significance of Ang-(1-7) indicated that this peptide was selective for subpopulations of Ang II binding sites. In this report we review recent findings and present new evidence that the actions of Ang-(1-7) in brain may be mediated by a specific subset(s) of Ang II receptors.

ANGIOTENSIN RECEPTORS IN THE DORSOMEDIAL MEDULLA OBLONGATA ASSOCIATED WITH THE VAGAL SENSORY AND MOTOR SYSTEM

Our interest in the modulatory effects of angiotensin peptides on baroreceptor reflex function led to the finding that Ang II receptors are present in the nucleus of the solitary tract (nTS) and vagal motor nucleus (dmnX)(1,2). These nuclei contain the first synapse of the baroreceptor vagal sensory afferent fibers and are the site of origin of vagal efferent motor neurons, respectively. In addition, we discovered that Ang II receptors in these brain regions are reduced following nodose ganglionectomy or cervical vagotomy, indicating that Ang II binding sites in the dorsomedial medulla are associated with the central components of both

Recent Advances in Cellular and Molecular Aspects of Angiotensin Receptors
Edited by Mohan K. Raizada et al., Plenum Press, New York, 1996

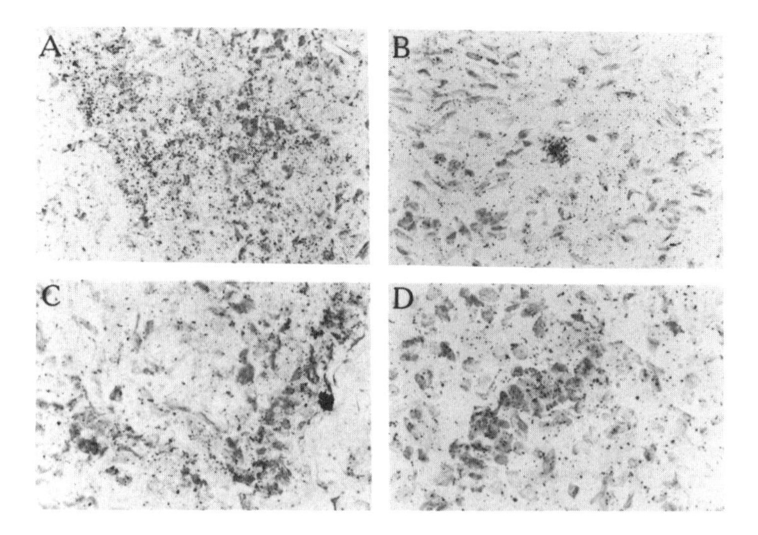

Figure 1. High power photomicrographs of emulsion autoradiography of ^{125}I-Ang II binding in the nodose ganglion (A and B) and the superior cervical ganglion (C and D). Exposed silver grains representing localization of the ^{125}I-Ang II binding sites can be seen as the black dots overlying the cresyl violet stained section. Note that the binding sites are associated with both the cell bodies and the surrounding tissue, consistent with our previous observations of binding associated with the peripheral process of the vagus nerve. Magnification is 500X for panels A and C and 625X for panels B and D.

the vagal sensory and motor systems (2). Use of high resolution emulsion autoradiography confirmed Ang II binding sites overlying cell bodies of the nTS and dmnX (3). Although the distribution of Ang II receptors in the dorsomedial medulla is more widespread than the projection of exclusively cardiovascular-related fibers in the vagal system, it closely parallels substance P cells and pathways and monoamine cell groups within the region (4). Further studies showed that specific, high affinity Ang II binding sites are present in the nodose ganglion and cervical vagus nerve (2). These binding sites undergo bi-directional transport in the peripheral vagus nerve, as indicated by the twenty-four hour accumulation above and below a ligature (5). These data are consistent with synthesis of Ang II receptors in cell bodies of the nodose ganglion and dmnX as well as receptor transport to the nerve processes centrally and peripherally. Indeed, we present new data for Ang II binding sites overlying cells and fibers in the nodose ganglion as well as the superior cervical ganglion in Figure 1.

ANGIOTENSIN RECEPTORS IN THE DORSOMEDIAL MEDULLA OBLONGATA MEDIATE IMPORTANT CARDIOVASCULAR ACTIONS

Localization of Ang II binding in the dorsal medulla prompted us to investigate the hemodynamic effects of the peptide in this brain region. Ang II binding sites in the dorsomedial medulla, the nodose ganglion and the vagus nerve of dogs or rats exhibit an affinity of ~0.8 nM, which is characteristic of presumed functional receptors for the peptide. Ang II lowered blood pressure and heart rate after injection of low (fmol) doses in the nTS or dmnX (6-10). Production of acute depressor and bradycardic responses to injections of Ang II was associated with both activation of vagal outflow and inhibition of sympathetic outflow (6,9). These hemodynamic effects were not dependent on intact aortic baroreceptor input, and thus may be due to postsynaptic actions (8). In addition, studies by us and others provided direct evidence that receptors in the nTS also influence the baroreceptor reflex. Microinjections of exogenous Ang II attenuated the baroreceptor reflex (11). More significantly, bilateral nTS injections of the classic Ang II antagonist Sar^1Thr^8-Ang II facilitated baroreflex control of heart rate, revealing tonic effects of endogenous angiotensin peptides in normotensive rats (12).

The above hemodynamic effects can be matched with actions of Ang II at presynaptic vagal afferent fibers or direct effects on postsynaptic cells within the nTS on the basis of anatomic, electrophysiologic and transmitter release data. For example, in addition to the above studies showing binding sites overlying cells within the nTS and dmnX, recent electrophysiologic evidence by Dr. K.L. Barnes revealed presynaptic as well as postsynaptic interactions of Ang II in the dorsomedial medulla (13). Angiotensin peptides directly excited neurons in this brain region in both rats and dogs (14-16), suggesting that the binding sites described above represent functional receptors. Synaptic blockade inhibited 50% of the cellular responses to Ang II further suggesting presynaptic actions (13). Consistent with presynaptic actions, Ang II and substance P excited the same neurons in a majority of cases and the latency for the onset of the Ang II tended to be longer than that for substance P (17). Ang II stimulated basal substance P immunoreactivity release from medullary slices, but tended to reduce potassium-evoked substance P release (17,18). Others reported that excitation of sensory neurons in the nodose ganglion by Ang II is followed by a partial depolarization that would tend to reduce responsiveness (19,20). Thus, our working hypothesis is that Ang II releases substance P from postsynaptic cells to mediate the acute hypotensive and bradycardic effects of the peptide in the dorsomedial medulla. In contrast, when vagal afferent fibers are stimulated (i.e., activation of the baroreceptor reflex), Ang II attenuates substance P release from these afferents as a mechanism for the Ang II attenuation of the reflex. Recent data support the first part of this hypothesis since the substance P antagonist ([Leu^{11}, ψCH_2NH10-11]-substance P) blocked the hemodynamic effects of Ang II injected into the nTS but had no effect on the response to glutamate injected at the same site (21).

Our observation that binding sites occurred in the nodose ganglion and peripheral vagus nerve process (containing both sensory and motor fibers) and were bi-directional transported in the nerve suggested that Ang II may act peripherally to alter sensory and motor functions, in addition to the above described central nervous system effects (5,22). Indeed, Ang II reduced the gain of the baroreceptor reflex in response to intravenous or cerebroventricular infusions of the peptide and during renal hypertension (23). Although Ang II inhibits presynaptic release of acetylcholine (24), its action on vagal afferent activity was largely unknown. New studies showed excitation of rat nodose ganglion neurons by Ang II (19,20), the functional significance of which is not yet established. Large doses of Ang II given in an isolated carotid sinus sensitize baroreceptor activity (25). We suggested that Ang II binding sites may exist in the SA or AV nodes or in peripheral sites containing sensory nerve endings (carotid sinus, carotid body, aortic arch, aortic depressor or carotid sinus nerve) (2,22), a concept supported by reports of Ang II binding sites within the conduction system (SA and AV nodes) of the rat heart (26). Thus, circulating Ang II may act at multiple sites to modify baro- or chemoreceptor afferent nerve discharges.

ANGIOTENSIN RECEPTOR HETEROGENEITY AS DEFINED BY DIFFERENTIAL PROPERTIES OF ANG-(1-7) AND ANG II IN THE DORSOMEDIAL MEDULLA OBLONGATA

We found that Ang-(1-7) is generated in homogenates of the dorsomedial medulla and hypothalamus (27) and is present endogenously (28,29). Ang-(1-7) was previously thought to be inactive since it exhibited no dipsogenic activity and had weak pressor effects (30). However, Ang-(1-7) released vasopressin from isolated hypothalamo-neurohypophysial explants (31), altered arterial pressure and heart rate after dorsomedial medulla microinjections (7,8), and was excitatory to neurons of the paraventricular hypothalamus (16) and dorsomedial medulla (15). In initial studies carried out prior to the development of subtype selective antagonists, the effects of both peptides were inhibited by [Sar^1Thr^8]-Ang II (8,31). In subsequent studies, it was apparent that Ang-(1-7) was working at a subset of angiotensin receptors. For example, while Ang-(1-7) was not as potent as Ang II in releasing substance P from dorsomedial medulla slices, it was effective in increasing substance P release from the hypothalamus (18). Unilateral sino-aortic denervation, associated with a 12% reduction in Ang II binding sites in the nTS, had no effect on responses to Ang II but potentiated responses to Ang-(1-7) (8). Campagnole-Santos et al. (32) recently reported that intraventricular infusion of Ang-(1-7) facilitates, whereas Ang II inhibited, the baroreceptor reflex. Ang-(1-7) was devoid of excitatory actions in the nodose ganglion (19), but did excite neurons of brain of both dogs and rats (15,16). Thus, Ang-(1-7) may participate in selective angiotensin-related cardiovascular actions in the central nervous system.

On the basis of the differential responses to Ang II and Ang-(1-7), subnuclei within the canine dorsomedial medulla were examined for the ability of Ang-(1-7) to compete for ^{125}I-Ang II binding using previously published conditions (2,5,33). The ^{125}I-Ang II binding extends roughly from 0.5 mm caudal to 3-4 mm rostral to obex in the medial and dorsal subnuclei of the nTS, in the dmnX and in the area postrema. Ang II competed for the ^{125}I-Ang

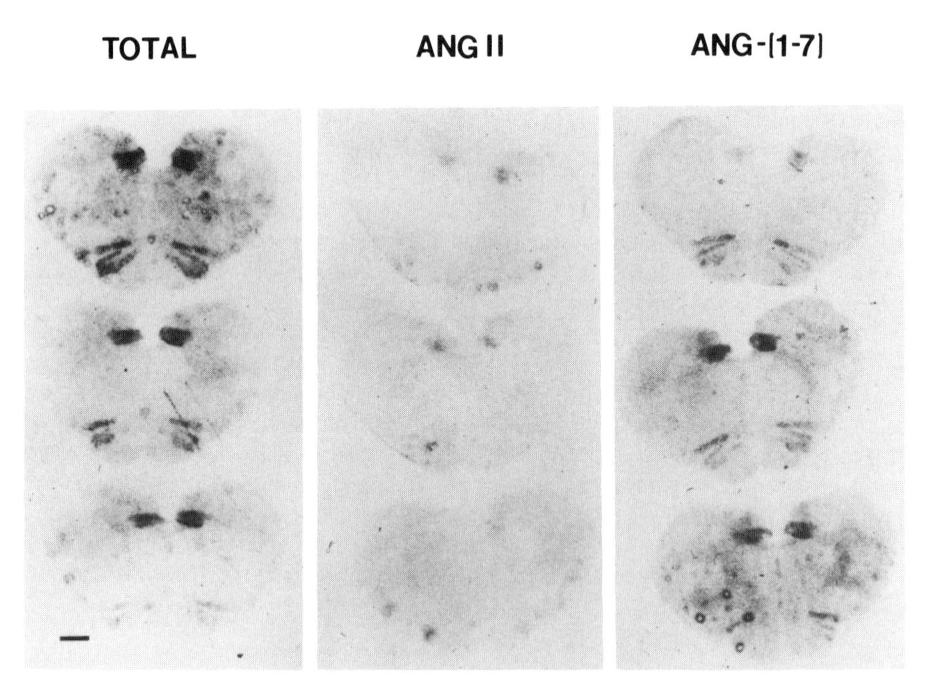

Figure 2. Displayed here are three sets of adjacent 14 μm sections from three different levels of the dorsomedial medulla of a single dog showing ^{125}I-Ang II binding in the area postrema, nTS and dmnX. The bottom sections are approximately 0.5 mm rostral to the landmark obex, the middle is 1.5 mm rostral and the top is 2.5 mm rostral to obex (scale bar = 2 mm). Total binding (~0.4 nM ^{125}I-Ang II without competitors) is shown in the first panel. The second panel shows that excess unlabeled Ang II (1 μM) competed for binding at all three levels. In contrast, the third panel reveals that 1 μM Ang-(1-7) competed for binding as completely as Ang II in the rostral area only, as illustrated in the top section.

II binding at all levels (see Figure 2). In contrast, only in the most rostral portion of the nTS and dmnX (2.5-3.5 mm rostral to obex) did Ang-(1-7) appear to compete for Ang II binding to the same extent as Ang II (Figure 2). Similar observations have been made in rat dorsomedial medulla (34), although the rostro-caudal differences are not as distinct in the rat since the extent of ^{125}I-Ang II binding is only 2.5 mm rather than over 4 mm in the dog.

Dorsal Motor Nucleus of the Vagus

Competition with Ang Peptides

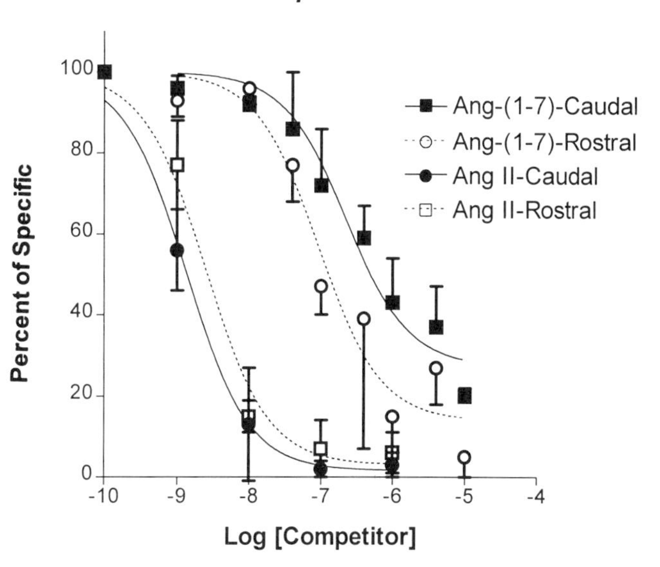

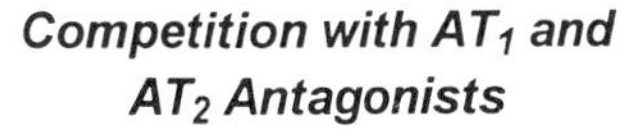

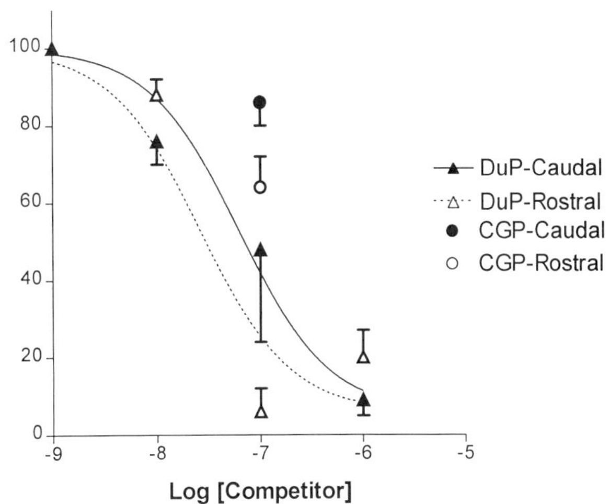

Figure 3. Upper panel: Competition curves obtained for Ang II and Ang-(1-7) in the caudal (0.5 mm caudal to obex to 2.5 mm rostral to obex) and rostral (2.5 to 4 mm rostral to obex) dorsal motor nucleus of the vagus of 13 male dogs. Total binding was determined at a concentration of ~0.3 nM ^{125}I-Ang II and specific binding as that competed for by 10 μM Ang II. Each data point represents the mean ± SEM of values from 2-6 dogs. Data were analyzed and curves fit using GraphPad Prizm (GraphPad Software Inc., version 1.03, 1994). Lower panel: Competition obtained with losartan (DuP; 0.001-1 μM) and CGP 42112A (CGP; 0.1 μM) in the caudal (0.5 mm caudal to obex to 2.5 mm rostral to obex) and rostral (2.5 to 4 mm rostral to obex) dorsal motor nucleus of the vagus of 6 male dogs. Total binding was determined at a concentration of ~0.3 nM ^{125}I-Ang II and specific binding as that competed for by 1 μM Ang II. Each data point represents the mean ± SEM of values from 2-4 dogs. Data for DuP were analyzed and curves fit using GraphPad Prizm.

Once this observation was made, we performed more detailed competition studies using several concentrations of Ang-(1-7). Binding was assessed in the caudal (~0-2.5 mm rostral to obex) and rostral (~2.5 to 4 mm rostral to obex) aspects of the nTS and dmnX. Ang II competed for sites in the rostral nTS (data not shown) and dmnX (Figure 3, upper panel) of the dorsomedial medulla with an affinity of 2 - 3 nM and in the caudal areas with an affinity of 1 nM. The IC_{50} for competition with Ang-(1-7) tended to be lower in the rostral (80-90 nM) versus caudal (120-220 nM) region of the nTS (data not shown) and dmnX (Figure 3, upper panel). These findings suggest that the relative potency of Ang-(1-7) is only ~30-40 fold less than Ang II in the rostral, but ~100-200 fold less potent in the caudal dorsomedial medulla. Another interesting finding was that Ang-(1-7) competed for 85 ± 9% and 95 ± 5% of the sites in the rostral dmnX at 1 and 10 μM which is comparable to the maximal competition by Ang II. However, only 57 ± 11 % and 80 ± 2 % competition occurred with Ang-(1-7) at 1 and 10 μM in the caudal dmnX ($p < 0.05$, compared with rostral area at 10 μM). Similar findings were again observed in the nTS. Thus, it appears that Ang-(1-7) is able to effectively compete for almost all of the specific Ang II binding in the rostral but not caudal aspects of the nuclei of the dorsal medulla.

The development of new subtype selective antagonists allowed us to further define the pharmacological characteristics of Ang II binding sites in the dorsal medulla. Competition studies were carried out with subtype selective Ang II receptor antagonists in dogs using *in vitro* receptor autoradiography (35,36). Losartan was used to identify the AT_1 sensitive sites and CGP 42112A was used for the AT_2 sites. As shown in the competition curves for the dmnX in lower panel of Figure 3 and the dorsomedial medullary brain sections in Figure 4, the AT_1 selective losartan was a potent competitor in the canine dorsomedial medulla at all rostro-caudal levels. Maximal competition with losartan (1 μM) was 80-90% both in the rostral and caudal medulla. Thus, AT_1 receptors represent at least 80-90% of the binding sites, consistent with previous reports in rat dorsomedial medulla (37-39). Similar to the findings with Ang-(1-7), rostro-caudal differences existed with respect to the amount of competition with AT_2 antagonist CGP42112A at doses that were selective and produced maximal competition at AT_2 sites in other brain regions (39) and the periphery (35,36). In the caudal nTS and dmnX the AT_2 antagonist CGP42112A (0.1 μM) competed for 17 ± 6% of the specific binding. This indicates that a small population (10-20%) of AT_2 sites exists in caudal dorsal medulla of the dog and that competition with AT_1 and AT_2 antagonists accounts for roughly 100% of the specific Ang II binding as reported in rat brainstem (40). In contrast, in the rostral nTS and dmnX the AT_2 antagonist (0.1 μM) competed for a greater proportion of receptors (caudal: 17 ± 6%; rostral: 40 ± 5%; $p < 0.05$), again suggesting differences in the characteristics of the rostral Ang II binding sites as was indicated with the Ang-(1-7) studies. Furthermore, since losartan also competed for 80 ± 7% of the binding in the rostral dorsomedial medulla, the greater than 100% competition indicates that binding sites in this region may not be as selective for the AT_1 and AT_2 antagonists. These findings are consistent with recent reports by others in rat brain (39,41). Although not discussed, these earlier studies reported greater than 100% competition in the dorsal medulla as compared with other brain areas using the same concentration of competitors. Our preliminary studies of the rostral dorsomedial medulla of rats similarly show greater than 100% competition with both AT_1 and AT_2 compounds, but the separation in rostral and caudal areas is not as distinct as for the dog (34). Further studies are required with detailed competition curves using the more selective PD123319 or PD123177 AT_2 antagonists and combined use of the AT_1 and AT_2 competitors. However, the above preliminary evidence suggests that in brain there is receptor that is not as selective for the two classes of compounds. In fact, studies by Dr. Tallant revealed similar findings in astrocytes cultured from several brain regions (see chapter by Tallant et al.), raising the intriguing possibility that the AT_2-Ang-(1-7) sensitive sites seen in the rostral medulla are present on astrocytes. A similar receptor that recognized both AT_1 and AT_2 compounds has been identified in the kidney that also showed 40 nM affinity for Ang-(1-7) (42). As reported by Speth (43), the AT_2 binding sites in brain may also exhibit different characteristics than those in the periphery.

To further substantiate the relative difference in binding in the rostral and caudal medulla we compared the pattern of ^{125}I-Ang II binding with that of 2 nM ^{125}I-Ang-(1-7). Consistent with the competition studies, specific ^{125}I-Ang-(1-7) binding defined by competition with 10 μM Ang-(1-7) was detected in the rostral but not caudal dorsomedial medulla (Figure 5). Again we have similar results with ^{125}I-Ang-(1-7) binding in the dorsomedial medulla of the rat (unpublished observations).

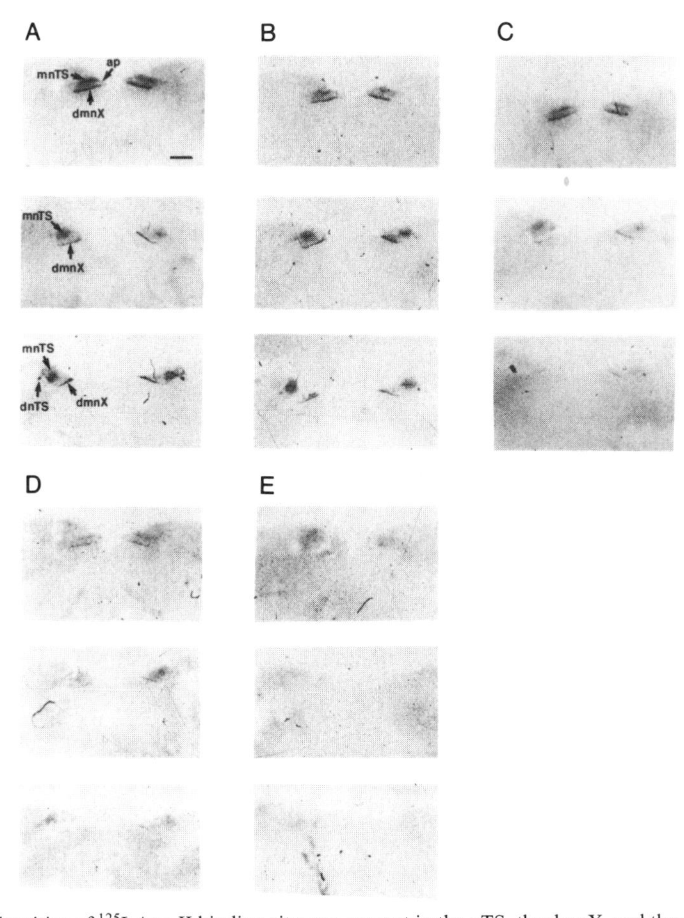

Figure 4. High densities of ^{125}I-Ang II binding sites are present in the nTS, the dmnX, and the area postrema (ap) of the canine dorsomedial medulla at three rostro-caudal levels (top - 0.5 mm rostral to obex; middle - 2 mm rostral to obex; bottom - 3 mm rostral to obex). Five sets of consecutive adjacent 14 μm sections are shown (scale bar = 1 mm): A) Total binding in the absence of any competitor at the three rostro-caudal levels; B) Competition with 0.1 μM Ang-(1-7) illustrating partial competition at the rostral sites (lowest panel) but not caudal sites (note the lower concentration than in Fig. 2); C) CGP 42112A also produced greater competition at a 0.1 μM concentration in the rostral compared with caudal areas; D) Losartan was a potent competitor of the ^{125}I-Ang II binding at 1 μM at all rostro-caudal levels; E) Nonspecific binding in the presence of 1 μM Ang II showing almost total competition at all levels.

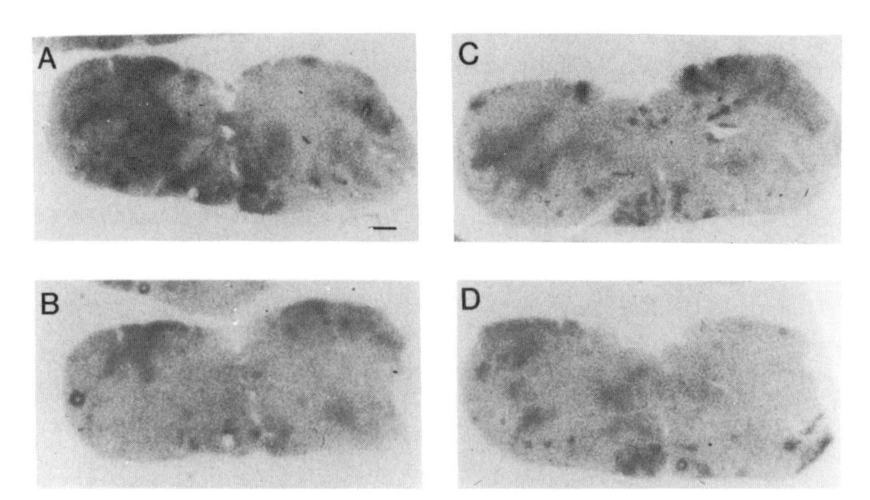

Figure 5. Using ~2 nM ^{125}I-Ang-(1-7), specific binding was detected only in the rostral (Panel C) but not caudal (Panel A) dorsomedial medulla in adjacent sections of the same dog (scale bar = 1 mm). Non-specific binding in the presence of 1 μM unlabeled Ang-(1-7) is shown in Figures 3B and 3D.

CONCLUSIONS

We clearly show that the majority of Ang II receptors in dog caudal dorsomedial medulla are classified as AT_1, although a smaller population of AT_2 sites (10-20%) may also be present. These data are consistent with findings in the brainstem of the rat (39-41). We also observed that a greater percentage of the Ang II receptors in the rostral part of the dorsomedial medulla of the dog appeared to recognize the AT_2 antagonists, as well as Ang-(1-7), in contrast to the caudal aspects of these nuclei. In addition, Ang-(1-7) was approximately 7 times more potent in the rostral area compared with the caudal aspects of the dorsomedial medulla. Thus, our research supports the contention that subpopulations of AT_1 receptors exist, possibly situated on different parts of the neuronal circuits comprising the baroreceptor reflex arc. Moreover, the finding of a rostro-caudal difference in the ability of Ang-(1-7) to bind to Ang II receptors in this brain region may also explain the differential effects of the two peptides on substance P release, after sino-aortic denervation, and on the baroreceptor reflex in this brain region (8,18).

In addition to revealing a rostro-caudal difference in binding to endogenous angiotensin peptides, there was a similar pattern for recognition by the AT_2 competitor. We have not used Ang-(1-7) to compete for the ^{125}I-Ang II binding in the presence of AT_1 or AT_2 antagonists, which would definitively demonstrate actions of Ang-(1-7) at one or the other sites. Nonetheless, the findings support our previous observations that responses to Ang-(1-7) may be more sensitive to inhibition by AT_2 antagonists. For example, prostaglandin release in response to Ang-(1-7) in human astrocytes was blocked by AT_2 antagonists but higher concentration of AT_1 antagonists were required to inhibit responses (44-46). In contrast, the Ang II-dependent prostaglandin release in astrocytes was attenuated by lower doses of DuP 753 (losartan) or L-158,809 (44,45,47), indicating the predominant involvement of classic AT_1 receptors in the Ang II response. More recently, vasopressin release in response to Ang II was blocked by the AT_2 selective antagonists PD 123177 or CGP 42112A as well as losartan (48). While the effect of these inhibitors on vasopressin release in response to Ang-(1-7) has not been tested, Ang-(1-7)-induced neuronal excitation in the paraventricular nucleus was inhibited by CGP 42112A (16). New findings reported by Santos et al. (49) and Felix et al. (50) suggest that a D-Ala^7 analog of Ang-(1-7) acts in brain as a selective Ang-(1-7) antagonist (i.e., this compound did not block the actions of Ang II). Since our previous data suggest overlapping actions of the two peptides, it remains to be seen whether there are non-Ang II receptor actions associated with Ang-(1-7) in brain, as we have seen in the periphery (51).

Finally, Harding et al. (52) showed that another shorter fragment of the angiotensin peptide system, Ang-(3-8) was active and has its own distinct binding site (Ang IV receptor). Characterization of this binding site indicated that it is neither AT_1 nor AT_2, nor is it competed for by sarcosine substituted analogs of Ang II or antagonists (52). Consistent with the binding characteristics, the cognitive and cerebrovascular actions of the peptide were not blocked by [Sar^1Ile^8]-Ang II or subtype selective antagonists(53,54). The Ang-(3-8) receptor exhibited high affinity for Ang-(3-7), raising the possibility that shorter analogs of Ang-(1-7) may activate this receptor. However, the responses observed with Ang-(1-7) are a subset of those seen with Ang II in most cases and are blocked completely by [Sar^1Thr^8]-Ang II. Thus it is unlikely that the actions of Ang-(1-7) are mediated by the Ang IV receptor. Moreover, in contrast to the major vasodilator component seen with Ang-(1-7), Ang-(3-7) displayed only pressor effects in our study in the pithed rat (51). Others have also shown vasoconstrictor effects of Ang-(3-8) in different vascular beds that are mediated by Ang II AT_1 receptors (55,56). Again, these findings reveal differences in peripheral versus central angiotensin receptors and their responses to endogenous angiotensin peptides.

In summary, initial studies suggested an anatomical relationship of Ang II binding sites to neurons and pathways in the dorsomedial medulla containing transmitters such as substance P or serotonin, or synthesizing enzymes such as tyrosine hydroxylase at the light microscopic level (4,57,58). A functional relationship was subsequently demonstrated in the dorsomedial medulla and hypothalamus for Ang II. However, the finding of differential actions of Ang-(1-7) with respect to release of substance P and norepinephrine and the baroreceptor reflex (17,18) led to the additional characterization of Ang II receptor binding in this brain region. These studies suggest that Ang-(1-7) is somewhat selective for Ang II receptors recognizing AT_2 antagonists and strengthen the concept for differential actions of the two angiotensin peptides. Moreover, our findings emphasize the physiologic importance of the enzymatic processing pathways involved in the formation of Ang II and Ang-(1-7) since endogenous levels of the two peptides vary in different tissues endogenously, and high levels of Ang-(1-7) exist in the plasma of hypertensive patients and animals treated with converting enzyme inhibitors (59-63). Significantly, we have recent data to suggest that sensitivity to the depressor effects of Ang-(1-7) is enhanced in two models of hypertension (64,65). Thus, the mechanisms behind these enhanced effects are now an important aspect of future studies.

ACKNOWLEDGMENTS

This work was supported by grants HL-38535 (DID) and HL-51952 (DID, CMF) from the National Institutes of Health. Part of the work was carried out during the tenure of an Established Investigatorship to DID from the American Heart Association. We thank Susan M. Bosch for excellent technical assistance in these studies. We wish to thank Drs. Pieter Timmermans and Ronald Smith at DuPont for supplying the losartan (DuP 753) and Dr. Marc de Gasparo at CIBA-GEIGY for supplying the CGP 42112A.

REFERENCES

1. Speth, R. C., Wamsley, J. K., Gehlert, D. R., Chernicky, C. L., Barnes, K. L., and Ferrario, C. M. (1985) Brain Res. *326*:137-143
2. Diz, D. I., Barnes, K. L., and Ferrario, C. M. (1986) Brain Res. Bull. *17*:497-505
3. Szigethy, E. M., Barnes, K. L., and Diz, D. I. (1992) Brain Res. Bull. *29*:813-819
4. Diz, D. I., Block, C. H., Barnes, K. L., and Ferrario, C. M. (1986) J. Hypertension *4(suppl 6)*:S468-S471
5. Diz, D. I. and Ferrario, C. M. (1988) Hypertension *11(suppl I)*:I-139-I-143
6. Diz, D. I., Barnes, K. L., and Ferrario, C. M. (1984) J. Hypertension *2*:53-56
7. Campagnole-Santos, M. J., Diz, D. I., Santos, R. A. S., Khosla, M. C., Brosnihan, K. B., and Ferrario, C. M. (1989) Am. J. Physiol. *257*:H324-H329
8. Campagnole-Santos, M. J., Diz, D. I., and Ferrario, C. M. (1990) Hypertension *15:(Suppl. I)*:I-34-I-39
9. Casto, R. and Phillips, M. I. (1985) Am. J. Physiol. *249*:R341-R347
10. Rettig, R., Healy, D. P., and Printz, M. P. (1986) Brain Res. *364*:233-240
11. Casto, R. and Phillips, M. I. (1986) Am. J. Physiol. *250*:R193-R198
12. Campagnole-Santos, M. J., Diz, D. I., and Ferrario, C. M. (1988) Hypertension *11(suppl I)*:I-167-I-171
13. Barnes, K. L., Qu, L., and McQueeney, A. J. (1992) Soc. Neurosci. *18*:755(Abstract)
14. Barnes, K. L., Knowles, W. D., and Ferrario, C. M. (1988) Hypertension *11*:680-684
15. Barnes, K. L., Knowles, W. D., and Ferrario, C. M. (1990) Brain Res. Bull. *24*:275-280
16. Felix, D., Khosla, M. C., Barnes, K. L., Imboden, H., Montani, B., and Ferrario, C. M. (1991) Hypertension *17*:1111-1114
17. Barnes, K. L., Diz, D. I., and Ferrario, C. M. (1991) Hypertension *17*:1121-1126
18. Diz, D. I. and Pirro, N. (1992) Hypertension *19(Suppl II)*:II-41-II-48
19. Widdop, R. E., Krstew, E., and Jarrott, B. (1992) Clinical & Experimental Hypertension - Part A, Theory & Practice *14*:597-613
20. Widdop, R. E., Krstew, E., and Jarrott, B. (1990) European J. Pharmacol *185*:107-111
21. Diz, D. I., Fantz, D. L., Benter, I. F., and Bosch, S. M. (1995) Hypertension (Abstract)
22. Diz, D. I., Barnes, K. L., and Ferrario, C. M. (1987) Fed. Proc. *46*:30-35
23. Ferrario, C. M., Ueno, Y., Diz, D. I., and Barnes, K. L. (1986) in Pathophysiology of Hypertension — Regulatory Mechanisms (Zanchetti, A. and Tarazi, R. C. eds) pp. 431-454, Elsevier, Amsterdam
24. Potter, E. K. (1982) Br. J. Pharmacol. *75*:9-11
25. Schmid, P. G., Guo, G. B., and Abboud, F. M. (1985) Fed. Proc. *44*:2388-2392
26. Saito, K., Gutkind, J. S., and Saavedra, J. M. (1987) Am. J. Physiol. *253*:H1618-H1622
27. Santos, R. A. S., Brosnihan, K. B., Chappell, M. C., Pesquero, J., Chernicky, C. L., Greene, L. J., and Ferrario, C. M. (1988) Hypertension *11(Suppl1)*:153-157
28. Chappell, M. C., Brosnihan, K. B., Diz, D. I., and Ferrario, C. M. (1989) J. Biol. Chem. *264*:16518-16521
29. Block, C. H., Santos, R. A. S., Brosnihan, K. B., and Ferrario, C. M. (1989) Peptides *9*:1395-1401
30. Khosla, M. C., Smeby, R. R., and Bumpus, F. M. (1974) in Angiotensin (Page, I. H. and Bumpus, F. M. eds) pp. 126-161, Springer-Verlag, Berlin
31. Schiavone, M. T., Santos, R. A. S., Brosnihan, K. B., Khosla, M. C., and Ferrario, C. M. (1988) Proc. Natl. Acad. Sci. USA *85*:4095-4098
32. Campagnole-Santos, M. J., Heringer, S. B., Batista, E. N., Khosla, M. C., and Santos, R. A. (1992) Amer J Physiol *263*:R89-94
33. Diz, D. I. and Ferrario, C. M. (1989) Hypertension *14*:342(Abstract)
34. Diz, D. I., Bosch, S. M., Moriguchi, A., Ganten, D., and Ferrario, C. M. (1994) J Hypertension *12*:S71(Abstract)
35. Goldfarb, D. A., Diz, D. I., Tubbs, R. R., Ferrario, C. M., and Novick, A. C. (1994) J.Urology *151*:206-213
36. Chappell, M. C., Diz, D. I., and Jacobsen, D. W. (1992) Peptides *13*:313-318
37. Tsutsumi, K. and Saavedra, M. (1991) J. Neurochem. *56*:348-351
38. Rowe, B. P., Grove, K. L., Saylor, D. L., and Speth, R. C. (1991) Regul. Pept. *33*:45-53
39. Gehlert, D. R., Gackenheimer, S. L., and Schober, D. A. (1991) Neurosci *44*:501-514
40. Obermuller, N., Unger, T., Culman, J., Gohlke, P., deGasparo, M., and Bottari, S. P. (1991) Neurosci. Lett. *132*:11-15
41. Wamsley, J. K., Herblin, W. F., Alburges, M. E., and Hunt, M. (1990) Brain Res. Bull. *25*:397-400
42. Ernsberger, P., Zhou, J., Damon, T. H., and Douglas, J. G. (1992) Amer J Physiol *263*:F411-F416
43. Speth, R. C. (1993) Regulatory Peptides *44*:189-197
44. Jaiswal, N., Diz, D. I., Tallant, E. A., Khosla, M. C., and Ferrario, C. M. (1991) Am. J. Physiol. *260*:R1000-R1006

45. Jaiswal, N., Tallant, E. A., Diz, D. I., Khosla, M. C., and Ferrario, C. M. (1991) Hypertension *17*:1115-1120
46. Tallant, E. A., Jaiswal, N., Diz, D. I., and Ferrario, C. M. (1991) Hypertension *18*:32-39
47. Leung, K. H., Chang, R. S. L., Lotti, V. J., Roscoe, A., Smith, R. D., Timmermans, P. B. M. W. M., and Chiu, A. T. (1992) Am. J. Hypertens. *5*:648-656
48. Hogarty, D. C., Speakman, E. A., Puig, V., and Phillips, M. I. (1992) Brain Res *586*:289-294
49. Santos, R. A. S., Campagnole-Santos, M. J., Baracho, N. C. V., Fontes, M. A. P., Silva, L. C. S., Neves, L. A. A., Oliveira, D. R., Caligiorne, S. M., Rodrigues, A. R. V., Gropen, C. ,Jr., Carvalho, W. S., Silva, C. S. E., and Khosla, M. C. (1994) Brain Res. Bull. *35*:293-298
50. Ambuhl, P., Felix, D., and Khosla, M. C. (1994) Brain Res. Bull. *35*:289-291
51. Benter, I. F., Diz, D. I., and Ferrario, C. M. (1993) Peptides *14*:679-684
52. Harding, J. W., Cook, V. I., Miller-Wing, A. V., Hanesworth, J. M., Sardinia, M. F., Hall, K. L., Stobb, J. W., Swanson, G. N., Coleman, J. K., Wright, J. W., and et al (1992) Brain Res *583*:340-3
53. Naveri, L., Stromberg, C., and Saavedra, J. M. (1994) Journal of Cerebral Blood Flow & Metabolism *14*:1096-1099
54. Wright, J. W., Miller-Wing, A. V., Shaffer, M. J., Higginson, C., Wright, D. E., Hanesworth, J. M., and Harding, J. W. (1993) Brain Res Bull *32*:497-502
55. Nossaman, B. D., Feng, C. J., Kaye, A. D., and Kadowitz, P. J. (1995) Am. J. Physiol. *268*:L302-L308
56. Gardiner, S. M., Kemp, P. A., March, J. E., and Bennett, T. (1993) Br. J. Pharmacol. *110*:159-162
57. Block, C. H. (1987) in Brain Peptides and Catecholamines in Cardiovascular Regulation (Buckley, J. P. and Ferrario, C. M. eds) pp. 109-124, Raven Press, New York
58. Block, C. H., Barnes, K. L., and Ferrario, C. M. (1987) J. Cardiovasc. Pharmacol. *10(suppl 12)*:230-234
59. Ferrario, C. M., Barnes, K. L., Block, C. H., Brosnihan, K. B., Diz, D. I., Khosla, M. C., and Santos, R. A. S. (1990) Hypertension *15 (Suppl I)*:I13-I19
60. Ferrario, C. M., Jaiswal, N., Yamamoto, K., Diz, D. I., and Schiavone, M. T. (1991) Clin. Cardiol. *14 IV*:56-62
61. Kohara, K., Brosnihan, K. B., and Ferrario, C. M. (1993) Peptides *14*:883-891
62. Campbell, D. J., Lawrence, A. C., Towrie, A., Kladis, A., and Valentijn, A. J. (1991) Hypertension *18*:763-773
63. Kohara, K., Brosnihan, K. B., Chappell, M. C., Khosla, M. C., and Ferrario, C. M. (1991) Hypertension *17*:131-138
64. Benter, I. F., Ferrario, C. M., Morris, M., and Diz, D. (1995) Am. J. Physiol. (In press)
65. Moriguchi, A., Matsumura, K., Tallant, E. A., Ganten, D., and Ferrario, C. M. (1995) Hypertension (In press)

25

ATYPICAL (NON-AT_1, NON-AT_2) ANGIOTENSIN RECEPTORS

Roger D. Smith

Department of Clinical Biochemistry
Addenbrooke's Hospital
Cambridge CB2 2QQ, United Kingdom

INTRODUCTION

Angiotensin II (Ang II), the principal active component of the renin-angiotensin system (RAS), plays a major role in the physiology of the cardiovascular system by affecting diverse target tissues such as vascular smooth muscle (to stimulate vasoconstriction), myocardium (to increase heart rate and contractile force), adrenal cortex (to release aldosterone) and medulla (to release catecholamines), pituitary (to regulate release of vasopressin and other hormones), kidney (to promote sodium reabsorption) and brain (to increase thirst and drinking, and to stimulate a centrally-mediated pressor response) (reviewed in 1-3). The peptide also promotes cardiac hypertrophy and contributes to the neointimal proliferation of vascular smooth muscle following arterial injury (4,5). The potential role of Ang II as a growth factor also implicates it in the pathogenesis of hypertension and atherosclerosis (6). In addition to these effects on the cardiovascular system, Ang II may also be active in non-cardiovascular tissues such as the ovary (7,8).

Multiple mechanisms of signal transduction have been demonstrated for Ang II. For example, depending on the target cell or tissue, Ang II stimulates phosphoinositide turnover (with elevation of intracellular Ca^{2+} ($[Ca^{2+}]_i$) and activation of protein kinase C), inhibits adenylate cyclase, activates guanylate cyclase, releases prostaglandins and regulates Ca^{2+} channels (reviewed in 9-11). In view of these pleiotrophic actions of Ang II, multiple angiotensin receptor subtypes have been postulated for some years (12-15). However, their existence has only been confirmed in recent years with the introduction of highly-selective non-peptide angiotensin receptor antagonists (reviewed in 16-18) which have allowed a classification of angiotensin receptors into AT_1 and AT_2 subtypes (19). AT_1 receptors are defined by their sensitivity to biphenylimidazole antagonists typified by DuP 753, whereas AT_2 receptors are sensitive to tetrahydroimidazopyridine antagonists such as PD 123177 and PD 123319 (19). The peptide, CGP 42112A, has also been used as an AT_2-specific ligand, although this agent displays both agonistic and antagonistic properties at AT_2 receptors (depending on cell type), and also binds a non-angiotensin site in brain and spleen (15,20). However, despite differential sensitivities to these antagonists, both AT_1 and AT_2 receptors

Recent Advances in Cellular and Molecular Aspects of Angiotensin Receptors
Edited by Mohan K. Raizada et al., Plenum Press, New York, 1996

have high affinity for Ang II, angiotensin III (Ang III) and the peptide antagonists, [Sar^1,Ile^8]-Ang II and saralasin, but only low affinity for angiotensin I (Ang I) (19).

AT_1 and AT_2 receptors have been cloned and sequenced and are members of the seven transmembrane domain superfamily of receptors (21-24). However, the AT_2 receptor has only 32-34% amino acid sequence homology with the AT_1 receptor and exhibits different properties. For example, ligand binding to AT_1 receptors is sensitive to the reducing agent, dithiothreitol (DTT), and to non-hydrolysable GTP analogues such as GTPγS (12-15). In contrast, DTT enhances ligand binding to the AT_2 site (12-15). Furthermore, this site is unaffected by GTPγS which suggests either that it lacks (or has inefficient) G protein coupling, or that it couples to a unique class of G proteins (23,24). The AT_1 receptor activates phosphoinositide hydrolysis and/or inhibits adenylate cyclase (9-11,19) whereas the signalling pathway(s) activated by the AT_2 receptor is less clearly defined. However, the latter receptor has been reported to decrease cyclic GMP (25,26), to stimulate protein tyrosine phosphatase activity (26) and to increase outward K^+ currents (27).

However, the availability of subtype-specific antagonists which have facilitated the identification and characterisation of AT_1 and AT_2 receptors does not preclude the possibility that additional angiotensin receptor subtypes may also exist for which selective antagonists (or agonists) are not yet available. Indeed, there have been several recent reports of high affinity angiotensin binding sites (on a range of cells and tissues) that are insensitive to both AT_1 and AT_2 receptor antagonists. These 'atypical' angiotensin binding sites, some of which appear to activate intracellular signalling pathways (and may therefore be tentatively termed AT_n receptors), are the subject of this short review.

BINDING SITES ON MYCOPLASMA

Specific high-affinity binding sites for ^{125}I-Ang II have been described on two species of *Mycoplasma* which commonly infect eukaryotic cell cultures, namely *Mycoplasma hyorhinis* and *Acholeplasma laidlawii* (28,29). However, this phenomenon does not appear to be a general feature of *Mycoplasmataceae*, since angiotensin binding sites were not present on two strains of *M. hominis* (28).

Binding sites were detectable both on *M. hyorhinis*-infected OK opossum kidney cells as well as on monocultures of *M. hyorhinis* itself (but not on uninfected OK cells) (28). Although the sites have high affinity for Ang II (K_D 5.1 nM), they differ markedly from AT_1 and AT_2 receptors in also having high affinity for Ang I (K_D 1.6 nM), but only very low affinity for Ang III (K_D ~330 μM) and saralasin. The sites also have moderate affinity (K_D 57 nM) for Ang (1-7), but are distinct from both AT_1 and AT_2 receptors since ^{125}I-Ang II binding was unaffected by micromolar concentrations of DuP 753 or CGP 42112A.

Curiously, the binding of ^{125}I-Ang II to these sites was sensitive to micromolar concentrations of the unrelated peptides, aprotinin and bacitracin, and this was a specific effect since chymostatin, pepstatin and leupeptin each had no effect at concentrations up to 100 μM. DTT appeared to exert a biphasic effect on ^{125}I-Ang II binding: concentrations in the range 0.3 - 30 mM increased ^{125}I-Ang binding (to 150% of the control value), whereas the highest concentration of DTT tested (100 mM) reduced ^{125}I-Ang II binding by ~70% .

Specific ^{125}I-Ang II binding sites were also found on *A. laidlawii*-infected rat aortic smooth muscle (SMC) and glomerular mesangial cells (GMCs) which appear to be very similar to those present on *M. hyorhinis*-infected OK cells (29). Like *M. hyorhinis*-infected OK cells, the sites on *A. laidlawii*-infected SMCs and GMCs have high affinity for Ang II (K_D 0.75 nM) and Ang I (K_D 0.72 nM), very low affinity for Ang III (K_D 31 μM) and are insensitive to both AT_1 and AT_2 specific antagonists (DuP 753 and PD 123319 respectively). However the *A. laidlawii* sites appear to differ from the *M. hyorhinis* sites since they have

200-fold higher affinity for saralasin (K_D 123 nM) and 40-fold higher affinity for Ang (1-7) (K_D 1.3 nM). Furthermore, in contrast to its effect on the *M. hyorhinis* sites, DTT partially decreased (by ~30%) ^{125}I-Ang II binding to the *A. laidlawii* sites with a half-maximal effect at 0.6 mM. GTPγS had no effect.

At present it is unclear whether the observed differences between the ^{125}I-Ang II binding sites on *M. hyorhinis*-infected OK cells and *A. laidlawii*-infected SMCs and GMCs represent real differences between these two binding sites, or merely arise from variations in experimental design between the two studies. In the case of *A. laidlawii*, it is also unclear whether the binding site resides on the microorganism's plasma membrane, or whether infection of eukaryotic cells stimulates the expression or activation of a latent binding site encoded by the host cell genome. However, the presence of ^{125}I-Ang II binding sites on monocultures of *M. hyorhinis* with binding characteristics similar to those of *M. hyorhinis*-infected OK cells suggests that the microorganisms themselves express angiotensin binding sites (28).

The function(s), if any, of such microorganism-encoded ^{125}I-Ang II binding sites is unclear. However, Ang II had no effect on $[Ca^{2+}]_i$, ^{45}Ca uptake or cyclic AMP production in *M. hyorhinis*-infected OK cells, and no functional responses to added Ang II of *A. laidlawii*-infected SMCs or GMCs, or of *M. hyorhinis* mono-cultures, have been reported. However, the ability of *M. hyorhinis*-infected OK cells to internalise (at 37°C) and subsequently degrade ^{125}I-Ang II (28) suggests that the site may be have some function.

Between 5-35% of all cell cultures are estimated to be infected with various species of *Mycoplasma*, with *M. hyorhinis* and *A. laidlawii* being particularly common contaminants (30). Furthermore, since many species of *Mycoplasma* exert no cytopathogenic effects on host cells, such infection may be difficult to detect. It is therefore essential to exclude *Mycoplasma* contamination when evaluating the significance of putative atypical angiotensin binding sites on cultured eukaryotic cells.

XENOPUS RECEPTORS

High (K_D 1.6 nM) and low (K_D 22 nM) affinity angiotensin binding sites that are insensitive to both DuP 753 and PD 123319 have been identified on *Xenopus* cardiac membranes (31), and similar sites are also present on *Xenopus* follicular oocytes (32). The cardiac sites have high affinity for Ang II, Ang III and [Sar^1, Ile^8]-Ang II, moderate affinity for Ang I and saralasin, low affinity for CGP 42112A, and are sensitive to DTT (32,33). Two distinct but closely related cardiac receptors termed xAT have been cloned and sequenced (33,34): they have 89% amino acid sequence homology with each other and 60-63% homology to the AT$_1$ receptor. Transcripts encoding the receptor were detected in *Xenopus* lung, liver, kidney, spleen and heart, but not in the adrenal, intestine or smooth muscle (34).

When transfected into defolliculated *Xenopus* oocytes, xAT mediated an increase in $[Ca^{2+}]_i$ in response to added Ang II. Unusually, [Sar^1, Ile^8]-Ang II, saralasin and CGP 42112A were also agonists at this site (33,34). In view of its functional similarity and relatively high homology to the AT$_1$ receptor, it therefore seems likely that xAT represents an amphibian counterpart of the mammalian AT$_1$ receptor (which has acquired its unusual pharmacological properties as a result of evolutionary divergence) (33).

Recently, the structural requirements underlying the pharmacological differences between xAT and AT$_1$ receptors have been investigated (35). Specific sites in the AT$_1$ receptor that determine DuP 753 binding were identified using mutant rat AT$_1$ receptors (transiently expressed in COS-7 cells) in which non-conserved amino acids were replaced by the corresponding *Xenopus* residues. Analysis of a bank of single point and combined mutants revealed that amino acid substitutions at Val^{108}, Ala^{163}, Thr^{198}, Ser^{252}, Leu^{300} and Phe^{301}

resulted in marked attenuation of DuP 753 binding (35). Interestingly, each of these residues are located within the membrane-spanning regions of the receptor and are distinct from the site at which peptide antagonists bind.

AVIAN RECEPTORS

In the chicken, Ang II causes a unique biphasic blood pressure response: an initial, rapid depressor response followed by a smaller secondary pressor response (36). The pressor response is believed to be indirect, being mediated via the release of catecholamines. However, the initial depressor response occurs via vascular angiotensin receptors which mediate vasodilatation (37). However, these chicken receptors appear to differ from AT_1 and AT_2 receptors since they have only moderate affinity for [Sar^1, Ile^8]-Ang II (37).

Ang II stimulates angiogenesis in the chick embryo chorioallantoic membrane (38) via high affinity (K_D 2.7 nM) angiotensin receptors which are expressed in this tissue. These receptors are clearly distinct from AT_1 or AT_2 receptors (since they are insensitive to DuP 753 and PD 123319) and have high affinity for Ang I but only low affinity for Ang III. Like the chicken vascular receptors, they also have only moderate affinity for [Sar^1, Ile^8]-Ang II, although they are sensitive to micromolar concentrations of CGP 42112A (38).

Using oligonucleotide primers based on the AT_1 receptor, an angiotensin receptor has been cloned and sequenced from the turkey adrenal gland (39,40). Although this receptor has ~75% overall sequence homology to the AT_1 receptor, it is insensitive to both DuP 753 and PD 123319. It has high affinity for Ang II (K_D 0.17 nM) and [Sar^1, Ile^8]-Ang II, but low affinity for Ang I and CGP 42112A. Transcripts encoding the receptor were abundant in the adrenal cortex, but were undetectable in any other turkey tissues (39). Since angiotensin binding sites are present in a variety of turkey tissues, the adrenal receptor may therefore either be distinct from the receptors in other turkey tissues, or the stability of its mRNA may be much lower than those in the other tissues (39).

Ang II treatment of COS-7 cells transfected with the turkey adrenal receptor stimulated inositol phosphate production which was unaffected by the presence of an excess (30 μM) of DuP 753 (40). In view of its functional similarity and high homology to the AT_1 receptor, it therefore seems likely that the turkey adrenal receptor represents an avian counterpart of the AT_1 receptor (analogous to the *Xenopus* receptor).

MAMMALIAN RECEPTORS

Neuro-2A Murine Neuroblastoma Cells

Chaki and Inagami have described the presence of specific, moderately high affinity (K_D 12 nM) ^{125}I-Ang II binding sites that are insensitive to both DuP 753 and PD 123319 on Neuro-2A cells (41-43). Although the affinity of these sites for Ang I was not reported, they exhibited low affinity for Ang III, and were partially sensitive to DTT (with a half-maximal effect at ~1 mM). Interestingly, differentiation of Neuro-2A cells with prostaglandin E_1 resulted in a 10-fold increase in the number of these binding sites, without any significant change in their affinity. The sites were insensitive to GTPγS and were unaffected by ion channel blockers such as nifedipine, diltiazem and veratridine.

Although Ang II had no effect on phosphoinositide hydrolysis or cyclic AMP formation, the peptide stimulated a rapid (30 s) and dose-dependent increase in cyclic GMP formation in Neuro-2A cells, with a maximal (8.3-fold) increase at 1 μM. The increase in cyclic GMP formation appeared to be mediated by soluble guanylate cyclase, via a mechanism involving activation of nitric oxide synthase secondary to influx of extracellular Ca^{2+} (43). This effect was completely inhibited by co-incubation of the cells with [Sar^1, Ile^8]-Ang II (1 μM), but was unaffected by the presence of equimolar concentrations of either DuP 753 or PD 123319. The authors therefore concluded that the stimulated increase of cyclic GMP formation was mediated via atypical (non-AT_1, non-AT_2) angiotensin receptors which they termed AT_3 (44).

However, since testing of the Neuro-2A cultures used in these experiments for *Mycoplasma* contamination was not reported (41-43), it is important to establish whether or not the findings are reproducible in *Mycoplasma*-free cultures. If uninfected Neuro-2A cells do not express atypical angiotensin receptors, deliberate infection of such cultures with *M. hyorhinis* or *A. laidlawii* could be used to investigate the mechanism by which angiotensin receptors expressed on infecting microorganisms are able to activate a host cell signalling pathway (elevation of cyclic GMP), or by which *Mycoplasma* induces the expression or activation of host-encoded angiotensin receptors.

IEC-18 Rat Intestinal Epithelial Cells

Smith *et al* recently described the presence of AT_1 angiotensin receptors that couple to phosphoinositide hydrolysis on a cultured rat intestinal epithelial (RIE-1) cell line (45,46). Two related rat intestinal epithelial cell lines, IEC-6 and IEC-18 (47), were also found to express specific high affinity angiotensin binding sites (48). However, although the sites on IEC-6 cells were (like RIE-1 cells) the AT_1 subtype, the sites on IEC-18 cells were atypical, being insensitive to micromolar concentrations of either DuP 753 or PD 123319 (48). These cultures were free of *Mycoplasma* contamination, as determined by staining with the DNA binding fluorochrome, 4′-6-diamidino-2′-phenylindole (30,49,50).

The IEC-18 sites have high affinity for Ang I (K_D 3.4 nM) and Ang II, moderate affinity for Ang (1-7) and the unrelated peptide, bombesin, low affinity for Ang (3-8) and very low affinity for Ang III. ^{125}I-Ang I binding was partially inhibited by DTT but GTPγS had no effect. Treatment of IEC-18 cells with Ang II (100 nM) stimulated a rapid (30 s), but transient elevation of cyclic GMP that was unaffected by the presence of an excess (10 μM) of either DuP 753 or PD 123319 (R. D. Smith, manuscript in preparation). The IEC-18 sites therefore appear to be similar to the sites on Neuro-2A cells. However, whereas the latter sites appear to have relatively high affinity for [Sar^1, Ile^8]-Ang II (since the cyclic GMP response of Neuro-2A cells to 1 μM-Ang II was completely inhibited by an equimolar concentration of [Sar^1, Ile^8]-Ang II) (42), the sites on IEC-18 cells had only low affinity for [Sar^1, Ile^8]-Ang II.

The *mas* Oncogene Product

The *mas* oncogene encodes a putative seven transmembrane domain receptor which has been claimed to be an angiotensin receptor on the basis of a functional response (elevation of $[Ca^{2+}]_i$) to Ang II of *mas*-injected *Xenopus* oocytes and *mas*-transfected NG 115-401L neural cells (51). Like AT_1 receptors, the *mas*-encoded sites have high affinity for Ang II and Ang III although the rank order of binding affinities (Ang III > Ang III) of these two peptides is reversed compared to their affinities for AT_1 receptors. They also have low affinity for Ang I and are sensitive to DTT. However, the sites are insensitive to DuP 753 and the AT_2-specific antagonist, Exp 655 (52). Uniquely amongst angiotensin receptors, however, they are sensitive to the broad spectrum peptide receptor antagonist, [D-Arg^1, D-Pro^2, D-$Trp^{7,9}$, Leu^{11}]-substance P (53), but are unable to recognise mono-, or di-iodinated derivatives of Ang II or Ang III (52). Furthermore, substituted octapeptide Ang II antagonists behave as partial agonists at the *mas*-encoded sites.

However, objections have been raised to the claim that *mas* encodes an angiotensin receptor (54). For example, *mas* exhibits less than 10% overall amino acid sequence homology to AT_1 receptors (51) and no increased binding of ^{125}I-Ang II was reported for *mas*-injected or transfected cells. *Xenopus* oocytes themselves have also subsequently been found to express endogenous angiotensin receptors (present predominantly on their surrounding follicular cells) which are difficult to remove completely by routine methods of defolliculation (55). Since these follicular receptors are able to mediate an increase in oocyte $[Ca^{2+}]_i$ by signal transfer through gap junctions (55), they may have been responsible for some or all of the Ca^{2+} responses to Ang II of *mas*-injected oocytes.

RECEPTORS FOR METABOLIC FRAGMENTS OF ANG II

Although Ang II (and, to a lesser extent, Ang III) have historically been considered to be the principal active components of the RAS, recent studies have indicated that some smaller fragments of Ang II, particularly Ang (1-7) and Ang (3-8), are also biologically active. Furthermore, these fragments appear to act via receptors that are distinct from AT_1 and AT_2 angiotensin receptors.

Ang (3-8) Hexapeptide (Ang IV)

Ang (3-8), which has also been termed Ang IV, has low affinity for classical (AT_1 and AT_2) angiotensin receptors (56). Nevertheless, this hexapeptide binds specifically and with high affinity (K_D 0.6-1.85 nM) to sites present in a range of tissues including adrenal, kidney, heart, liver, lung, uterus, blood vessels and a variety of brain nuclei (57-62). The sites, which have been termed AT_4 (62), are distinct from AT_1 and AT_2 receptors since they are insensitive to DuP 753, PD 123177 and CGP 42112A. They also have low affinity for Ang II, Ang III and [Sar^1, Ile^8]-Ang II, and are unaffected by GTPγS (57-61).

Ang (3-8) possesses unique biological properties that may be important in cardiovascular physiology. For example, the peptide produces endothelium-dependent vasodilatation when applied topically (in the presence of L-arginine) to rabbit cerebral arterioles (63), increases renal cortical blood flow in the rat (57) and decreases renin release following systemic infusion in humans (64). In cultured chick cardiocytes, Ang (3-8) antagonises Ang II-stimulated protein and RNA synthesis (65). Ang (3-8) is also active in rat brain, intracerebroventricular infusion of the peptide improving memory acquisition, retention and retrieval (66).

These findings suggest that the AT_4 site represents a functional, but atypical angiotensin receptor which mediates specific cellular responses. Indeed, the vasodilatory effect of Ang (3-8) on cerebral arterioles was inhibited by methylene blue, suggesting that this response (in contrast to responses elicited at AT_1 and AT_2 receptors) is mediated via soluble guanylate cyclase (63). Furthermore, although both Ang (3-8) and Ang II stimulate inositol phosphate production and increase $[Ca^{2+}]_i$ in vascular smooth muscle cells, the kinetics of the calcium response and the pattern of inositol phosphates production was different, suggesting different mechanisms of signal transduction in response to each peptide (67).

Ang (1-7) Heptapeptide

Ang (1-7) is a biologically active peptide present in the brain (68,69) which may have reciprocal actions to those of Ang II in the central control of blood pressure. Although Ang (1-7) is as potent as Ang II in stimulating the secretion of vasopressin (70), the heptapeptide produces vasodepressor and bradycardic responses when injected into the dorsomedial medulla of rats (71,72). Cultured vascular endothelial cells also mediate the production of Ang (1-7) from added Ang I (73), and the peptide has a selective coronary vasoconstrictor effect in the hamster heart (74). Ang (1-7) enhances the production of prostaglandins by several cells including astrocytes (75) and endothelial cells (76).

Ang (1-7)-stimulated prostaglandin release from astrocytes was inhibited by CGP 42112A but not by DuP 753 (75), whereas PD 123177, but not CGP 42112A or DuP 753, attenuated Ang (1-7)-induced prostaglandin release from endothelial cells (76). CGP 42112A inhibited the increase in activity of individual paraventricular neurons stimulated by Ang (1-7) (77). However, in contrast to these findings, the cardiovascular effects of Ang (1-7) were unaffected by CGP 42112A, DuP 753 or PD 123319, but were inhibited by [Sar^1, Thr^8]-Ang II, suggesting that these effects are mediated via a receptor distinct from either AT_1 or AT_2 receptors (78). Consistent with this hypothesis, the central pressor effect of Ang (1-7) was unaffected by either DuP 753 or CGP 42112A (79).

It therefore seems likely that a specific receptor (distinct from AT_1 and AT_2 receptors) exists for Ang (1-7) which mediates at least some of the actions of the heptapeptide. However, despite reports of functional antagonism of the actions of Ang (1-7), there are no reports of binding studies using ^{125}I-labelled Ang (1-7). However, the availability of a specific antagonist (termed A-779) which selectively blocked the pressor response to Ang (1-7), without affecting AT_1-mediated pressor responses to Ang II (79), should facilitate characterisation of this putative receptor.

Concluding Remarks

In view of their homology to AT_1 receptors and coupling to phosphoinositide hydrolysis (with elevation of $[Ca^{2+}]_i$), the *Xenopus* and avian receptors described above probably represent non-mammalian counterparts of the mammalian AT_1 receptor. However, the existence of additional non-AT_1, non-AT_2 mammalian angiotensin receptor subtypes that have unique pharmacological profiles, but for which specific antagonists are not yet available, seems very likely. Indeed, this would not be surprising since families comprising multiple subtypes of seven transmembrane domain receptors are known to exist for several ligands. For example, the muscarinic and serotonin families each comprise five, the somatostatin family four and the opioid family three principal receptor subtypes (80). However, the unequivocal existence of additional (AT_n) angiotensin receptor subtypes will only be established by isolating and sequencing cDNAs that encode these putative receptors.

REFERENCES

1. Peach, M. J. (1977) Physiol. Rev. *57*: 313-370
2. Saavedra, J. M. (1992) Endocr. Rev. *13*: 329-380
3. Sealey, J. E. and Laragh, J. H. (1990) in Hypertension: Pathophysiology, Diagnosis and Management (eds. J. H. Laragh, B. M. Brenner), Raven Press, New York, pp 1287-1317
4. Dzau, V. J., Gibbons, G. H. and Pratt, R. E. (1991) Hypertension *18* (suppl II): II-100-II-105
5. Schelling, P., Fischer, H. and Ganten, D. (1991) J. Hypertens. *9*: 3-15
6. Krieger, J. E. and Dzau, V. J. (1991) Hypertension *18* (suppl I): I-3-I-17
7. Stirling, D., Magness, R. R., Stone, R., Waterman, M. R. and Simpson, E. R. (1990) J. Biol. Chem. *265*: 5-8
8. Pucell, A. G., Bumpus, F. M. and Husain, A. (1987) J. Biol. Chem. *262*: 7076-7080
9. Catt, K. J., Sandberg, K. and Balla, T. (1993) in Cellular and Molecular Biology of the Renin-Angiotensin System (eds M. K. Raizada, M. I. Phillips, C. Sumners), CRC Press, Boca Raton, Florida, pp 307-356.
10. Heemskerk, F. M. J. and Saavedra, J. M. (1994) in Angiotensin Receptors (eds J. M. Saavedra, P. B. M. W. M. Timmermans), Plenum Press, New York, pp 177-191.

11. Bottari, S. P., de Gasparo, M., Stecklings, M. and Levens, N. R. (1993) Front. Neuroendocrinol. *14*: 123-171
12. Gunther, S. (1984) J. Biol. Chem. *259*: 7622-7629
13. Douglas, J. G. (1987) Am. J. Physiol. *253*: F1-F7
14. Speth, R. C. and Kim, K. H. (1990) Biochem. Biophys. Res. Commun. *169*: 997-1006
15. Whitebread, S., Mele, M., Kamber, B. and DeGasparo, M. (1989) Biochem. Biophys. Res. Commun. *163*: 284-291
16. Smith, R. D., Chiu, A. T., Wong, P. C., Herblin, W. F. and Timmermans, P. B. M. W. M. (1992) Annu. Rev. Pharmacol. Toxicol. *32*: 135-165
17. Timmermans, P. B. M. W. M., Wong, P. C., Chiu, A. T., and Herblin, W. F. (1991) Trends Pharmacol. Sci. *12*: 55-62
18. Timmermans, P. B. M. W. M., Carini, D. J., Chiu, A. T., Duncia, J. V., Price, W. A., Wells, G. J., Wong, P. C., Johnson, A. L. and Wexler, R. R. (1991) Am. J. Hypertens. *4*: 275S-281S
19. Bumpus, F. M., Catt, K. J., Chiu, A. T., DeGasparo, M., Goodfriend, T., Husain, A., Peach, M. J., Taylor, D. G. and P. B. M. W. M. Timmermans (1991) Hypertension *17*: 720-721
20. de Oliveira, A. M., Viswanathan, M., Heemskerk, F. M. J., Correa, F. M. A. and Saavedra, J. M. (1994) Biochem. Biophys. Res. Commun. *200*: 1049-1058
21. Sasaki, K., Yamano, Y., Bardhan, S., Iwai, N., Murray, J. J., Hasegawa, M., Matsuda, Y. and Inagami, T. (1991) Nature *351*: 230-233
22. Murphy, T. J., Alexander, R. W., Griendling, K. K., Runge, M. S. and Bernstein, K. E. (1991) Nature *351*: 233-236
23. Mukoyama, M., Nakajima, M., Horiuchi, M., Sasamura, H., Pratt, R. E. and Dzau, V. J. (1993) J. Biol. Chem. *268*: 24539-24542
24. Kambayashi, Y., Bardhan, S., Takahashi, K., Tsuzuki, S., Inui, H., Hamakubo, T. and Inagami, T. (1993) J. Biol. Chem. *268*: 24543-24546
25. Sumners, C. and Myers, L. M. (1991) Am. J. Physiol. *260*: C79-C87
26. Bottari, S. P., King, I. N., Reichlin, S., Dahlstroem, I., Lydon, N. and DeGasparo, M. (1992) Biochem. Biophys. Res. Commun. *183*: 206-211
27. Kang, J., Posner, P. and Sumners, C. (1994) Am. J. Physiol. *267*: C1389-C1397
28. Bergwitz, C., Madoff, S., Abou-Samra, A.-B. and Juppner, H. (1991) Biochem. Biophys. Res. Commun. *179*: 1391-1399
29. Whitbread, S., Pfeilschifter, J., Ramjoue, H. and DeGasparo, M. (1993) Reg. Peptides *44*: 233-238
30. Hay, R. J., Macy, M. L. and Chen, T. R. (1989) Nature *339*: 487-488
31. Sandberg, K., Ji, H., Millan, M. A. and Catt, K. J. (1991) FEBS Lett. *284*: 281-284
32. Ji, H., Sandberg, K. and Catt, K. J. (1990) Mol. Pharmacol. *39*: 120-123
33. Bergsma, D. J., Ellis, C., Nuthulaganti, P. R., Nambi, P., Scaife, K., Kumar, C. and Aiyar, N. (1993) Mol. Pharmacol. *44*: 277-284
34. Ji, H., Sandberg, K., Zhang, Y. and Catt, K. J. (1993) Biochem. Biophys. Res. Commun. *194*: 756-762
35. Ji, H., Leung, M., Zhang, Y., Catt, K. J. and Sandberg, K. (1994) J. Biol. Chem. *269*: 16533-16536
36. Nishimura, H., Nakamura, Y., Sumner, R. P. and Khosla, M. C. (1982) Am. J. Physiol. *242*: H314-H324
37. Stallone, J. N., Nishimura, H. and Khosla, M. C. (1989) J. Pharmacol. Exp. Therapeut. *251*: 1076-1082
38. Le Noble, F. A. C., Schreurs, N. H. J. S., van Straaten, H. W. M., Slaaf, D. W., Smits, J. F. M., Rogg, H. and Struijker-Boudier, H. A. J. (1993) Am. J. Physiol. *264*: R460-R465
39. Carsia, R. V., McIlroy, P. J., Kowalski, K. I. and Tilly, J. L. (1993) Biochem. Biophys. Res. Commun. *191*: 1073-1080
40. Murphy, T. J., Nakamura, Y., Takeuchi, K. and Alexander, R. W. (1993) Mol. Pharmacol. *44*: 1-7
41. Chaki, S. and Inagami, T. (1992) Biochem. Biophys. Res. Commun. *182*: 388-394
42. Chaki, S. and Inagami, T. (1992) Eur. J. Pharmacol. *225*: 355-356
43. Chaki, S. and Inagami, T. (1993) Mol. Pharmacol. *43*: 603-608
44. Inagami, T, Iwai, N., Sasaki, K., Yamamo, Y., Bardhan, S., Chaki, S., Guo, D.-F. and Furuta, H. (1992) J. Hypertens. *10*: 713-716
45. Blay, J. and Brown, K. D. (1984) Cell Biol. Int. Rep. *8*: 551-560
46. Smith, R. D., Corps, A. N., Hadfield, K. M., Vaughan, T. J. and Brown, K. D. (1994) Biochem. J. *302*: 791-800
47. Quaroni, A., Wands, J., Trelstad, R. L. and Isselbacher, K. J. (1979) J. Cell Biol. *80*: 248-265
48. Smith, R. D. (1994) Cell Biol. Int. *18*: 737-745
49. Russell, W. C., Newman, C. and Williamson, D. H. (1975) Nature *253*: 461-462
50. Freshney, R. I. (1994) in Culture of Animal Cells (R. I. Freshney), Wiley-Liss, New York, pp 243-252
51. Jackson, T. R., Blair, L. A. C., Marshall, J., Goedert, M. and Hanley, M. R. (1988) Nature *335*: 437-440

52. Hanley, M. R. (1991) J. Cardiovasc. Pharmacol. *18* (Suppl. 2): S7-S13
53. Jensen, R. T., Jones, S. W., Folkers, K. and Gardner, J. D. (1984) Nature *309*: 61-63
54. Catt, K. J. and Abbott, A. (1991) Trends Pharmacol. Sci. *12*: 279-281
55. Sandberg, K., Bor, M., Ji, H., Markwick, A., Millan, M. A. and Catt, K. J. (1990) Science *249*: 298-301
56. Glossman, H., Baukal, A. J. and Catt, K. J. (1974) J. Biol. Chem. *249*: 825-834
57. Swanson, G. N., Hanesworth, J. M., Sardinia, M. F., Coleman, J. K. M., Wright, J. W., Hall, K. L., Miller-Wing, A. V., Stobb, J. W., Cook, V. I., Harding, E. C. and Harding, J. W. (1992) Regul. Pept. *40*: 409-419
58. Harding, J. W., Cook, V. I., Miller-Wing, A. V., Hanesworth, J. M., Sardinia, M. F., Hall, K. L., Stobb,, J. W., Swanson, G. N., Coleman, J. K. M., Wright, J. W. and Harding, E. C. (1992) Brain Res. *583*: 340-343
59. Hall, K. L., Hanesworth, J. M., Ball, A. E., Felgenhauer, G. P., Hosick, H. L. and Harding, J. W. (1993) Regul. Pept. *44*: 225-232
60. Hanesworth, J. M., Sardinia, M. F., Krebs, L. T., Hall, K. L. and Harding, J. W. (1993) J. Pharmacol. Exp. Therapeut. *266*: 1036-1042
61. Miller-Wing, A. V., Hanesworth, J. M., Sardinia, M. F., Hall, K. L., Wright, J. W., Speth, R. C. Grove, K. L. and Harding, J. W. (1993) J. Pharmacol. Exp. Therapeut. *266*: 1718-1726
62. Sardinia, M. F., Hanesworth, J. M., Krishnan, F. and Harding, J. W. (1994) Peptides *15*: 1399-1406
63. Haberl, R. L., Decker, P. J. and Einhaupl, K. M. (1991) Circ. Res. *68*: 1621-1627
64. Kono, T., Ikeda, F., Tamiguchi, A., Imura, H., Oseko, F., Yoshioka, M. and Khosla, (1985) Acta Endocrinol. *109*: 249-253
65. Baker, K. M. and Aceto, J. F. (1990) Am. J. Physiol. *259*: H610-H618
66. Braszko, J. J., Kupryszewski, G., Witczuk, B. and Wisniewski, K. (1988) Neuroscience *27*: 777-783
67. Dostal, D. E., Murahashi, T. and Peach, M. J. (1990) Hypertension *15*: 815-822
68. Chappell, M. C., Brosnihan, K. B., Diz, D. I. and Ferrario, C. M. (1989) J. Biol. Chem. *264*: 16518-16523
69. Ferrario, C. M., Barnes, K. L., Block, C. H., Brosnihan, K. B., Diz, D. I., Khosla, M. C. and Santos, R. A. S. (1990) Hypertension *15* (suppl I): I-13-I-19
70. Schiavone, M. T., Santos, R. A. S., Brosnihan, K. B., Khosla, M. C. and Ferrario, C. M. (1988) Proc. Natl. Acad. Sci. USA *85*: 4095-4098
71. Campagnole-Santos, M. J., Diz, D. I., Santos, R. A. S., Khosla, M. C., Brosnihan, K. B. and Ferrario, C. M. (1989) Am. J. Physiol. *257*: H324-H329
72. Silva, L. C. S., Fontes, M. A. P., Campagnole-Santos, M. J., Khosla, M. C., Campos, R. R., Guertzenstein, P. G. and Santos, R. A. S. (1993) Brain Res. *613*: 321-325
73. Santos, R. A. S., Brosnihan, K. B., Jacobsen, D. W., DiCorleto, P. E. and Ferrario, C. M. (1992) Hypertension *19* (suppl II): II-56-II-61
74. Kumagai, H., Khosla, M., Ferrario, C. M. and Fouad-Tarazi, F. M. (1990) Hypertension *15* (suppl I): I-29-I-33
75. Jaiswal, N., Tallant, E. A., Diz, D. I., Khosla, M. C. and Ferrario, C. M. (1991) Hypertension *17*: 1115-1120
76. Jaiswal, N., Diz, D. I., Chappell, M. C., Khosla, M. C. and Ferrario, C. M. (1992) Hypertension *19* (suppl II): II-49-II-55
77. Felix, D., Khosla, M. C., Barnes, K. L., Imboden, H., Montani, B. and Ferrario, C. M. (1991) Hypertension *17*: 1111-1114
78. Benter, I. F., Diz, D. I. and Ferrario, C. M. (1993) Peptides *14*: 679-684
79. Santos, R. A. S., Campagnole-Santos, M. J., Baracho, N. C. V., Fontes, M. A. P., Silva, L. C. S., Neves, L. A. A., Oliveira, D. R., Caligiorne, S. M., Rodrigues, A. R. V., Gropen, C., Carvalho, W. S., Silva, A. C. S. E. and Khosla, M. C. (1994) Brain Res. Bull. *35*: 293-298
80. Watson, S. and Arkinstall, S. (1994) The G-Protein Linked Receptor Factsbook, Academic Press, London

26

BRAIN ANGIOTENSIN II AND RELATED RECEPTORS: NEW DEVELOPMENTS

J. M. Saavedra, A. M. de Oliveira, O. Jöhren, and M. Viswanathan

Section on Pharmacology, Laboratory of Clinical Science
National Institute of Mental Health
10 Center Drive MSC 1514
Bldg 10, Room 2D-45
Bethesda, Maryland 20892-1514

INTRODUCTION

Our research has recently been focused on the characterization and role of brain angiotensin II receptor subtypes (1). In a manner similar to that of the periphery (2), brain angiotensin II receptors were initially characterized in two subtypes, AT1 and AT2 (1,3).

Brain AT1 receptors are, for the most part, involved in the classical central actions of angiotensin II, such as regulation of blood pressure, drinking, salt appetite, sympathetic activity and pituitary function (1,4).

Little is known, on the other hand, about the functions of the AT2 receptor subtype. Our developmental studies, which revealed a very high expression of AT2 receptors in certain brain areas, point to a crucial role of this receptor subtype in the maturation of sensory and motor function in the rat (3). In addition, AT2 receptors are selectively expressed in cerebral blood vessels (5) and their stimulation modulates cerebrovascular autoregulation (6).

There are indications of further heterogeneity for both AT1 and AT2 receptor subtypes. At least in the rat, AT1 receptors have been further classified into AT1A and AT1B subtypes, with selective distribution and regulation in the brain and peripheral tissues (7,8). We have also proposed, in the brain, further heterogeneity for the AT2 receptors, (AT2A and AT2B, possibly coupled and uncoupled to G-proteins, respectively) (9).

Our recent work revealed further complexity for the brain angiotensin receptors, and indicated the existence of other, non-angiotensin but probably related receptors, of potential physiological importance. We will emphasize, in this chapter, our recent discovery of "atypical" angiotensin II receptors in the gerbil brain, and the unexpected finding of non-angiotensin macrophage/microglial receptors associated to brain injury.

Recent Advances in Cellular and Molecular Aspects of Angiotensin Receptors
Edited by Mohan K. Raizada et al., Plenum Press, New York, 1996

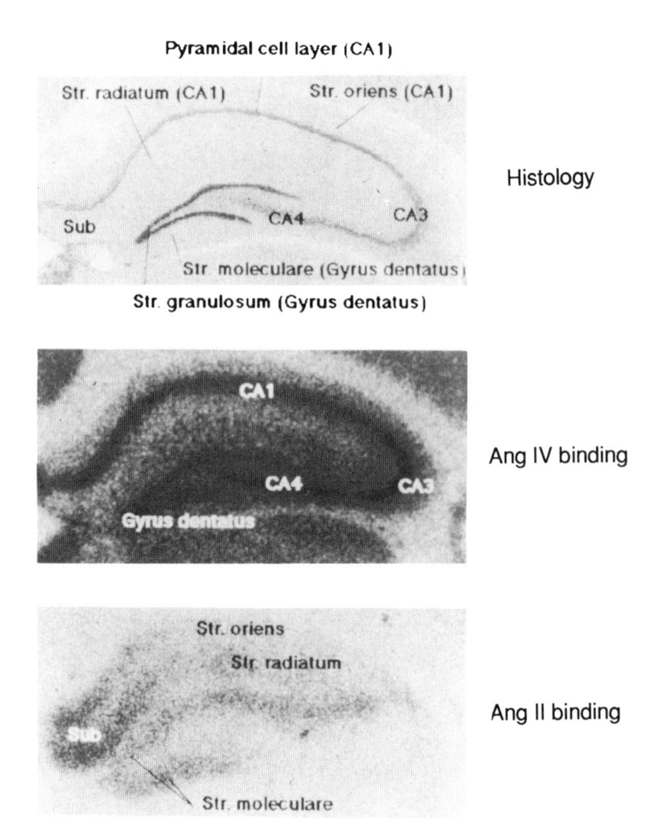

Figure 1. Autoradiographic localization of Angiotensin II and Angiotensin IV binding in the gerbil hippocampus. Top figure: histology, toluidine blue. Sub: subiculum. Middle figure: Angiotensin IV, total binding. Lower figure: Angiotensin II, total binding.

RESULTS AND DISCUSSION

After our initial discovery of the presence of angiotensin II AT_2 receptors in rat cerebral arteries (5) we asked the question whether their stimulation or blockade could play a role in the prevention or treatment of stroke. We decided to initiate studies on gerbils, a widely used model of stroke (10). In gerbils, temporary closure of both carotid arteries results in very selective neuronal death (11). Because the anterior communicating arteries in the gerbil are incompletely developed, permanent occlusion of one carotid results in a unilateral and massive stroke (12). The gerbil brain, however, was reported to contain very few angiotensin II receptors (13).

We conducted studies to first determine if the gerbil expressed any kind of angiotensin receptors in the cerebral arteries, and we could not find any (Saavedra, de Oliveira and Viswanathan, unpublished results). We saw, however, a high expression of angiotensin II binding in the hippocampus. The largest number of sites was located in the antero-medial hippocampus, and was restricted to the subiculum, stratum oriens, radiatum, the lacunar molecular layer of the CA1 subfield, and to the molecular layer of the gyrus dentatus (Figure

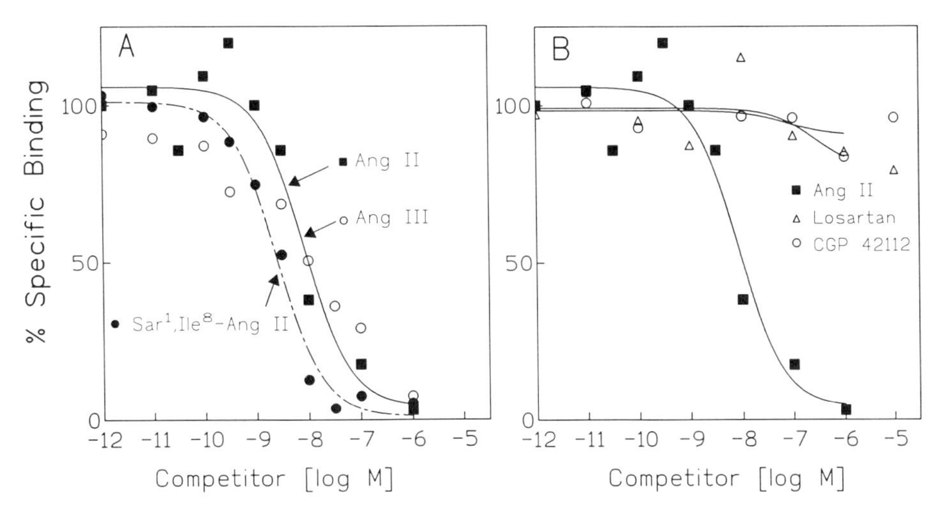

Figure 2. Competition studies of [^{125}I]Sar^{1}-Ang II binding to gerbil hippocampus. Consecutive sections were incubated in the presence of 5 x 10^{-10} M [^{125}I]Sar^{1}-Ang II and increasing concentrations of competitors. Results are expressed as % specific binding, after substraction of nonspecific binding, obtained by incubation in the presence of 5 x 10^{-6} M Ang II, and represent one typical example. A: competition by Ang II, Ang III, and Sar^{1}, Ile^{8}-Ang II. B: competition by Ang II, losartan and CGP 42112.

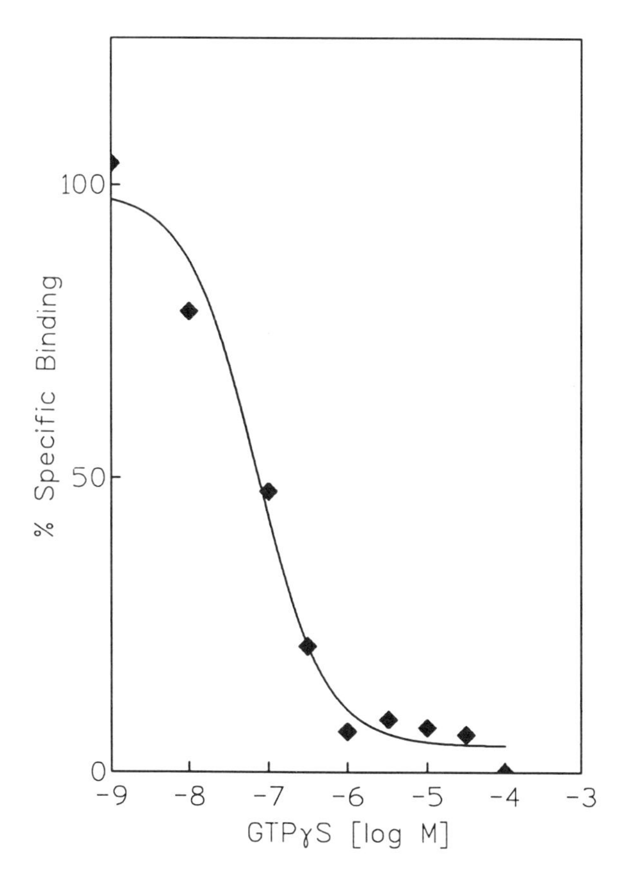

Figure 3. Effect of GTPγS on gerbil hippocampal [^{125}I]Sar^{1}-Ang II binding. Consecutive brain sections were incubated with 5 x 10^{-10} M [^{125}I]Sar^{1}-Ang II and increasing concentrations of GTPγS. Results are expressed as % of specific binding, and represent a typical experiment.

1). This binding was intercalated with binding for angiotensin IV (Ang IV) as revealed with [^{125}I]Ang IV, which was present in the pyramidal layer of the CA1 and CA3 subfields and the granular cell layer of the gyrus dentatus (Figure 1). In turn, the areas rich in Ang IV binding were devoid of Ang II binding (Figure 1).

We further characterized the gerbil hippocampal Ang II binding by selective displacement with Ang II related peptides and receptor subtype competitors, and by binding in the presence of guanine nucleotides. The gerbil hippocampus expressed a single class site, with high affinity for Ang II, Ang III, and Sar^1,Ile^8-Ang II, (Figure 2A) which was totally inhibited by incubation with increasing concentrations of guanine nucleotides (Figure 3). Conversely, this site had very low affinity for the selective AT_1 blocker losartan, for the selective AT_2 displacer CGP 42112 (Figure 2B) and could not be displaced by Ang IV (results not shown).

Our results indicate that the gerbil hippocampal binding sites are Ang II sites, but pharmacologically different from the AT_1 and AT_2 receptor subtypes. They are not similar to any of the recently discovered Ang II receptors, that of amphibians (14), turkey adrenal gland (15), the proposed rat AT_3 receptor (16), or the receptor recently described by Chaki and Inagami (17) and certainly not the guinea pig hippocampal Ang IV receptor

(18). For this reason, and according to the accepted preliminary standard nomenclature for Ang II receptors (19), the gerbil hippocampal receptor, since it represents Ang II binding not displaced with high affinity by the AT_1 or AT_2 selective ligands, should be considered an "atypical" Ang II receptor.

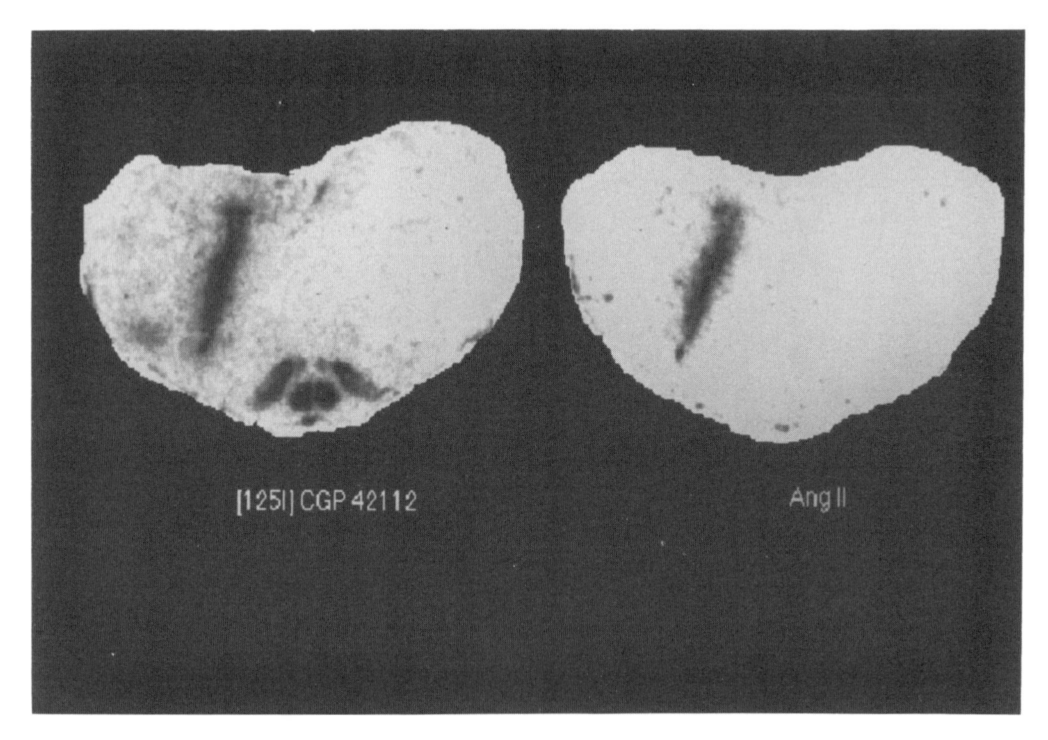

Figure 4. Binding of [^{125}I]CGP 42112 to rat brain stem. Consecutive sections were incubated with [^{125}I]CGP 42112 alone (left figure) or in the presence of 5 x 10^{-6} M unlabeled Ang II. Note that the left figure shows binding to the inferior olivary complex as well as to the brain lesion, represented by an oblique cut through the brainstem. In the right figure, unlabeled Ang II was able to displace all AT_2 receptors from the inferior olive, but was not able to displace the [^{125}I]CGP 42112 binding to the brain lesion.

Nothing is presently known regarding the functional role of the gerbil Ang II "atypical" hippocampal receptors. Because of their distribution, however, it is tempting to speculate that they could play roles in memory mechanisms, in the central control of emotional behavior, and be selectively affected by alterations in cerebral circulation such as those occurring during stroke.

During the course of our attempts to elucidate the role of rat brain AT_2 receptors, we unexpectedly found still another new binding site. We wished to determine the origin of the AT_2 receptors located in the inferior olivary complex and the cerebellar cortex of the immature rat (3). We were intrigued by the massive decline (about 90 %) in AT_2 receptor expression in both regions after the second week of life (3), and we speculated that these receptors played a role in the synaptic organization and development of the inferior olivary-cerebellar pathway.

To clarify whether the cerebellar receptors could have an origin in the inferior olive, a knife cut was performed at the level of the brainstem in two-week old rats. We chose, for this experiment, to reveal the AT_2 receptors using [^{125}I]CGP 42112, a ligand originally developed as selective for this receptor subtype (20,21). We found that the brainstem lesion did not result in any change in the number of AT_2 receptors, indicating that the inferior olivary and the cerebellar receptors had different anatomical origins, and there was probably no transport of AT_2 receptors between the two areas (unpublished results). We were, however, rapidly distracted from the original experimental hypothesis. We observed a very intense expression of binding alongside the lesion, a phenomenon totally unexpected. More importantly, we discovered that unlabeled angiotensin II, which was used to determine non-specific binding in adjacent sections, and which displaced all binding detected by [^{125}I]CGP 42112 in the inferior olive or the cerebellum, had no affinity for [^{125}I]CGP 42112 binding alongside the lesion (Figure 4).

We realized that the [^{125}I]CGP 42112 binding associated with the brain lesion was unrelated to angiotensin II, and we immediately set ourselves up to study and characterize this novel binding site. We quickly found that in the rat, very few tissues expressed non-angiotensin CGP 42112 binding sites under normal conditions. The red pulp of the spleen was the tissue showing the most remarkable amount of binding (22). Very low binding, about 5 % of that found in the spleen, was expressed in the lung and thymus (22), and all other tissues, including the normal rat brain, were negative for CGP 42112 binding.

We later confirmed that the CGP 42112 binding was associated to spleen macrophages (22) and that it was expressed in the brain only after lesions, where it was associated to macrophages-activated microglia (23,24).

We continue our work on the assumption of the existence of an ever increasing number of "angiotensin" and related receptors, some of which are better characterized than others, and some which may bind to endogenous ligands still to be characterized.

REFERENCES

1. Saavedra, J.M. (1992) Endocr. Rev. *13*:329-380.
2. Timmermans, P.B.M.W.M., Wong, P.C., Chiu, A.T., Herblin, W.F., Benfield, P., Carini, D.J., Lee, R.J., Wexler, R.R., Saye, J.A.M. and Smith, R.D., 1993, Pharmacol. Rev. *45*:205-251.
3. Tsutsumi, K. and Saavedra, J.M. (1991). Am. J. Physiol. *261*:R209-R216.
4. Tsutsumi, K. and Saavedra, J.M. (1991). Endocrinology *129*:3001-3008.
5. Tsutsumi, K. and Saavedra, J.M. (1991). Am. J. Physiol. *261*:H663-H670.
6. Strömberg, C., Naveri, L. and Saavedra, J.M. (1993) J. Cereb. Blood Flow Metab. *13*:298-303.
7. Iwai, N. and Inagami, T. (1992). FEBS *298*:257-260.
8. Yoshida, H., Kukuchi, J., Guo, D.F., Furuta, H., Iwai, N., van del Meer-de Jong, R., Inagami, T. and Ichikawa, I. (1992). Biochem. Biophys. Res. Commun. *186*: 1042-1049.

9. Tsutsumi, K. and Saavedra, J.M. (1992). Mol. Pharmacol. *41*:290-297.
10. Allen, K.L., Busza, A.L., Proctor, E., King, M.D., Williams, S.R., Crockard, H.A. and Gadian, D.G. (1993). NMR in Biomedicine, *6*:181-186.
11. Crain, B.J., Westerkam, W.D., Harrison, A.H. and Nadler, J.V. (1988). Neuroscience *27*:387-402.
12. Berry, K., Wisniewski, H.M., Svarzbein L. and Baez, S. (1975). J. Neurol. Sci. *25*:75-92.
13. Harding, J.W., Stone, L.P. and Wright, J.W. (1981). Brain Res. *205*:265-274.
14. Ji, H., Sandberg, D. and Catt, K.J. (1991). Mol. Pharmacol. *39*:120-123.
15. Murphy, T.J., Nakamura, Y., Takeuchi, K. and Alexander, R.W. (1993). Mol. Pharmacol. *44*:1-7.
16. Sanberg, K., Ji, H., Clark, A.J.L., Shapira, H. and Catt, K.J. (1992). J. Biol. Chem. *267*:9455-9458.
17. Chaki, S. and Inagami, T. (1992). Biochem. Biophys. Res. Commun. *182*:388-394.
18. Harding, J.W., Cook, V.I., Miller-Wing, A.V., Hanes, J.M., Sardinia, M.F., Hall, K.L., Stobb, J.W., Swanson, G.N., Coleman, J.K.M., Wright, J.W. and Harding, E.C. (1992). Brain Res. *583*:340-343.
19. Bumpus, F.M., Catt, K.J., Chiu, A.T., de Gasparo, M., Goodfriend, M., Husain, T., Peach, A., Taylor, M.J.Jr. and Timmermans, P.B.M.W.M. (1991). Hypertension *17*:720-721.
20. Speth, R.C. (1993). Regul. Pept. *44*:189-197.
21. Heemskerk, F.M.J., Zorad, S., Seltzer, A. and Saavedra, J.M. (1993). NeuroReport *4*:103-105.
22. de Oliveira, A.M., Viswanathan, M., Heemskerk, F.M.J., Correa, F.M.A. and Saavedra, J.M. (1994). Biochem. Biophys. Res. Commun. *200*: 1049-1058.
23. Viswanathan, M., de Oliveira, A.M., Correa, F.M.A. and Saavedra, J.M. (1994). Brain Res. *658*:265-270.
24. Viswanathan, M., de Oliveira, A.M., Wu, R.M., Chiueh, C.C. and Saavedra, J.M. (1994). Cell. Molec. Neurobiol. *14*:99-104.

27

RECEPTORS FOR (3-8) ANGIOTENSIN IN BRAIN CELLS

AngIV Binding in Brain Cells

Conrad Sernia, Bruce Wyse, Siok-Keen Tey, and Su-Lin Leong

Neuroendocrine Laboratory
Department of Physiology and Pharmacology
University of Queensland
Queensland 4072, Australia

INTRODUCTION

Angiotensin II (AngII) is the peptide effector hormone of the blood renin-angiotensin system (RAS). Central to the role of the RAS as a hormonal regulator of cardiovascular function are the actions of AngII on vascular tone and on fluid and electrolyte homeostasis (1). It is commonly accepted that the functions of the blood RAS are supported by local angiotensin-generating systems in various tissues, including the heart, kidney, adrenal, blood vessels and brain (2-4). Hence, production of AngII in particular brain areas can lead to hypertensive changes in the periphery.(3,5,6).

However AngII is unlikely to be the sole effector peptide of the RAS; smaller N-terminal truncated peptides, (2-8)AngII, (3-8)AngII, and the C-terminal truncated (1-7)AngII are all potential contenders as additional and alternative active peptides to AngII (3,6,7). Joe Harding and his colleagues (8-15) have been instrumental in providing evidence that (3-8)AngII (also termed AngIV) is biologically active and probably acts via a receptor with distinctly different ligand-binding profiles from those of the AT_1 and AT_2 subtypes of the AngII receptor (15-18). This putative AngIV receptor has been found in the brain, heart, kidney, vasculature, liver, lung and adrenal cortex (8-15,19,20). It binds AngII and AngIII with low affinity and does not bind losartan, a selective AT_1 receptor antagonist or PD123177 and CGP42112A, both selective for the AT_2 receptor (17). The AngIV receptor has been implicated in a diversity of physiological functions, ranging from the control of blood flow in the kidney (8) and brain (21) to enhancing memory retrieval and retention (11,22). There is also evidence that AngIV antagonises the hypertrophic action of AngII in cultured chick myocytes(23).

Miller-Wing et al (24) investigated the distribution of brain AngIV receptors in the guinea pig and found widespread binding in various brain regions, with particular high densities in the neocortex, hippocampus, thalamus, inferior and superior colliculi and cerebellum (9,24). These data were obtained from radioreceptor assays of tissue homogenates and from autoradiography of tissue sections. Such methods are appropiate for determining the brain areas which exhibit abundant binding but they are inappropiate for the identification of the types of cells involved. Thus, it is unclear whether the binding is restricted to either neurones or neuroglia, or is found in both. The identification of the cell types which bind AngIV is of key significance to the elucidation of AngIV function since glia and neurones have distinct functions. Furthermore, the use of a heterogenous source of receptors, as in an homogenate of brain tissue, limits the usefulness of data on ligand-specificity or of the physicochemical nature of AngIV receptors, since these properties could differ between cell types. For these reasons we have pursued the culture of defined cells as experimental models for the investigation of AngIV receptors. We searched for suitable cell lines and found that the rat C6 glioma expressed abundant AngIV binding activity. We describe here the AngIV receptors in these cells, as well as the partial characterization of AngIV sites in RT4-D6 (a glial cell line) (25) and primary cultures of astrocytes and neurones. Receptors were characterized by competitive radioreceptor assay, autoradiography and high pressure liquid chromatography (HPLC).

MATERIALS AND METHODS

Chemicals

(3-8)Angii and (4-8)Angii Were Obtained from Peninsula (Belmont, California). Angi, Angii, Bestatin, and Bsa (Bovine Serum Albumin) Were Purchased from Sigma-Aldrich (N.S.W., Aust.). Plummer's Inhibitor Was from Calbiochem (Alexandria, N.S.W., Aust.). Losartan and Pd 123319 Were Gifts from Dupont Merck Pharmaceuticals (Wilmington, Delaware) and Parke-Davis Pharmaceuticals (Michigan), Respectively. All Tissue Culture Media, Fetal Calf Serum (Fcs), Trypsin, Hepes and Other Tissue Culture Solutions Were Purchased from Gibco (Glen Waverly, Vic., Aust.). Poly-D-Lysine, Insulin and Transferrin Were Obtained from Sigma-Aldrich (N.S.W., Aust.).

Cell Culture

Rat C6 (American Type Culture Collection CC107) and RT4-D6 (25) glial cell lines were maintained in an atmosphere of 97.5% air/2.5% CO_2, at 37°C, in HEPES-buffered DMEM containing 10% fetal calf serum and 100U/ml penicillin and 100U/ml streptomycin. All experiments were performed at confluency.

Astrocyte cultures were prepared from the hypothalamic/thalamic region of neonate rats, as described previously (26). Cells were dissociated with trypsin and resuspended in sorbitol-DMEM (glucose-free)+10% FCS to remove any oligodendrites. After 3 days the cells were grown in normal glucose-DMEM+10%FCS until they reached confluency. They were then subcultured into 6-well plates and regrown to confluency (10-12 days). These cultures consist of approximately 98% astrocytes (26). Neurones were prepared from gestational day 16 rat fetuses as described previously (26).

Radioreceptor Assay

All binding assays were performed in triplicate using 6 well plates. Wells were washed once with 1ml of Dulbecco's PBS pH 7.2 at 37°C followed by the addition of 0.5 mls of radioreceptor assay buffer (Dulbecco's PBS at pH 7.2, 0.6% bovine serum albumin, 50μM Plummers inhibitor, 10μM bestatin) containing 50,000 cpm ^{125}I-AngIV and unlabelled AngIV at concentrations of 10,50,100 and 500 pmol; 1,5,50,and 500 nM; and 10 μM for non-specific binding. After 20 mins at 22°C, bound and free were separated by 5 x 2ml washes with cold Dulbecco's PBS, 0.6% BSA and cells digested with 0.5mls of 1M NaOH followed by 0.5ml of water. The radioactivity was counted (γ-counter,LKB Model 2174) and used to determine the affinity constant (Ka) and receptor density (Ro) by the computer program, LIGAND (Biosoft, Cambridge, UK). Receptor densities were expressed relative to protein content, as measured by the Bradford method. Competitive displacement curves were generated by 20 mins incubations at 22°C with each of the following ligands: losartan (an AT1 receptor ligand), PD123319 (AT2 receptor ligand), Sar^1,Ile^8-AngII, AngII, AngIII and AngIV and (4-8)AngII. Binding data were analysed and expressed graphically by the computer program, PRISM (Graphpad Software, San Diego, California). The observed rate of association (k_{obs}) was calculated from binding experiments performed at 22°C in the presence of 0.3nM ^{125}I-AngIV over a period of 1 hour. Non-specific binding at corresponding time points was determined with 10μM unlabelled AngIV. The dissociation rate constant (k_{-1}) was calculated from experiments at 22°C, consisting of the pre-incubation of ^{125}I-AngIV for 20 mins followed by the dissociation of bound ligand over 60 mins with the addition of 10 μM AngIV. The apparent first order association constant (k_{obs}) and the dissociation rate constant (k_{-1}) were calculated with INPLOT (Graphpad Software, San Diego, California) and the association rate constant (k_{+1}) from the equation $(k_{+1}) = (k_{obs} - k_{-1})/[L]$, where L is the ligand concentration.

Iodination of Angiotensin IV

AngIV was iodinated by a 2 min incubation of 5μg AngIV, 20μg chloramine T, and 500μCi Na^{125}I in a total volume of 50 μl potassium phosphate buffer (0.3M, pH 7.6). The ^{125}I-AngIV was eluted isocratically on a reverse phase high pressure liquid chromatography column (C18, maxil 10) using 20% acetonitrile in potassium-triethylamine buffer (20mM, pH 3.0). The radiodinated AngIV (≈ 2000 Ci/mmol) was stored at -20°C.

Degradation of Ligand

The amount of AngIV which was metabolized during the incubation time for radioimmunoassays was examined by HPLC. Radioreceptor assay buffer containing 10^6 cpm/ml of ^{125}I-AngIV was incubated with cells for 20 mins at 22°C. The buffer and cells were collected and treated with 10% trichloroacetic acid for 30 mins and then centrifuged at 1500 x g for 20 mins to remove denatured proteins. Peptides in the supernatants were removed by C18 mini columns which had been pre-conditioned with 10 ml of 0.1% triflouroacetic acid (TFA) solution. The minicolumns were washed with 4 ml of 0.1% TFA, 4 ml 10% acetonitrile (ACN) in 0.1% TFA and finally the peptides were eluted with 4 ml of a 90:10 mixture of 10% ACN to 0.1% TFA. The extract was dried, reconstituted in 50 μl of a mixture of 15:85 of 10% acetonitrilre and 0.1% TFA and loaded onto a LKB Ultrapac C18 reverse phase HPLC column. Peptides were separated with a gradient of 15 to 40 % TFA in ACN over 30mins at a flow rate of 1ml/min. Fractions were collected at minute intervals and the radioactivity of each fraction determined. The overall recovery of radioactivity from the extraction and separation procedure was about 90%. Radioiodinated AngIV which had not been incubated with cells but otherwise treated identically was used as a control.

Receptor Autoradiography

Cells were grown on 8-chamber slides and used at 80% confluency. Media was removed and cells were lightly prefixed with 2% formaldehyde at 4°C for 5 mins. After a

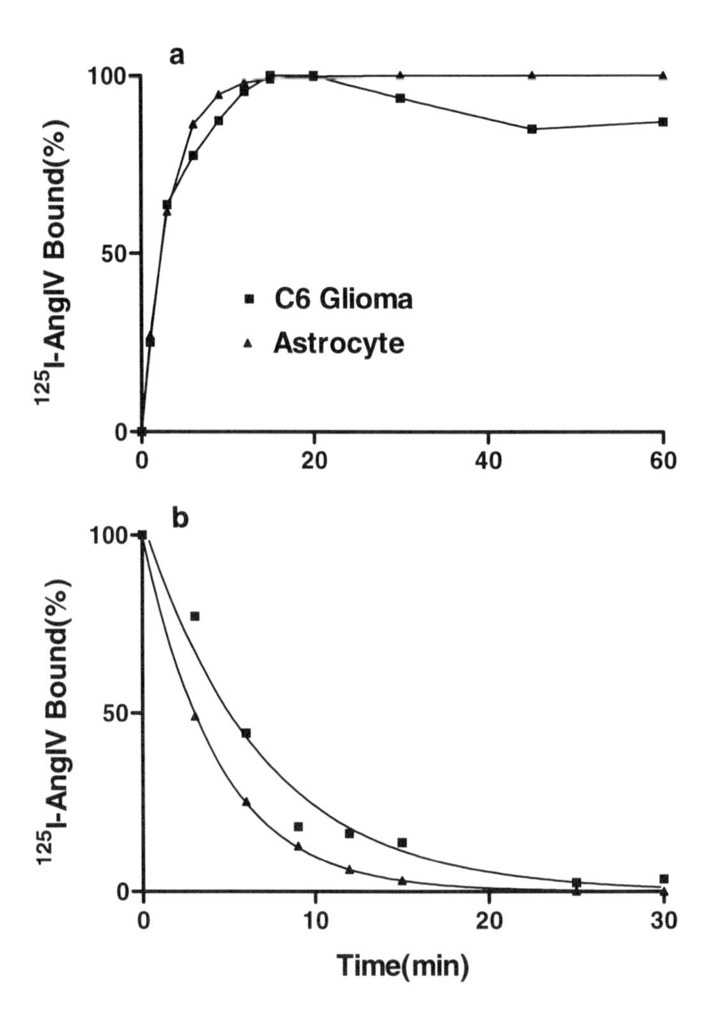

Figure 1. Time course for the association (a) and dissociation (b) of 0.3nM ^{125}IAngIV to C6 glioma and primary astrocyte cells. These results were used to calculate the observed association rates (k_{obs}) and dissociation rates (k_{-1}) by non-linear regression using the INPLOT software package on the assumption that the binding approximated first order kinetics. The data are summarized in Table 1. The decrease in specific binding observed in C6 cells at longer time intervals suggests some degradation of ligand was present. Data points represent the means of 3-4 experiments.

wash with radioreceptor buffer, the cells were incubated for 20 mins with 30pM ^{125}I-AngIV alone or together with 1μM AngIV (sufficient to saturate specific binding sites), AngII, Sar^1,Ile^8-AngII, losartan or PD123319. The supernatant was then removed and the cells washed 3 times with radioreceptor assay buffer at 4°C. The slides were dried for 3 hours with air and fixed overnight with formaldehyde vapours at 50°C. They were then dipped in emulsion (K.2, Kodak), placed in a light-free box and left at -20°C for 2 weeks. The location of bound ^{125}I-AngIV was indicated by dark silver grains in the photographic emulsion.

Table 1. Summary of kinetic data shown in Figure 1 and the association constants calculated from k_{+1}/k_{-1}

Cell type	On-rate (k_{+1}) min^{-1} nM^{-1}	Off-rate (k_{-1}) min^{-1}	Assoc. constant (Ka) nM^{-1}
C6 glioma	0.303	0.146	2.08
astrocytes	0.307	0.221	1.39

RESULTS

The time-course for the specific binding of 0.3nM ^{125}I-AngIV to C6 glioma and primary glial cells at 22°C is shown in Figure 1a. Maximal binding was reached between 15 and 20 min and maintained up to 60min for astrocytes. Some decrease was evident in C6 glioma cells after 30 min, suggesting a degradation of the ligand. Based on these results, the optimal conditions for radioreceptor assays were assumed to occur at an incubation time of 20 min and hence all binding assays requiring equilibrium conditions were done at a standard incubation time of 20 min. The rates of association for C6 and astrocytes were determined from the specific binding over the first 20 min. Non-linear regression and pseudo first-order kinetic analysis of the data was performed with INPLOT to obtain observed association rates, k_{obs}. The rate of dissociation of bound radioligand was determined in the presence of 10μM AngIV over a 30 min period, as shown in Fig 1b.The off rates, k_{-1} were calculated with INPLOT. Using the equation: $(k_{+1}) = (k_{obs} - k_{-1})/[L]$ on-rates, k_{+1} were calculated. The summary of these results and of the corresponding association constants ($Ka = k_{+1}/k_{-1}$) is shown in Table 1.

Competitive displacement of ^{125}I-AngIV by AngIV in C6, RT4-D6 and astrocytes is shown in Fig 2. All the ANGIV displacement curves show a clear point of inflection, typically obtained when two binding sites with different affinities are present. The best fit to the data was obtained by a two binding-site analysis. Curvilinear Scatchard plots confirmed the presence of two sites, as shown in Fig 3a. For comparison with AngII receptors, the binding of the non-selective AngII antagonist Sar^{1}Ile^{8}AngII in C6 cells was also examined, as shown in Figure 3b. Only one binding site was present with an affinity (Ka) of 5.8nM^{-1} and a density of 4.8 fmole/mg protein. The mean affinities (K_1, K_2;) and densities (R_1, R_2) of the two sites for AngIV are summarized in Table 2. The different cells lines show similar high affinity K1 sites but a wide range of receptor density. It is notable that the values of K1 (the high affinity site) for C6 and astrocytes are in close agreement with those calculated from the kinetic data shown in Table 1. A comparison of our results with published values for bovine coronary (BCVEC) and aortic (BAEC) endothelial cells (obtained at 37^{0}C) show higher affinities but densities comparable to our cell lines (Table 2).

The specificity of the AngIV receptor was examined by competition displacement assays in C6, astrocytes and neurones; the results for C6 only are shown in Fig 4. The most effective competitors were AngIV and AngIII, both of which saturated the specific binding sites at 1μM and displayed inflection points between 80-85% binding; (4-8)AngII was also able to compete with ^{125}I-AngIV to a concentration of 0.01μM for about 20% of the total

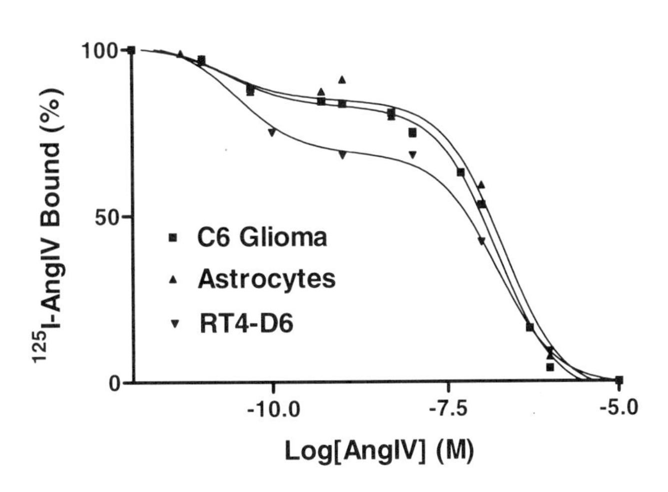

Figure 2. Competitive displacement of radiolabelled ^{125}I-AngIV by increasing concentrations of unlabelled AngIV in glia cells lines C6 and RT4-D6 and primary astrocyte cultures from neonatal rat brains. All three curves show points of inflection typical of binding to two sites of different affinities

binding sites. Higher concentrations of (4-8)AngII were ineffective in displacing radioligand further. The non-selective angiotensin receptor antagonist, Sar[1],Ile[8]-AngII, the AT1-selective antagonist, losartan, and the AT2-selective ligand, PD123319, were all ineffective at competing for AngIV receptors. The displacement curves for AngIV, AngIII and (4-8)AngII were best fitted (regression coefficient $R^2 \geq 0.97$) by a two-site model. A similar hierachy of ligand specificity (AngIV>AngIII>>(4-8)AngII>>losartan= PD123319=Sar[1],Ile[8]-AngII) was obtained for astrocytes and neurones. Ligand competition was not tested in RT4-D6 cells.

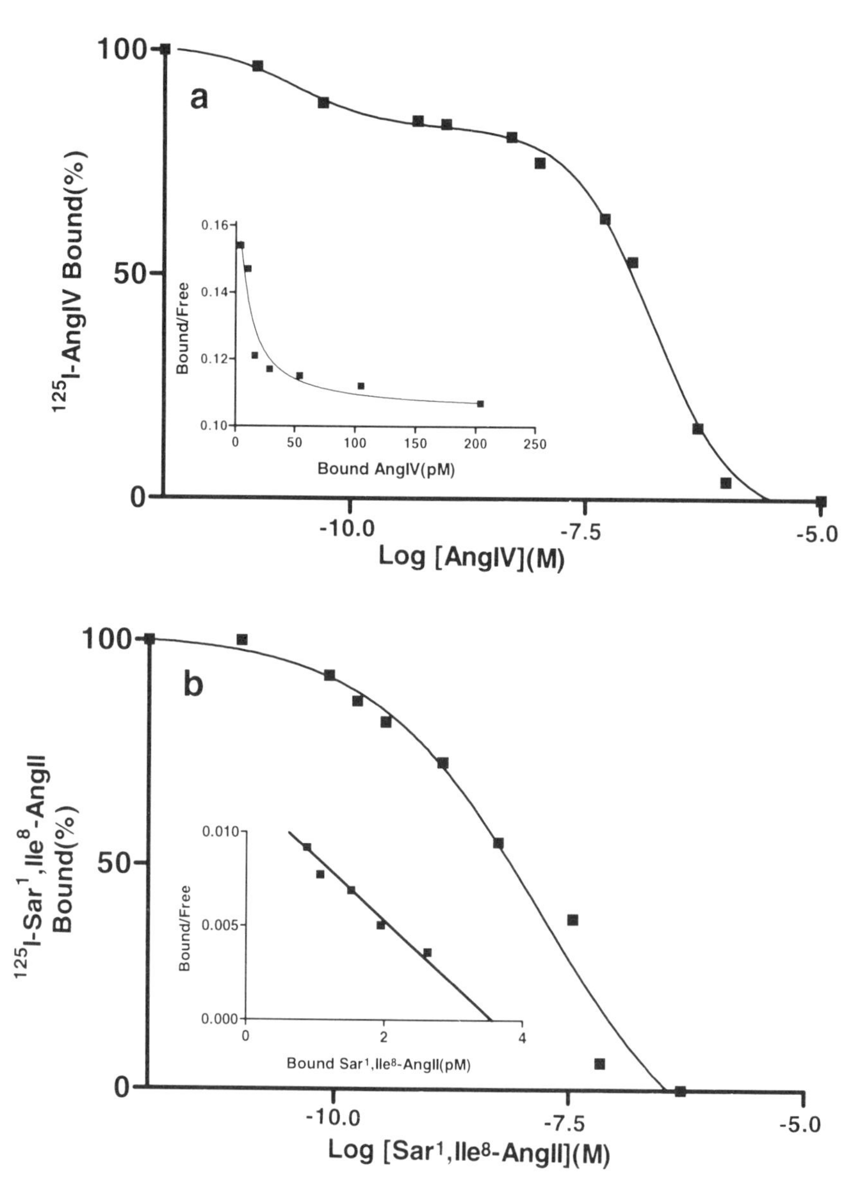

Figure 3. Competition displacement curves for AngIV (a) and Sar[1] Ile[8]-AngII (b) in C6 cells with the corresponding Scatchard plots shown as insets. Note the different binding characteristics of the two ligands to the same cell type.

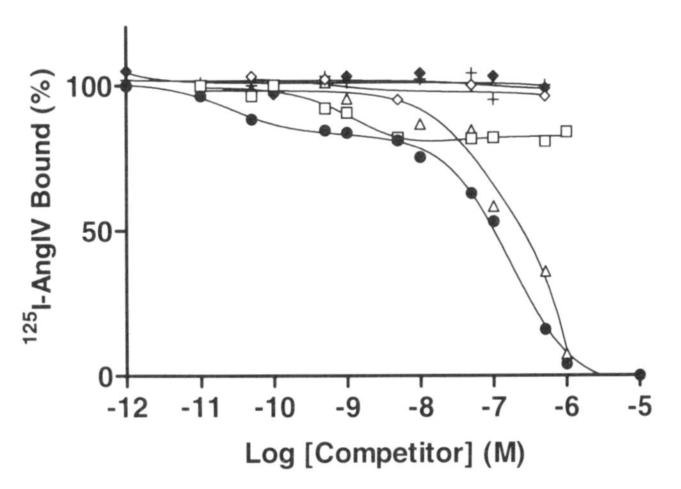

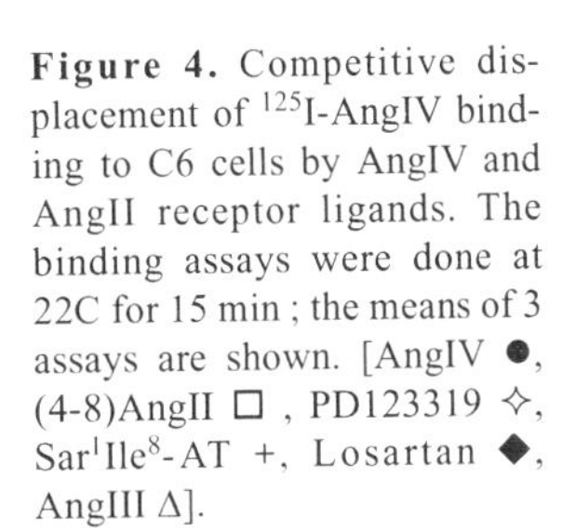

Figure 4. Competitive displacement of ^{125}I-AngIV binding to C6 cells by AngIV and AngII receptor ligands. The binding assays were done at 22C for 15 min ; the means of 3 assays are shown. [AngIV ●, (4-8)AngII □ , PD123319 ✧, Sar^{1}Ile^8-AT +, Losartan ◆, AngIII Δ].

The peptides in extracts of C6 and astrocyte cells after a 15 min incubation with $1x10^6$cpm ^{125}I-AngIV were separated by HPLC to determine the extent of ligand degradation (Fig 5) The purity of the ligand before its incubation with cells was confirmed by its elution as a single peak at fractions 20-23. The results show that a large fraction of the radioactivity eluted in the expected fractions for pure ^{125}I-AngIV, although a second smaller peak of degraded product was also present at fractions 16-18.

The cellular localization of ^{125}I-AngIV binding was examined by radioreceptor autoradiography in C6 cells , astrocytes and neurones (Figure 6). Dark grains were observed over cell bodies and down the branches of cells incubated with radioligand alone but few grains were observed in the presence of 1 μM AngIV. For C6 cells, the binding sites appeared to be distributed over most cells and not restricted to a subpopulation. In constrast, primary astrocytes and neurones consisted of subpopulations of cells expressing receptors. The addition of 1 μM AngII, AngIII, Sar^{1}Ile^{8}AngII, losartan or PD123319 did not abolish binding of ^{125}I-AngIV (not shown), thereby confirming the selectivity for AngIV.

Table 2. Affinity constants (K1,K2) and respective receptor densities (R1,R2) for AngIV binding in various cell types. Data represent mean values from Scatchard analysis of radioreceptor assays by LIGAND. *Data obtained at 37^{0}C by Hall et al (13); [BCVEC=bovine coronary venular endothelial cells, BAEC=bovine aortic endothelial cells]

Cell type	K_1 nM^{-1}	R_1fmol/mg prot	K_2nM^{-1}	R_2fmol/mg prot
C6 glioma	2.49	34	0.18	563
RT4-D6	6.05	6	0.07	756
Neurones	2.72	109	0.07	1723
Astrocytes	5.71	192	0.28	1425
BCVEC*	68.5	6	0.72	594
BAEC*	37.2	10	0.23	434

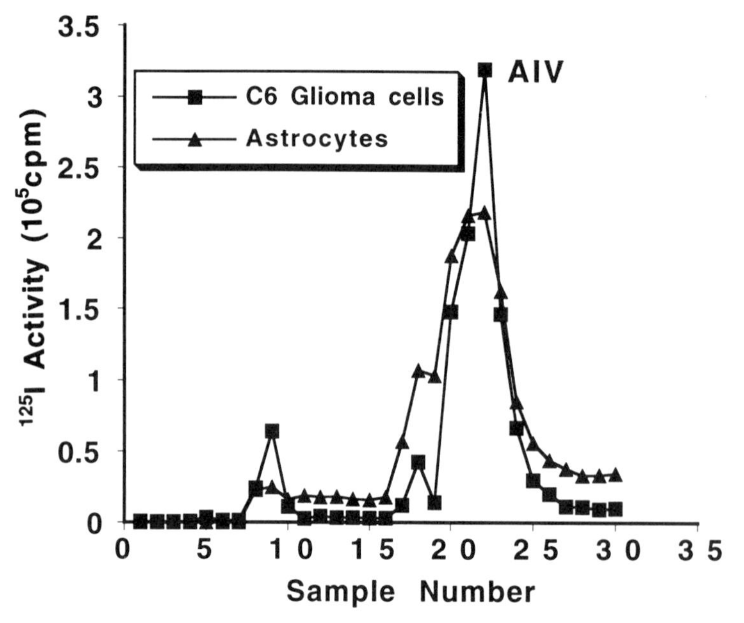

Figure 5. HPLC separation of degraded fragments of AngIV formed during radioreceptor assays. C6 and astrocyte cells were incubated with 10^6 cpm ^{125}IAngIV for 15 min at 22°C, the cells treated with TCA to precipitate proteins, and the peptides in the supernatant extracted as described in Methods. Peptides were eluted with a 15-40% TFA gradient in acetonitrile and compared with the elution profile of pure TCA-treated ^{125}IAngIV, which eluted as a peak at fractions 20-23 (not shown). Data points represent the means of 3-6 experiments.

DISCUSSION

In this study, we have shown that cell lines of glial cells (C6 and RT4-D6) as well as primary cultures of astrocytes and neurones have binding sites for AngIV. These sites were shown to be distinct from AT_1 and AT_2 receptors by the failure of the non-selective angiotensin receptor antagonist, Sar^{1}Ile^8-AngII, the AT_1-selective antagonist, losartan, and the AT_2-selective ligand, PD123319, to displace AngIV in competition assays and in autoradiography. Some displacement activity was found with (4-8)AngII, which is in agreement with recent studies showing that deletion or substitution of amino acids at the N-terminus dramatically decreases the affinity of the peptide for the AngIV receptor (12). AngIII also showed competitive displacement of Ang IV in our radioreceptor assays, although it was not so noticeable in our receptor autoradiography. It is likely that enzymatic conversion to AngIV is occurring in the radioreceptor assays, as reported previously (13). The binding to the AngIV receptors was saturable and reversible, reaching equilibrium conditions in 20 minutes at 22°C. The ligand being bound was verified by HPLC as being mostly AngIV and not some degraded fragment. Since C6 cells have AT receptors as well as AngIV sites, we compared the binding by the non-selective AT receptor ligand ^{125}I-Sar^{1}Ile^8-AngII with that of ^{125}I-AngIV. In contrast to the AngIV sites, ^{125}I-Sar^{1}Ile^8-AngII bound to a single class of binding sites which were present at a lower density (9.8fmol/mg protein) than the high affinity AngIV site (34 fmol/mg protein; Table 2). Swanson et al (10) also showed that , in tissues which have both AngIV and AT receptors, the density of AngIV receptors is generally greater.

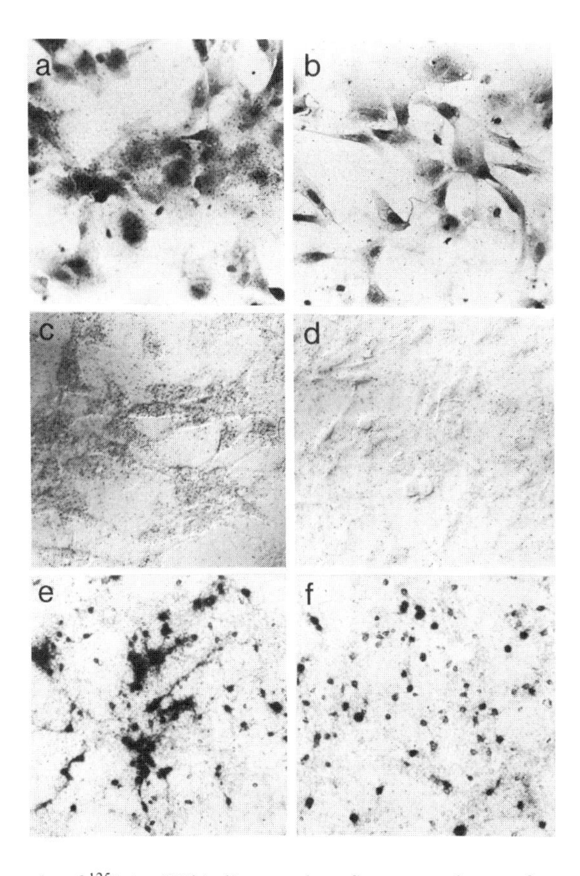

Figure 6. Autoradiographs of ^{125}I-AngIV binding to subconfluent monolayers of astrocytes (panels a,b), C6 glioma (panels c,d), and neurones (panels e,f). Panels a,c and e show a heavy and even cellular distribution of bound ^{125}I-AngIV (dark grains) which could be displaced by excess non-radiolabelled 1μM AngIV, shown in panels b,d and f. Dark grains were seen in most C6 cells but only in subpopulations of astrocytes and neurones. Note the high density of grains over neurite extensions as well as perikarya in panel e. Astrocytes were lightly stained with toluidine blue to show cell bodies; C6 and neurones were left unstained. Magnification: X100 for neurones and X200 for others.

High-affinity sites with a Ka of around 1-3 nM^{-1} were obtained from Scatchard plots and from kinetic analyses. These values are like those reported for the brain, heart and venular endothelial cells (10). In addition to a high affinity site, we observed a second site of lower affinity (K2 = 0.07 to 0.28 nM^{-1};Table 2) and higher capacity, as suggested by the curvilinear Scatchard and by the inflection points at 75-85% in the ligand competition assays. In general, previous studies report only one binding site. A reason for this difference may lie in the species investigated; the rat in our study and the rabbit, bovine and guinea pig tissues in previous studies (8-15,19,20). However in some cases, two binding sites have been found; for example in vasclular endothelium (see Table 2). There are also experimental conditions which could lead to a curvilinear Scatchard plot and consequently, the apparent presence of receptor heterogeneity. In particular, the degradation of AngIV ligand, non-equilibrium conditions at low ligand concentrations and receptor internalization are factors which influence the determination of binding isotherms (19,27) Thus at present our observation of a second binding site is to be accepted with caution until further elucidation of the nature of the AngIV receptor. In any event, it is highly unlikely that the lower affinity site represents AT receptors since these are present at a lower density than the high affinity AngIV sites (10).

The status and significance of AngIV receptors are presently contentious issues due to the absence of firm evidence that AngIV is the endogenous ligand and that the AngIV receptor is a distinct receptor molecule. Nevertheless, evidence exists which supports a role of AngIV in cardiac hypertrophy (23) but not contractility (14); in cerebral and renal blood flow (10,21) and in the enhancement of memory (11,22). The widespread distribution of AngIV receptors in the guinea-pig brain supports possible central actions of AngIV (15). Furthermore, the predominance of these sites in cognitive and sensorimotor areas, rather than in those cardiovascular and osmoregulatory areas which express AT receptors (18), dimishes the possibility that the effects of ANGIV are in fact being mediated by AT receptors. Finally, angiotensinogen, the essential substrate for the generation of angiotensins, is present in areas of the brain with high AngIV receptor density (28); and so the local production of AngIV is possible.

Our results show that these ANGIV sites are present in transformed glial cells of brain (C6) and peripheral (RT4-D6) origin, as well as in primary astrocyte cells. Glial cells are not directly involved in neurotransmission and therefore are unlikely to be directly involved in cognitive effects observed by Braszko et al (22) and Wright et al (11). However primary neurones also have AngIV receptors over the perikarya and extending down the neurites. These receptors could mediate a neuromodulatory or neurotransmitter function.of AngIV. Finally, the AngIV receptor ligand properties of the transformed and primary glial cells were similar, which suggests that these cell lines could be used as models of AngIV action as well as sources of AngIV receptors for the physicochemical characterization and identification of this new putative "angiotensin" receptor.

ACKNOWLEDGMENTS

This work was supported by grants to CS from the National Health and Medical Research Council of Australia and from the Brain Research Fund (University of Queensland). Losartan and PD123319 were generously provided by Dupont Merck Pharmaceuticals (Wilmington, Delaware) and Parke-Davis Pharmaceuticals (Ann Arbor, Michigan) respectively. Ms Tey and Leong were recipients of University of Queensland Vacation Scholarships. The assistance of R.Mann, Dr. Tang Zeng and K. Greenland is gratefully acknowledged

REFERENCES

1. Eberhardt, R.T., Kevak, R.M., Kang, P.M. and Frishman, W.H. (1993). J. Clin. Pharmacol. *33*:1023-1028.
2. Campbell, D.J. (1987). J. Cardiovasc. Pharmacol. *10*:S1-S8
3. Phillips, M.I., Speakman, E.A. and Kimura, B. (1993). Regul. Peptides *43*:1-20.
4. Brown, L. and Sernia, C. (1994). Exp. Physiol. Pharmacol. *21*:811-818.
5. Bunnemann, B., Fuxe, K. and Ganten, D. (1993). Regul. Peptides *46*: 487-509.
6. Saavedra, J.M. (1992). Endocr. Revs. *13*:329-386
7. Ferrario, C. M., Brosnihan, K. B., Diz, D. I., Jaiswal, N., Khosla, M. C., Milsted, A. and Tallant, E. A. (1991). Hypertension *18:* (Suppl. III): S126-S133.
8. Coleman, J.K.M., Krebbs, L., Ong, B., Hanesworth, J.M., Lawrence, K.A., Harding, J.W. and Wright, J.W.(1992). FASEB. J. *6:* A437.
9. Harding, J.W., Cook, V.I. Miller-Wing, A.V., Hanesworth, J.M., Sardinia, M.F., Hall, K.L., Stobb, J.W., Swanson, G.N., Coleman, J.K.M., Wright, J.W. and Harding, E.C. (1992). Brain. Res. *583*: 340-343
10. Swanson, G.N., Hanesworth, J.M., Sardinia, M.F., Coleman, J.K.M., Wright, J.W., Hall, K.L., Miller-Wing, A.V., Stobb, J.W., Cook, V.I., Harding, E.C. and Harding, J.W.(1992) Regul. Peptides *40*:409-419.

11. Wright, J.W., Miller-Wing, A.V., Shaffer, M.J., Higginson, C., Wright, D.E., Hanesworth, J.M. and Harding, J.W.(1993).Brain Res. Bull. *32*: 497-502
12. Sardinia, M.F., Hanesworth, J.M., Krebs, L.T. and Harding J.W. (1993). Peptides *14*: 949-954.
13. Hall, K.L., Hanesworth, J.M., Ball, A.E., Felgenhaur, G.P., Hosick, H.L. and Harding, J.W.(1993). Regul. Peptides *44*: 225-232.
14. Hanesworth, J.M. , Sardinia, MF., Krebs, L. T., Hall, K.L. & Harding, J.W.(1993). J. Pharmacol. Exper. Ther. *266*: 1036-1042.
15. Wright, J.W. and Harding, J.W.(1994). Neurosci. Biobehav. Rev.*18*: 21-53.
16. Whitebread, S., Mele, M., Kamber, B. and deGasparo, M.(1989). Biochem. Biophys. Res. Commun. *163*: 284-291.
17. Timmermans, P.B., Wong, P.C., Chiu, A.T. and Harblin, W.F.(1991). Trends Pharmacol. Sci. *12*: 55-62.
18. Stecklings, U.M., Bottari, S.P. and Unger, T. (1992).TIPS. *13*: 365-368.
19. Bernier, S.G., Fournier, A. and Guillemette, G.(1994). Eur. J. Pharmacol. *271*: 55-63.
20. Jarvis, M.F., Gessner, G.W. and Ly, C.Q. (1992). Eur. J. Pharmacol. *219*: 319-322.
21. Haberl, R.L., Decker, P.J.and Einhaupl, K.M.(1991). Circ. Res. *68*: 1621-1627.
22. Braszko, J.J., Kupryszewski, G., Witczuk, B. and Wisniewski, K.(1988) Neurosci. 27: 777-783.
23. Baker, K.M. and Aceto, J.F. (1990). Am. J. Physiol. *259*: H610-H618.
24. Miller-Wing, A. V., Hanesworth, J. M., Sardinia, M. F., Hall, K. L., Wright, J. W., Speth, R. C., Grove, K. L. and Harding, J. W. (1993). J. Pharmacol. Exp. Ther. *266*: 1718-1726.
25. Hagiwara,N., Imada, S. & Sueoka,N. (1993). J. Neurosci Res 36,646-656.
26. Thomas, W.G., Greenland, K..J., Shinkel, T.A. and Sernia, C.(1992). Brain Res. *588*:191-200.
27. Beck, J.S. and Goren, H.J.(1983). J. Receptor Res. *3*: 561-577.
28. Sernia C.(1995). Regul. Peptides *57:1-18.*

INDEX